# NONEMISSIVE
# ELECTROOPTIC
# DISPLAYS

## Earlier Brown Boveri Symposia

Flow Research on Blading ● 1969
*Edited by L. S. Dzung*

Real-Time Control of Electric Power Systems ● 1971
*Edited by E. Handschin*

High-Temperature Materials in Gas Turbines ● 1973
*Edited by P. R. Sahm and M. O. Speidel*

# NONEMISSIVE ELECTROOPTIC DISPLAYS

*Edited by*

## A. R. Kmetz

*and*

## F. K. von Willisen

*BBC Brown, Boveri & Company Limited*

PLENUM PRESS • NEW YORK AND LONDON

Library of Congress Cataloging in Publication Data

Brown, Boveri Symposium on Nonemissive Electrooptic Displays, Baden, Switzerland,
1975.
   Nonemissive electrooptic displays.

   Includes indexes.
   1. Information display systems–Congresses. 2. Electrooptical devices–Congresses. I.
Kmetz, A. R. II. Willisen, F. K. von. III. Brown Boveri Research Center. IV. Title.
TK7882.I6B67 1975              621.38'0414                    76-17060
ISBN-13: 978-1-4613-4291-5     e-ISBN-13: 978-1-4613-4289-2
DOI: 10.1007/978-1-4613-4289-2

Proceedings of the Fourth Brown Boveri Symposium on
Nonemissive Electrooptic Displays held at the Brown
Boveri Research Center, Baden, Switzerland, September 29-30, 1975

© 1976 Plenum Press, New York
Softcover reprint of the hardcover 1st edition 1976

A Division of Plenum Publishing Corporation
227 West 17th Street, New York, N.Y. 10011

For the contribution by E. P. Raynes (pp. 25–41),
Copyright © Controller HBMSO London 1975.

# FOREWORD

Shortly after the inception of the Brown Boveri Research Center in 1966, plans were made to organize a series of biennial scientific symposia. A different subject was to be chosen for each symposium with the following requirements in mind:

- It should characterize a part of a scientific discipline; in other words, it should describe an area of scholarly study and research.
- It should be of current interest in the sense that important results have recently been obtained and considerable research effort is under way in the world's scientific community. In other words, there must be a good reason why the symposium should be held now, rather than five years earlier or five years later.
- It should bear some relation to the scientific and technological activity of Brown Boveri.

These symposia are intimately related to one of the basic concepts which govern the work of our Research Center: Close coupling between science and engineering. It is to this coupling that we owe the technical standard of our products and it is this coupling which we hope will be furthered by our symposia.

It is often said that the important technological innovations come from the basic sciences, and the transistor is taken as a brilliant example. Indeed, the transistor is based on quantum mechanics, a thoroughly fundamental scientific structure which in turn predicts the behavior of electrons in a crystal lattice and explains the existence of energy bands. Without knowledge of quantum mechanics, the invention of the transistor would have been impossible. This is a model case for the following sequence of events: first, scientific discovery; then, engineering invention.

It is good to note that this course, although frequent, is not the only path along which technological progress takes place. In fact, the opposite sequence is quite common: first, engineering invention; then, discovery of the underlying principles. Fluid dynamics, the subject of the first Brown Boveri symposium which took place six years ago, is almost a model for this process. The first steam turbines were designed and successfully built by

people who were only vaguely acquainted with the scientific laws underlying their machines, and, in fact, some of these laws were not even known
at the time. The fundamental concept of entropy was just being introduced
into the engineers' language, and much time elapsed before turbine designers
were willing to adopt this concept in their daily work. The fact that these
men were able to design very satisfactory turbines of 10 000 kilowatts in
this manner is an impressive illustration of their deep intuitive insight.

For further progress however, science was essential. Giant turbogener-
ators each having a rating of 1300 Megawatts are now probably the best
known among our products. The inventor still plays a key part in the design
process; but he must interact closely with the scientist, and he must frequently ask himself: "How far from the optimum is my invention? Am I ex-
pecting something that is against the fundamental laws of nature? If not,
what can be ultimately achieved with my idea?" Without the answer to these
questions, technical progress would be impossible beyond a certain point,
and this, of course, holds not only for turbines, but for all fields of technology.

Non-emissive displays exemplify a third course of development intermediate between the two foregoing cases. As little as five years ago, liquid crystal displays were little more than a laboratory curiosity; today watches and
calculators with such displays can be bought in stores almost throughout the
world. One can hardly find a better example to demonstrate how progress
can come about through the intimate coupling of science and technology —
or, to put it somewhat differently — through the simultaneous application
of insight gained through scientific analysis and knowledge obtained empiri-
cally.

The Brown Boveri interest in this field can be traced to 1969 when BBC
and Hoffmann-La Roche joined their R&D forces in an attempt to enter the
field of biomedical engineering. This joint program produced a large number
of exciting results, the best known of which was the discovery of the electrooptic properties of twisted nematic liquid crystal structures (the Schadt/
Helfrich Effect). After termination of the partnership, a research program in
the field of liquid crystal displays was continued and expanded at our
Research Center. Two years ago Brown Boveri decided to enter the field of
optoelectronic components when it started setting up a manufacturing plant
for liquid crystal displays. Needless to say the R&D effort has been further
intensified after the decision for large scale production of liquid crystal displays was reached. The result of this decision has been a very strong commit-
ment to the field of non-emissive displays, because we believe in their future
and in the market opportunities opened by the various technologies described in these Proceedings.

Since the establishment of the Brown Boveri Research Center, research
programs were started in most of the fields of the commercial activity of
Brown Boveri, including all three of our examples: semiconductor devices,
steam turbines and non-emissive displays. In the course of the years we were
surprised to find out how closely related these seemingly widely different

areas become if one moves toward the scientific end. We see one of the great challenges of today's industrial research in the task of looking for unanswered questions in widely different fields of technology and of seeking common answers in the sciences. This is only made possible by the organizational concept of corporate research which we have adopted, and of which these symposia are a manifestation.

The 1975 symposium on Non-Emissive Electrooptic Displays was attended by 96 participants from 6 European countries, the United States and Japan. It was both an honor and a pleasure to welcome scientists and engineers from so many parts of the world. Their willingness to travel to Baden and to spend two full days with us was a challenge as well as an obligation to us as organizers, and we sincerely hope that by the end of the meeting the expectations which prompted them to attend were fully met.

We should like to express our gratitude to all authors, who undertook the laborious task of preparing and delivering papers, and also to the discussion speakers, who contributed significantly to the success of the meeting.

To conclude, we should like to express our gratitude to every participant in the symposium. Special thanks are addressed to Drs. F. K. von Willisen, A. R. Kmetz and J. Nehring, who were responsible for the scientific part of the symposium, and also to Mr. E. Arn and Mrs. U. Richter for organizing and handling the administrative side of the meeting.

Owing to the limited space available and the desire to avoid an unwieldily large congress, it was not possible to extend invitations to a larger circle. With much regret, we had to disappoint a number of applicants who wished to participate. We hope that the present publication may be a partial compensation for those whom we did not have the pleasure of welcoming as our guest.

R. SCHÜPBACH                                      A. P. SPEISER
Director, Electronics Division                    Director of Research

PREFACE

Recent advances in microelectronics have greatly stimulated research and development on a new class of electrooptic displays which themselves emit no light and which therefore offer low-power, low-voltage operation. The technology of these non-emissive displays is a new field, growing rapidly in divergent directions. Researchers in this field who specialize in one of the many physical effects find it increasingly difficult to keep track of the nature and significance of developments in other areas and to maintain a realistic perspective on the relative importance of various aspects of their own work. Likewise reports of new developments, written for an audience of specialists, seldom fit the needs of potential users of non-emissive displays, who seek to evaluate and compare the various competing technologies with regard to a specific application. The evident need for a detailed technical overview of the non-emissive displays field from the viewpoint of application to practical display systems is the motivation for this book, and for the symposium out of which it grew.

Having determined to devote the fourth Brown Boveri Symposium to "Non-Emissive Electrooptic Displays", a program of topics was selected to cover the materials, phenomena and technology of this broad field. Leading scientists were then invited to prepare and present extensive review papers on these topics. Relevance to display applications was an explicit goal, and comparisons among various approaches were encouraged. The invited reviews were to be supplemented by short contributed papers and by technical discussions moderated by knowledgeable session chairmen.

The symposium took place on September 29 and 30, 1975 at the Brown Boveri Research Center in Baden, Switzerland. The entire first day was devoted to liquid crystals, reflecting their present dominance of non-emissive displays. Since the physics of the twisted nematic device is now well established, D. W. Berreman was able to use computer calculations to demonstrate the dependence of electrooptic performance on material and fabrication parameters. The principal display effects in cholesterics were reviewed by E. P. Raynes, and potentially major improvements from pleochroic dyes and hysteresis effects were cited. T. J. Scheffer presented a broad and rigorous survey of liquid crystal techniques for color displays. Relationships between molecular structure and display performance were discussed by D. Demus. Surface alignment techniques, including a method for weak anchoring, were reviewed by E. Guyon, who reported recent progress toward quantitative understanding of boundary effects.

The first three papers on the second day dealt with electrochromism. H. R. Zeller derived from basic principles some reasonable expectations for ultimate display performance. In surveying the possible materials and effects, I. F. Chang distinguished between electronic and electrochemical processes, finding the latter category much more promising for displays. Practical details of cell fabrication and driving methods were discussed by J. Bruinink. Later, material problems were identified by J. C. Lewis as the main obstacle to realizing the impressive potential of electrophoretic displays, and slim-loop PLZT emerged as the most promising candidate for a ferroelectric display in K. H. Härdtl's review. The symposium closed with a session on addressing techniques for non-emissive flat-panel displays. The compromises between optical performance and electrical complexity which limit matrix addressing of the various display effects were described and compared by A. R. Kmetz. Techniques for circumventing these limits by integrating electronic components into the display panel were reviewed by T. P. Brody, who reported striking progress with thin-film transistor arrays.

This book contains all invited review papers, as well as several short contributed papers, which were presented at the symposium. In addition the keynote paper by J. Kirton and concluding remarks by C. Hilsum provide useful information, as well as some stimulating opinions, concerning displays from a more general perspective. Discussion remarks, selected from written notes and tape recordings, are also included.

It is our pleasure to acknowledge the most gratifying cooperation of all authors with our editorial attempts to impose consistency of style and format on contributions from diverse scientific disciplines and nationalities. We are grateful for their responsive and considered support which made the timely completion of this book possible. We also acknowledge the indispensable efforts of Mrs. U. Richter who typed the entire camera-ready manuscript.

Baden, March 1976              A. R. Kmetz              F. K. von Willisen

# CONTENTS

### Session 1 — Display Engineering and Electrooptics
### Chairman: A. P. Speiser

REQUIREMENTS ON MODERN DISPLAYS

### Session 2 — Liquid Crystal Display Effects
### Chairman: W. Helfrich

ELECTRICAL AND OPTICAL PROPERTIES OF TWISTED NEMATIC
STRUCTURES

CHOLESTERIC TEXTURE AND PHASE CHANGE EFFECTS

LIQUID CRYSTAL COLOR DISPLAYS

Short Communication: SPECTRUM OF VOLTAGE CONTROLLABE
COLOR FORMATION WITH NEMATIC LIQUID CRYSTALS

### Session 3 — Liquid Crystal Materials and Technology
### Chairman: F. J. Kahn

CHEMICAL COMPOSITION AND DISPLAY PERFORMANCE

### Session 4 — Electrochromic Displays
### Chairman: C. J. Gerritsma

### Session 5 — Alternative Non-Emissive Displays
### Chairman: R. Engelbrecht

### Session 6 — Display Systems
### Chairman: S. Kobayashi

# LIST OF PARTICIPANTS

E. Arn
Brown Boveri Research Center,
5401 Baden, Switzerland

Dr. G. Baur
Institut für Angewandte Festkörperphysik der Fraunhofer-Gesellschaft,
7800 Freiburg i. Br., Eckerstr. 4,
Fed. Rep. Germany

Dr. D. W. Berreman
Bell Laboratories,
600 Mountain Avenue, Murray Hill,
N. J. 07974, USA

Dr. H. U. Beyeler
Brown Boveri Research Center,
5401 Baden, Switzerland

Dr. T. P. Brody
Research and Development Center,
Westinghouse Electric Corporation,
Beulah Road, Pittsburgh, Pa. 15235,
U.S.A.

Dr. J. Bruinink
N. V. Philips' Gloeilampenfabrieken,
Research Laboratories,
Eindhoven, Netherlands

Dr. R. A. Burmeister, Jr.
Hewlett-Packard, Solid State
Laboratory, 1501 Page Mill Road,
Palo Alto, Cal. 94304, USA

H. Cerutti
3042 Ortschwaben/BE, Switzerland

Dr. I. F. Chang
IBM Thomas J. Watson Research
Center, P. O. Box 218,
Yorktown Heights, N.Y. 10598, USA

J.-D. Cherix
Manufacture ROLEX, Haute Route 82,
2500 Bienne, Switzerland

Dr. J. Cognard
Ebauches S.A.,
2001 Neuchatel, 1, Faubourg
de l'Hopital, Switzerland

Dr. D. Demus
Martin-Luther-Universität Halle-
Wittenberg, 402 Halle (Saale),
Mühlpforte 1, German Dem. Rep.

Dr. P. Deverin
SSIH Management Services SA.,
96 Rue Stämpfli, 2500 Bienne,
Switzerland

H. A. Dorey
The Solartron Electronic Group,
Victoria Road, Farnborough,
Hants. England

Dr. R. Engelbrecht
RCA Laboratories Ltd.,
Badenerstr. 569, 8048 Zurich,
Switzerland

Dr. D. Erdmann
E. Merck, 61 Darmstadt 2,
Postfach 4119, Fed. Rep. Germany

Prof. Dr. A. G. Fischer
Universität Dortmund, 46 Dortmund-
Hombruch, August-Schmidt-Str. 23,
Fed. Rep. Germany

J. Fjortoft
Rank Research Laboratories,
P. O. Box 33, Great West Road,
Brentford, Middlesex, TW8 9AG, England

R. de Fluiter
Brown Boveri & Cie. AG., EKD,
5401 Baden, Switzerland

S. A. Foster
Kent Instruments Ltd., Biscot Rd.,
Luton, Beds., England

Dr. W. G. Freer
Rank Research Laboratories,
P. O. Box 33, Great West Road,
Brentford, Middlesex, TW8 9AG, England

P. George
CETEHOR, C. P. 1145, 25003 Besancon-
Cedex, France

Dr. C. J. Gerritsma
N. V. Philips' Gloeilampenfabrieken,
Research Laboratories,
Eindhoven, Netherlands

Dr. W. J. M. Gissane
ICI Corporate Laboratory,
P. O. Box 11, The Heath, Rumcorn,
Cheshire, England

Dr. T. Gladden
Centre Electronique Horloger S.A.,
2001 Neuchatel, rue Pourtales 13,
Switzerland

Dr. H. Gruler
Universität Ulm, Physikinstitut III,
79 Ulm, Oberer Eselsberg,
Fed. Rep. Germany

Dr. G. Guekos
Institut für Höhere Elektrotechnik,
ETH Zurich, 8006 Zurich,
Gloriastr. 35, Switzerland

Prof. Dr. E. Guyon
Universite de Paris-Sud, Centre d'Orsay,
Laboratoire de Physique des Solides,
91405 Orsay, France

Dr. K. H. Härdtl
Philips Forschungslaboratorium Aachen
GmbH., Postfach 1980, 5100 Aachen,
Fed. Rep. Germany

Dr. M. Hareng
Thomson-CSF, Laboratoire Central de
Recherches, Domaine de Corbeville,
C. P. 10, 91 Orsay, France

Prof. Dr. W. Helfrich
Freie Universität Berlin, 1000 Berlin 33,
Arnimallee 3, Fed. Rep. Germany

Dr. G. Heppke
Technische Universität Berlin,
1000 Berlin 12, Strasse des 17. Juni 135,
Fed. Rep. Germany

Dr. C. Hilsum
Royal Radar Establishment,
St. Andrews Road, Great Malvern,
Worcs. WR14 3PS, England

Dr. M. Hitchman
RCA Laboratories Ltd., Badenerstr. 569,
8048 Zurich, Switzerland

W. Hottinger
Brown Boveri Research Center,
5401 Baden, Switzerland

Dr. J. Hurault
LEP, Laboratoires d'Electronique et de
Physique Appliquee, 3, Avenue Descartes,
94 Limeil-Brevannes, France

Dr. R. Hurditch
Allen Clark Research Center,
The Plessey Company Ltd.,
Caswell, Towcester, Northants.,
England

K. Ichihara
WINGO Co. Ltd., 1-1108 Meishin-Guchi,
Toyonaka, Osaka 561, Japan

J. A. Jaccard
SSIH Management Services SA.,
96 Rue Stämpfli, 2500 Bienne,
Switzerland

Dr. J. Jouannic
Cie. Europeenne d'Horlogerie,
C. P. 1289, 25003 Besancon, France

Dr. F. J. Kahn
Hewlett-Packard, Solid State Laboratory,
1501 Page Mill Road,
Palo Alto, Cal. 94304, USA

Prof. Dr. H. Kelker
Farbwerke Hoechst AG.,
623 Frankfurt/M. 80, Postfach 800 320,
Fed. Rep. Germany

G. Keller
Brown Boveri & Cie. AG., EK,
5401 Baden, Switzerland

Dr. J. Kirton
Royal Radar Establishment,
St. Andrews Road, Great Malvern,
Worcs. WR14 3PS, England

Dr. A. R. Kmetz
Brown Boveri Research Center,
5401 Baden, Switzerland

Prof. Dr. S. Kobayashi
Faculty of Technology,
Tokyo University of Agriculture
and Technology, Koganei,
Tokyo 184, Japan

Dr. U. La Roche
Brown Boveri & Cie. AG., KLG,
5401 Baden, Switzerland

E. Leiba
Thomson-CSF, Laboratoire Central de
Recherches, Domaine de Corbeville,
C. P. 10, 91401 Orsay, France

Dr. J. C. Lewis
Allen Clark Research Center,
The Plessey Company Ltd., Caswell,
Towcester, Northants., England

Dr. M. McDermott
ICI Corporate Laboratory, P. O. Box 11,
The Heath, Rumcorn, Cheshire, England

Dr. G. Meier
Inst. für Angewandte Festkörperphysik
der Fraunhofer-Gesellschaft,
78 Freiburg i. Br., Eckerstr. 4,
Fed. Rep. Germany

Dr. Ch. Naly
Cie. Europeenne d'Horlogerie,
C. P. 1289, 25003 Besancon, France

Dr. J. Nehring
Brown Boveri Research Center,
5401 Baden, Switzerland

Dr. M. Osman
Brown Boveri & Cie. AG., ZLK,
5401 Baden, Switzerland

P. Peraud
Cie. Electro-Mecanique, 12, rue Portalis,
75383 Paris-Cedex 08, France

Dr. A. Perregaux
Brown Boveri & Cie. AG., EKD,
5401 Baden, Switzerland

Dr. C. von Planta
F. Hoffmann-La Roche & Co. AG.,
IV/Chem., 4002 Basel, Switzerland

Dr. L. Pohl
E. Merck, AZL,
61 Darmstadt, Frankfurterstr. 250,
Fed. Rep. Germany

Dr. A. de Quervain
Brown Boveri & Cie. AG., E,
5401 Baden, Switzerland

Dr. E. P. Raynes
Royal Radar Establishment,
St. Andrews Road, Great Malvern,
Worcs. WR14 3PS, England

Dr. J. Robert
CENG, Laboratoire d'Electronique et de
Technologie de l'Informatique,
38 Grenoble, France

Dr. D. Ross
RCA Laboratories,
David Sarnoff Research Center,
Princeton, N. J. 08540, USA

Dr. Y. Ruedin
Ebauches SA.,
2001 Neuchatel,
1, Faubourg de l'Hopital,
Switzerland

Y. Sasanuma
Citizen Watch Co. Ltd.,
Technical Laboratory Shimotomi,
Tokorozawa-Shi, Saitama Pref. 359,
Japan

Dr. E. Saurer
Ebauches SA.,
2001 Neuchatel, 1, Faubourg de
l'Hopital, Switzerland

Dr. M. Schadt
F. Hoffmann-La Roche & Co. AG.,
VI/Chem., 4002 Basel, Switzerland

H. Schaller
Bulova Watch Co., 44, Faubourg du
Jura, 2500 Bienne, Switzerland

Dr. A. Schauer
Siemens AG., Bauelemente und Grund-
lagenentwicklung, 8000 München 80,
Postfach 80 17 09, Fed. Rep. Germany

Dr. T. J. Scheffer
Brown Boveri Research Center,
5401 Baden, Switzerland

Dr. H. Scherrer
F. Hoffmann-La Roche & Co. AG.,
VI/Chem., 4002 Basel, Switzerland

M. F. Schiekel
AEG-Telefunken, Röhren-Abteilung,
79 Ulm, Söflingerstr. 100,
Fed. Rep. Germany

Prof. Dr. E. Schmidt
Brown Boveri & Cie. AG.,
5401 Baden, Switzerland

Dr. F. Schneider
Gesamthochschule Siegen,
593 Hüttental-Weidenau,
Paul-Bonatz-Str. 9-11,
Fed. Rep. Germany

Dr. C. Schüler
Brown Boveri Research Center,
5401 Baden, Switzerland

Dr. R. Schüpbach
Brown Boveri & Cie. AG., E,
5401 Baden, Switzerland

Dr. J. Schwarzmüller
Battelle Research Center,
1207 Carouge-Geneve,
7, Route de Drize, Switzerland

Dr. L. Slama
CERCEM-Laboratoires du Bourget,
49, Rue du Commandant Roland,
93 350 Le Bourget, France

Prof. Dr. A. P. Speiser
Brown Boveri Research Center,
5401 Baden, Switzerland

Dr. H. J. Stein
Brown Boveri & Cie. AG., EKD,
5401 Baden, Switzerland

Dr. S. Strässler
Brown Boveri Research Center,
5401 Baden, Switzerland

J. G. Suard
Timex Corporation,
1, Rue Denis Papin,
2500 Besancon, France

Dr. Y. Tokunaga
CITIZEN WATCH Co. Ltd.,
Shimotomi, Tokorozawa-Shi,
Saitama Pref. 359, Japan

Dr. K. Toriyama
Electron Tube Division,
Hitachi Ltd.,
Mobara Work, Mobara, Japan

Y. Tsuji
CITIZEN WATCH Co. Ltd.,
Technical Laboratory,
Shimotomi, Tokorozawa-Shi,
Saitama Pref. 359, Japan

R. Uriet
Schneider Electronique,
27-33 Rue d'Antony,
94 150 Rungis, France

Dr. L. Vannotti
Brown Boveri & Cie. AG., EKV,
5401 Baden, Switzerland

B. Volken
Zenith S.A., Billodes 34,
2400 Le Locle, Switzerland

Dr. R. Vuilleumier
Centre Electronique Horloger S.A.,
2001 Neuchatel, rue Pourtales 13,
Switzerland

Dr. G. Weibel
Solid State Research,
Zenith Radio Corporation,
1900 North Austin Avenue,
Chicago, Ill. 60639, USA

Dr. C. R. P. Wilcox
BDH Chemicals Ltd., Broom Road,
Poole BH12 4NN, England

P. Wild
Brown Boveri & Cie. AG., EKD,
5401 Baden, Switzerland

Dr. F. K. von Willisen
Brown Boveri Research Center,
5401 Baden, Switzerland

Dr. V. Wittwer
Inst. für Angewandte Festkörperphysik
der Fraunhofer-Gesellschaft,
7800 Freiburg i. Br., Eckerstr. 4,
Fed. Rep. Germany

Dr. H. J. Wullschleger
Brown Boveri & Cie. AG., EKD,
5401 Baden, Switzerland

Dr. B. Zega
Battelle Memorial Institute,
Centre de Recherche de Geneve,
1227 Geneve-Carouge,
7, Route de Drize, Switzerland

Dr. H. R. Zeller
Brown Boveri Research Center,
5401 Baden, Switzerland

# NONEMISSIVE
# ELECTROOPTIC
# DISPLAYS

# REQUIREMENTS ON MODERN DISPLAYS

JOHN KIRTON

Royal Radar Establishment, Malvern, Worcestershire, England

The purpose of this paper is to provide the foundation required to answer such questions as

"Is there an electrooptic technology which can be used to develop a display to perform a particular task?"

"Which display tasks can be carried out satisfactorily using a given electrooptic technology?"

A range of display tasks will be examined and a specification given for the performance that a user might expect. The specification will refer to the whole display (electrooptic transducer plus circuitry), and consideration must be given to the way in which the transducer is addressed before its required electrooptic performance can be determined. Long term aspects of display development must also be taken into account.

It is possible to identify the following classes of displays:

    (a) Indicators, both discrete and in arrays
    (b) Fixed messages and situation diagrams (mimics)
    (c) Extending bars and histograms
    (d) Single short rows of numerals
    (e) Alphanumeric terminals, oscilloscope displays
    (f) Radar displays
    (g) Command and control displays
    (h) TV monitors.

For the first four categories it is necessary to control brightness at up to 100 resolvable locations and in some cases to provide color coding. Category (e) requires up to 50 000 locations and occasionally color, and for (f) up to

500 000 elements and sometimes gray scale and color are needed. Command and control displays are needed in defense and for other complex control tasks. They are frequently used in conjunction with maps or other fixed information and require control of brightness and possibly color at up to $10^6$ locations. For all of these categories it is not necessary to change information in less than 0.25 seconds. For the present ultimate task, the TV monitor, 20 millisecond frame times are required with up to 500 000 elements and a full range of gray scale and color.

Sizes of display in all categories may vary from a few square centimetres to tens of square metres. For all categories an ideal display would be

- Easily legible in all lighting conditions.
  unambiguous and 'fail-safe'
- Cheap
- Compact
- Long lasting
- Usable over a wide temperature range
- Rugged
- Operated with very little power
- Safe

In order to convert display requirements into transducer characteristics it is necessary to consider addressing methods. Information is put into a display in a coded form which may depend on whether the source is a push button, keyboard, computer or other electronic or electrical device. Fewer than ten input conductors will normally be sufficient to carry the information which must then be decoded into a logical form which depends on how the transducer elements are activated. Often the voltage, power or frequency of the output from the decoding circuits is incompatible with the transducer and a matching stage has to be included. If the transducer is an analog device like a cathode ray tube, only one data connection may be needed but an amplifier would be required to step up power and voltage. The quest for flat displays and the increasing use of digital circuits has led to a move toward discrete element transducers. For these it is necessary to provide one connection per element or to use subtle methods such as multiplexing and binary shuttering in order to keep the number of connections within reasonable bounds. In addition voltage, frequency and power levels often have to be adjusted and thus the matching stage can dominate the whole display in terms of cost, weight and bulk, particularly when the display task is complex.

The following three addressing schemes are most commonly used in new display devices:

1) Individual connections can be made to each of a set of discrete elements usually provided with a common ground plane. Where the number of elements is small this is obviously the simplest method. Larger numbers can be controlled when it is possible to bond integrated distribution circuits directly to the lead-offs from the individual transducer elements. For even larger numbers of elements the electrooptic medium is deposited on a large scale integrated circuit which carries out all of the routing. Contact to one

side of the medium is made by pads on the IC and on the other by a single transparent ground plane electrode. So far this approach has been used, not to reduce interconnection difficulties, but to ease certain problems associated with matrix adressing[1]. However, it should be possible to produce wristwatch-sized complex displays with silicon technology and elsewhere in this volume T. P. Brody describes the use of thin film transistors to produce similar results over larger areas. The implications for the transducer are that it must be possible to interface it with the integrated circuit, but otherwise the demands are modest. For compatibility with silicon substrates, low voltage, low power driving is needed, but these requirements are eased for TFT substrates. In both cases, the shortest response time required for any application would be 20 ms.

2) The use of binary screen control is illustrated in Fig. 1. An area source projects electrons or light through a series of N screens, each of which

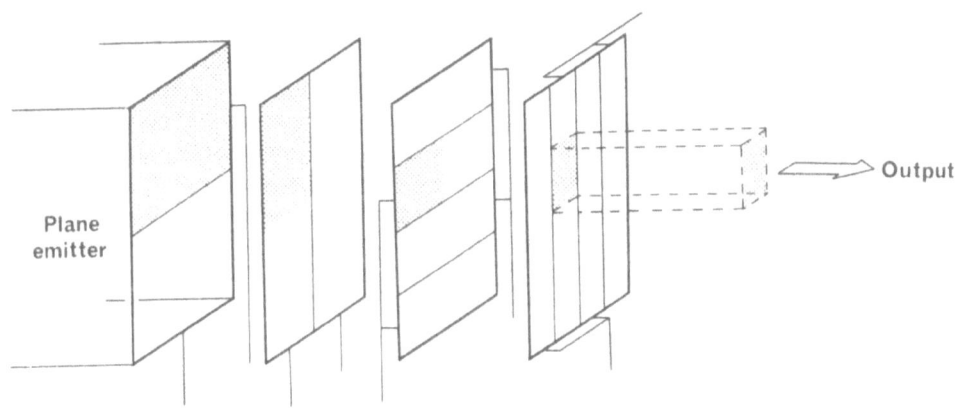

Fig. 1.   The Binary Screen Technique. Each screen halves the area of the incident light or electron beam so that $2^N$ discrete areas may be defined by N screens with up to 3N connections.

has two sets of apertures and 2 or 3 electrical connections. These apertures successively limit the original beam until a single area is selected from $2^N$ possible areas. In this way about 250 000 separately controllable elements may be selected (one at a time) using 18 screens and 36 or 54 electrical connections. The Northrop Digisplay[2], developed a few years ago, employed an area electron source in a flattened vacuum enclosure containing binary screens and using a conventional CRT phosphor. A more recent application of the technique has been the use of gas discharges by Sperry/Beckman[3]. A "pilot" or triggering glow is drawn through screens and induces a latched discharge in the front final section.

In both cases, a frame of information is written one element at a time, which means that the transducer must respond in 100 microseconds or less if a 10 000 element display is to have a frame time of 1 second. Where the transducer has no memory, the addressing scan must be continually repeated, and this implies the same considerable demands on speed and peak brightness which apply to the conventional cathode ray tube. The decay time must therefore be greater than the rise time by a factor equal to the number of elements. Failing this, the peak brightness or contrast must be considerably greater than the acceptable mean of say 100 cd/m$^2$ or 25 to 1 for emissive and non-emissive transducers respectively. Of course, where the transducer has inherent memory the response time ratio does not restrict the number of elements. The required rise time is simply the frame or page time divided by the number of elements. For a 10 000 element data display, this implies a 100 microsecond rise time; for TV, this figure reduces to about 100 nanoseconds. Additionally, the incorporation of a memory introduces difficulties in achieving gray scale. The use of an optical version of the technique seems attractive but many of the available cheap, large area modulators have response times which are too slow.

3) By using two dimensional matrix selection of discrete elements it is possible to make a more modest saving in the interconnections without demanding either such rapid and bright or latched performance from the transducer. An N×M matrix is equivalent to a multiplexed set of N characters each with M elements; both allow NM elements to be accessed through N+M connections. Although it is not possible to write all elements simultaneously, information can be introduced row by row. This imposes less stringent speed and brightness requirements but without a memory, refreshing is necessary. Addition of a memory again reduces required speed and brightness with the penalty of gray scale difficulties. (A particularly attractive invention would be a transducer with a multilevel memory.)

Matrix addressing makes a further demand on the transducer. It will be clear that simple matrix selection only works with a system which has a strongly non-linear electrooptic response to an increasing voltage. The discrimination ratio (response at the operating voltage divided by the response at half the operating voltage) is an important parameter if matrix operation is contemplated. A problem with non-emissive electrooptic effects is a tendency for slow response at voltages which give best discrimination ratio. At higher voltages the non-emissive effects saturate, and higher speed is gained at the expense of deteriorating discrimination ratio.

Clearly it is not possible to write a single list of ideal transducer specifications since these depend on the display task and on how the resolved elements are accessed. However, it is worth summarizing the electrooptic characteristics which influence display performance (Table 1). That simplicity aids cheapness is clear, but it is less obvious that versatile format also leads to lower cost. The aim would be to achieve cheapness by working with a transducer technology which is so adaptable that by simply changing masks and substrate dimensions it would be possible to produce a large number of display formats. The importance of legibility, life and temperature range is obvious. The entries under drive cost follow from the discussion on adressing.

## TABLE 1

### Important Transducer Characteristics

| | |
|---|---|
| 1. Type of Construction | Flatness |
| | Ruggedness |
| | Versatility of format |
| | Simplicity |
| 2. Legibility | Brightness |
| | Color |
| | Contrast over wide viewing sector |
| 3. Life | |
| 4. Temperature Range | |
| 5. Drive Cost | Operating voltage |
| | Operating power |
| | Discrimination ratio |
| | Speed |
| | Refreshed performance |
| | Memory |

Almost every transducer technology excels in at least one important respect, but none offers an ideal solution for any display task. For a particular display example, it is possible to imagine a set of ideal transducer characteristics but unfortunately no available technology can promise these. At present there is considerable research and development directed towards producing improved electrooptic performance, compatible silicon components and new electronic "accessories" like thin film transistor substrates, amorphous switches and photoconductors. In judging the relative merits of present technologies, the following points concerning their long-term viability should be considered in addition to the factors summarized in Table 1.

— At present, most transducer technologies can provide moderately good displays in one or more of the simple categories. Most also have potential for moving towards greater complexity, but none seems very likely to replace the cathode ray tube in domestic television receivers. The TV display field is therefore wide open to a completely new and more suitable approach to the problem.

— Although there are reasons for thinking that the next TV display may employ discrete elements, the problems of producing a competitively priced discrete element TV display so far seem insuperable. But what would be the possible consequences of success? If such a technology could solve the TV problem cheaply, it seems probable that it would be even more competitively priced for less complex tasks. This would be in contrast to the present situation where several technologies successfully coexist with the cathode ray tube. Perhaps a new technology which does not afford a prospect of providing a TV display should be either exploited rapidly or abandoned.

## REFERENCES

1   M. N. Ernsthoff, A. M. Leupp, M. J. Little and H. T. Peterson, IEEE Internat. Electron Devices Meet. Tech. Digest 73 CH 0781-5ED (1973) 548
2   L. A. Jefferies, S.I.D. Int. Symp. Digest 5 (1974) 96
3   C. D. Lustig, A. W. Baird, H. Vernon, J. B. Armstrong and G. Watts. S.I.D. Int. Symp. Digest 5 (1974) 128

## DISCUSSION

**H. A. Dorey (Solartron/Schlumberger)**
Your paper expresses well the viewpoint of the equipment supplier, but we must also consider user acceptance. Even if a display is legible, a customer will choose to buy and use it only if he finds its appearance preferable to competing technologies.

**J. C. Lewis (Plessey)**
User acceptance may be a time-dependent factor. Studies have found that error rates in reading seven-segment numerics are higher among naive subjects than among observers experienced in their use. Likewise there may be an educational time lag before the man-in-the-street will accept a new display.

**J. Kirton**
It is unfortunate that little data exists on this important factor; studies of user preferences are clearly needed.

**A. R. Kmetz (Brown Boveri)**
Would you elaborate on your statement to the effect that a miraculous solution to the flat TV problem would exterminate all competing technologies?

**J. Kirton**
Many technologies now coexist because each one offers a different compromise between advantages and disadvantages which makes it particularly suited to a specific range of display tasks. The problem of finding a discrete element replacement for the television picture tube is of a kind and magnitude that, if a solution can be found, it would drastically upset this coexistence. This is such a big "if" that I can understand your use of the word miraculous. The simple deduction from my argument is that if a given technology does not offer a potential solution to the TV monitor problem it should be exploited quickly.

**C. Hilsum (Royal Radar Establishment)**
I'm not sure I agree. After all, you don't find cathode ray tubes in wristwatches.

J. Kirton

That's just my point. The bulky, high-voltage CRT is no more an ideal display than any other now known. If a cheap, flat low-power, discrete element alternative were developed, it very likely would be applicable to the much simpler requirements of a watch.

# ELECTRICAL AND OPTICAL PROPERTIES OF

# TWISTED NEMATIC STRUCTURES

DWIGHT W. BERREMAN

Bell Laboratories, Murray Hill, New Jersey, USA

## SUMMARY

Key features observed in the optical behavior of Schadt-Helfrich nematic liquid crystal twist cells between polarizers following changes in an applied electric field are explained qualitatively. In addition, recently developed numerical methods for solving the hydrodynamic equations of Erickson and Leslie for planar flow and director reorientation in flat nematic layers with fixed boundaries following changes in field are described. Results of such computations together with numerical solutions of Maxwell's equations for light transmitted through a polarizer, twist cell and analyzer are illustrated. When known or reasonable estimated values of elastic, viscous, dielectric and optical parameters and of director orientation at the surfaces are used, computed optical behavior agrees well with experimental observations. Effects of adjusting cell thickness, applied field and other parameters are discussed.

## 1. INTRODUCTION

Numerical methods have recently been developed[1,2] to compute with high accuracy the optical and hydrodynamic behavior of twist cells[3,4] and other planar nematic or cholesteric liquid crystal devices. This behavior depends on the director orientation at the two plane surfaces (which is probably very nearly field-independent and hopefully uniform), the thickness of the liquid crystal layer, the electric (or magnetic) field applied across the liquid crystal, and the dielectric (or magnetic) anisotropy of the liquid crystal. It also depends on the wavelength and direction of the incident light, the optical anisotropy of the liquid crystal and its three elastic and five viscous parameters. If the liquid crystal is cholesteric the behavior also depends on the twist or pitch that would be assumed by an unstrained region of the material in the planar (Grandjean) texture.

Although there are a large number of parameters in the computation, most of them can be measured more or less independently. Measured or reasonable estimates of parameters yield agreement with observed optical behavior that is within the uncertainty of parameters. Some of the parameters can easily be adjusted to optimize the performance of a twist cell. We have used these numerical methods to obtain a better understanding of how the adjustment of such parameters affects the behavior of Schadt-Helfrich twist cells.

## 2. QUALITATIVE OPTICS OF TWISTED NEMATIC LAYERS

A typical twist cell device consists of a liquid crystal layer whose directors twist by a quarter turn between two parallel conducting electrodes, all sandwiched between two polarizers, one of which is backed by a reflector. The cell either reflects or absorbs light incident from the side opposite the reflector to a degree depending on the applied electric field, and, for a second or so, on the time after that field is changed. An optically equivalent system would be two twist cells that are mirror images of one another operating in series with transmitted light and without the reflector. For simplicity we shall describe the behavior of a single transmission cell. Such a cell is illustrated schematically in Fig. 1.

If there is no field across a quarter-turn twist cell and if polarizer and analyzer are parallel, almost no light will be transmitted because the slow twist of the liquid crystal director across the cell turns the plane of polarization by 90 degrees so that the light is stopped by the analyzer. If the polarizer and analyzer are crossed, nearly all the light transmitted by the polarizer is transmitted by the analyzer. The phenomenon of rotation of plane polarized light by a twisted structure that twists very litte in one optical wavelength was first explained by Mauguin[5].

When a strong electric field is applied across such a cell, the directors, after a time, are all oriented almost normal to the surface except in a region very near each surface which may become so thin at very high fields as to have no significant optical effect. In that case the liquid crystal layer behaves like a layer of uniaxial crystal with its optic axis normal to the surface. When viewed between parallel (or crossed) polarizers with highly divergent light a typical conoscopic figure in the form of a bright (or dark) Maltese cross appears (see Fig. 2). Rays incident in directions skew to the direction of polarizer and analyzer planes show interference fringes that are dependent on angle of incidence. Figure 3 shows detailed computations of the first interference fringe of the Maltese cross pattern at zero and 45° azimuth relative

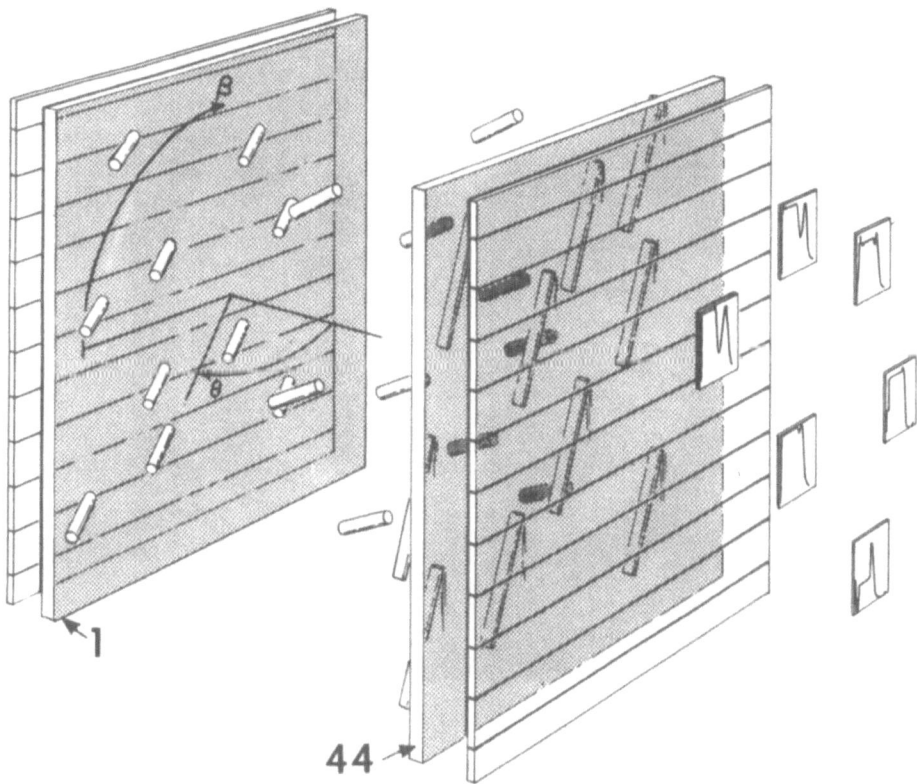

Fig. 1. Model of twist cell with polarizer and analyzer horizontal, enclosing twisted nematic with azimuth ($\beta$), tilt ($\alpha$) and twist in the directions assumed in the computations. Light enters at left. Exit ray directions are indicated by locations of plots in right foreground, which are reproduced on Fig. 7. Imaginary layers used in computations are numbered 1 to 44, as shown.

to the polarizer axis for angles of incidence from $0^\circ$ (normal) to $65^\circ$ for the cell of Baur et al.[6], assuming infinite applied field (homeotropic orientation).

The optical properties of quarter-turn twist cells fall between the limiting Maltese cross for extremely high fields and the almost complete 90 degrees rotation of plane polarized light that occurs when the field is off.

## 3. QUANTITATIVE METHODS FOR DYNAMICS AND OPTICS

Numerical methods to solve the hydrodynamic equations of Leslie[7] and Erickson[8] in twist cells are described by Van Doorn[1] and by Berreman[2]. The optical transmission through each instantaneous configuration of directors was computed using a numerical 4x4 matrix technique for solving Maxwell's equations that is correct to any desired accuracy[9,10].

Fig. 2. The Maltese cross conoscopic figure of a liquid crystal layer in homeotropic orientation, taken with polarizer and analyzer parallel using a very wide cone of light rays. (Photo courtesy of P. Cladis).

The equations of Leslie and Erickson for planar motion take on the form of the following four simultaneous differential equations. The first two represent balance of shear force, neglecting inertial effects that can be shown to be negligible:

$$\sigma_{zx} = T_{11}V'_x + T_{12}V'_y + T_{13}\dot{\theta} + T_{14}\dot{\beta} , \tag{1}$$

$$\sigma_{zy} = T_{21}V'_x + T_{22}V'_y + T_{23}\dot{\theta} + T_{24}\dot{\beta} . \tag{2}$$

Cartesian coordinates with layer thickness of the z-direction are assumed. The $\sigma$'s are constant shears independent of z to be evaluated at each instant of time. The primes represent differentials with respect to z and the dots represent differentials with respect to time. $V_x$ and $V_y$ are flow velocity components. The other two equations represent a balance of elastic, viscous, and electric torques in the absence of appreciable angular momentum change:

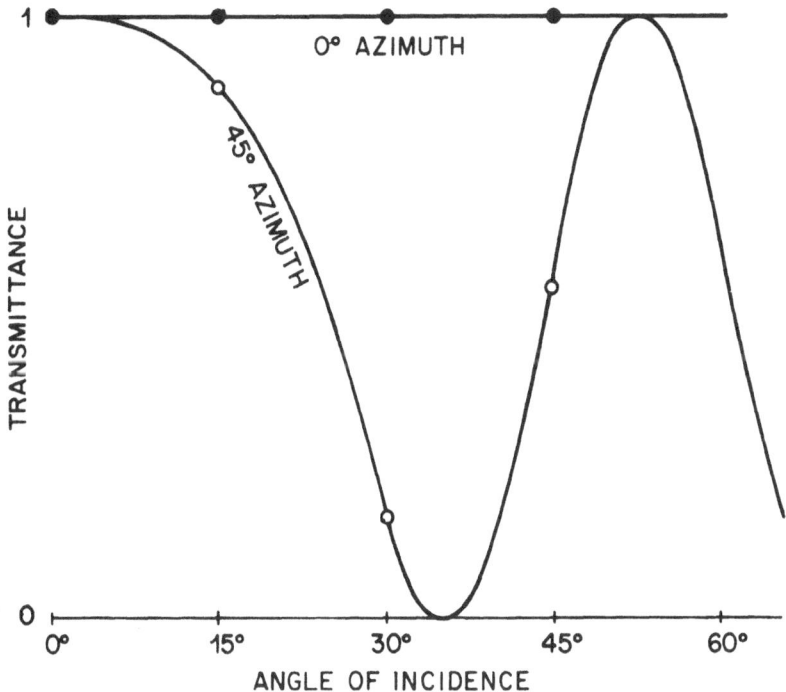

Fig. 3. Detailed plot of Maltese cross transmittance pattern versus angle of incidence for $V/V_c = \infty$, along horizontal or vertical azimuths (almost unity) and along either bisecting azimuth. $0^\circ$, $15^\circ$, $30^\circ$ and $45^\circ$ values, relating to succeeeding figures, are marked.

$$\lambda_1(z) = T_{31}V'_x + T_{32}V'_y + \gamma_1\dot{\theta} \qquad (3)$$

$$\lambda_2(z) = T_{41}V'_x + T_{42}V'_y + \gamma_1\dot{\beta} . \qquad (4)$$

All of the T's are functions of the director orientation angles $\theta$ and $\beta$, which depend on z at any instant and on the viscosities and elastic constants. They are given explicitly in Ref. 2. $\theta$ is the polar angle measured from the z-axis and $\beta$ is the azimuthal angle with respect to the x-axis. The $\lambda$'s are defined by

$$\lambda_1 = A_1\theta'' + A_2(\theta')^2 + A_3(\beta')^2 + A_4\beta'\beta'_0 + A_5(D_z^2/4\pi) ,$$

$$\lambda_2 = B_1\beta'' + B_2\theta'\beta' + B_3\theta'\beta'_0 .$$

The A's and B's are functions of $\theta$, $\beta$ and the elastic constants which are also given in Ref. 2. When the backflow is to be ignored, (1) and (2) are not used and the equations of motion are $\gamma_1 \dot{\theta} = \lambda_1(z)$ and $\gamma_1 \dot{\beta} = \lambda_2(z)$.

The $4 \times 4$ matrix optical method consists in computing a $4 \times 4$ matrix P such that

$$\begin{pmatrix} E_x \\ H_y \\ E_y \\ -H_x \end{pmatrix}_{\text{second surface}} = \underline{\underline{P}} \times \begin{pmatrix} E_x \\ H_y \\ E_y \\ -H_x \end{pmatrix}_{\text{first surface}}$$

where E and H are optical electric and magnetic field vectors. In first approximation, $\underline{\underline{P}} = \underline{\underline{P}}_n \cdots \underline{\underline{P}}_2 \underline{\underline{P}}_1$, where $\underline{\underline{P}}_i$ is a $4 \times 4$ matrix whose elements depend only on the wavelength and angle of incidence from vacuum of light, the thickness and the local optical dielectric tensor components in liquid crystal layer i which is thin enough that variation of the tensor is negligible. Layers 1 to n make up the whole liquid crystal layer[10].

## 4. NUMERICAL RESULTS FOR A TYPICAL CELL

We consider a quarter-turn twist cell with polarizer and analyzer oriented to transmit light with the electric vector nearly parallel to the plane of the directors adjacent to the first surface encountered by the light beam. Equilibrium and dynamic optical transmission by such a cell for light at some angles of incidence are described in a paper by Baur et al.[6] They determined that the physical parameters in Table 1 would best account for the optical properties they observed. Their liquid crystal was MBBA doped with two percent by weight of demethylaminobenzonitrile to give it positive dielectric anisotropy. Their interpretation was based on a simplified theory with only one viscosity parameter ($\gamma_1$) and without backflow[11]. We have used the five independent viscosity parameters of the Leslie-Erickson theory in proportions measured by Gähwiller[12], but scaled to give the value of $\gamma_1$ determined by Baur et al., as a basis for our computations.

Figure 4 shows the evolution of director configuration with time marked at 10 of 44 equally spaced levels within the cell, including the two fixed boundaries, when 17.5 volts is turned on (left) and then off (right), according to computations with parameters in Table 1.

Figure 5 shows transmittance of a single cell relative to that of the first polarizer alone, computed using the parameters in Table 1. The cell is initially at equilibrium with no applied field. A potential difference of 17.5 volts is applied for half a second, which is enough time to approach a new equilibrium configuration very closely, and then is turned off and transmission followed for another half second. The location of each plot in the figure represents the azimuth and angle of incidence of the light. Plots along

## TABLE 1

### Physical Parameters Used in Computations (CGS Units)

| | |
|---|---|
| $\theta_{ends}$ | $82^o$ |
| $\beta_{ends}$ | $2^o, 88^o$ |
| $\epsilon_1$ | 5.61 |
| $\epsilon_2$ | 5.33 |
| $k_{11}$ | $2.91 \times 10^{-7}$ |
| $k_{22}$ | $2.03 \times 10^{-7}$ |
| $k_{33}$ | $3.70 \times 10^{-7}$ |
| $\alpha_1$ | 0.0359 |
| $\alpha_2$ | -0.4283 |
| $\alpha_3$ | -0.0066 |
| $\alpha_4$ | 0.4598 |
| $\alpha_5$ | 0.2559 |
| thickness | 0.0014 |
| $n_{glass}$ | 1.5165 |
| $n_e$ | 1.743 |
| $n_0$ | 1.563 |
| wavelength | $6.328 \times 10^{-5}$ |

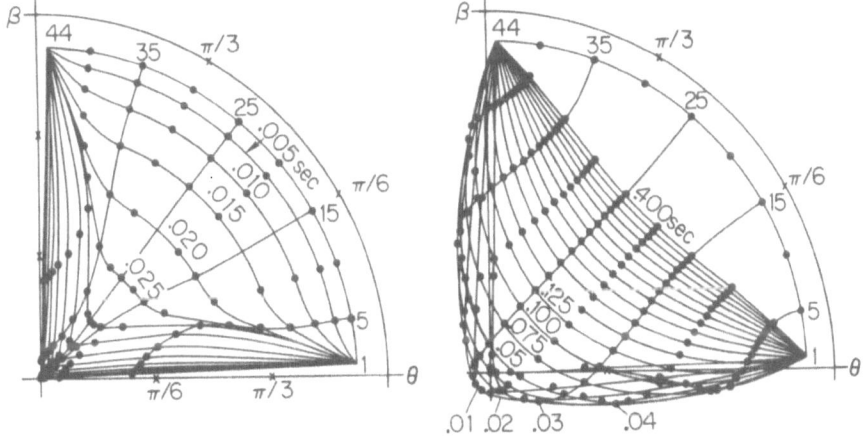

Fig. 4. Director orientation at various instants of time in the $\theta$, $\beta$ coordinate system, computed at 44 equally spaced levels using parameters in Table 1. The first level is at z = 0 and last at z = 14 $\mu$m. 17.5 V is switched on at t = 0 in the left figure, off from new equilibrium at t = 0 on the right.

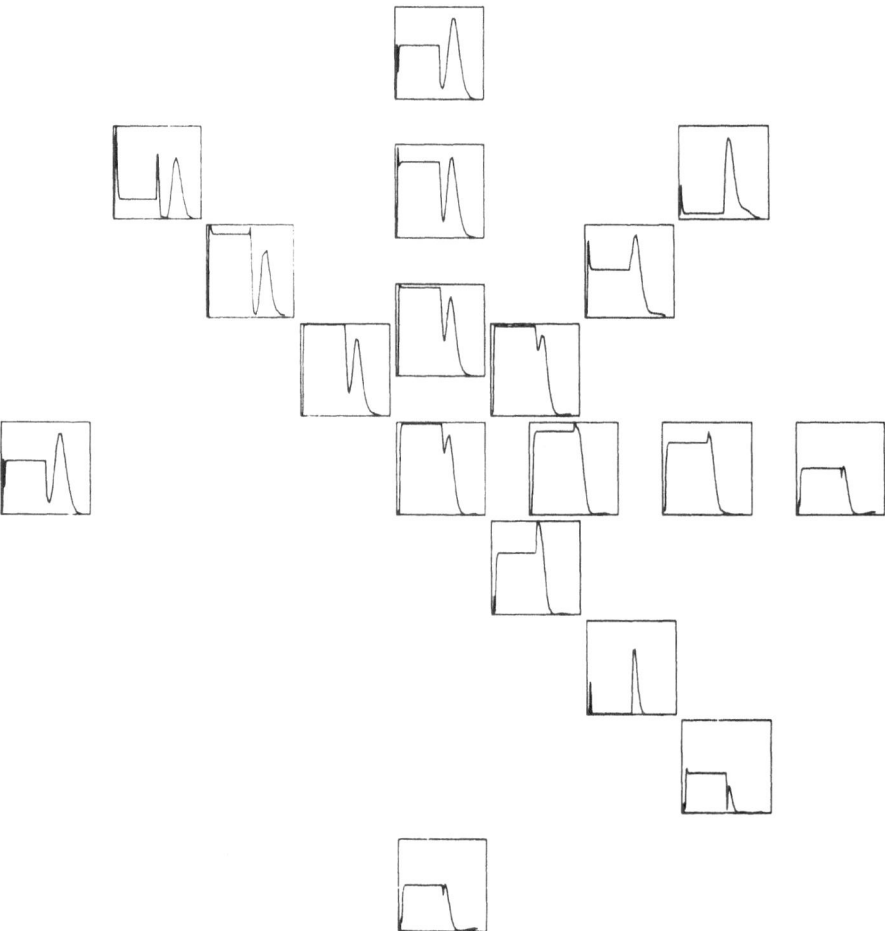

Fig. 5. Each curve shows transmittance as a function of time for a nematic twist cell with director orientation given in Fig. 4. The cell starts at equilibrium at zero volts, is switched on to 17.5 V for half a second, then off for half a second. Angles of incidence of 0°, 15°, 30° and 45° at various azimuths are indicated by placement of curves.

the vertical axis, for instance, are for light incident in the plane of polarization by the polarizer at angles of incidence -45°, 0°, 15°, 30° and 45° from normal.

There is very close symmetry in transmittance about an azimuthal plane bisecting the second and fourth quadrants. This symmetry can be explained by noting that transmittance is unchanged if the direction of propagation through the cell is reversed and that it depends little on whether the light is polarized parallel or perpendicular to the first directors it encounters, if the optical anisotropy is not very large. A beam incident at a given angle sees the same sequence of director orientations as a beam propagating in the reverse direction at the conjugate angle with respect to the symmetry plane but has opposite polarization relative to the first directors it encounters. Therefore the transmission at conjugate angles will be the same. This symmetry is not as exact if directors are tilted and twisted by differential amounts at equal distances from the two cell surfaces, or if the optical anisotropy is very large.

In these computations, we have adopted a fixed surface tilt angle of 8° and a slight azimuthal relaxation of 2°, so that the total twist is from 2° to 88° rather than 0° to 90°. These angles were found by Baur et al. to describe the properties of their cell most accurately. The relationship between tilt, twist and transmission directions is illustrated in Fig. 1.

17.5 volts is 5.2 times the critical voltage (3.36 V) below which tilt angles would relax to zero if the surface directors were exactly parallel to the surfaces. Even at this high voltage, there is large asymmetry in the Maltese cross optical transmittance pattern. Also, the principal axes do not remain fully transmitting with obliquely incident light. This is because the transition region from parallel to almost normal orientation of directors is not very short compared to the optical wavelength. In the "off" configuration the transmittance is always nearly zero, showing that the Mauguin approximation of slow twist is good then.

## 5. EFFECT OF ALTERING MATERIAL PARAMETERS

We now examine the effects of altering various parameters, one at a time. When the surface director tilt is assumed to be only two degrees rather than eight, it takes longer for the molecules to tilt upward when a field is applied (see Fig. 6). If there were no tilt and no random fluctuations there would be no torque, and rise time would theoretically be infinite. There is no appreciable change of the time required to return to the optical properties of the "off" configuration.

In Fig. 7 we show what would happen if the optical anisotropy (birefringence) were only half as great. The optical properties for rays at 30° from normal are quite similar to what would be obtained at about 20° with the anisotropy of MBBA. In particular, the interference fringes of the skew rays of the Maltese cross pattern occur at larger angles. This should be an advantage in most applications.

In Fig. 8 we halve the thickness rather than the optical anisotropy and observe almost the same optical effects. However, the times involved are all divided by four, which would also usually be advantageous. The equations of motion show that rise and decay times vary as the square of the thickness.

In Fig. 9 we show the effect of reducing $V/V_c$ from 5.2 to 2. Rise time is then greatly increased, and is roughly equal to the time required to return to the "off" configuration. Fall time is somewhat reduced because the director configuration in the cell is less distorted by the field. Optical properties in the "on" configuration are much more asymmetric and distorted from the ideal Maltese cross than with $V/V_c = 5.2$.

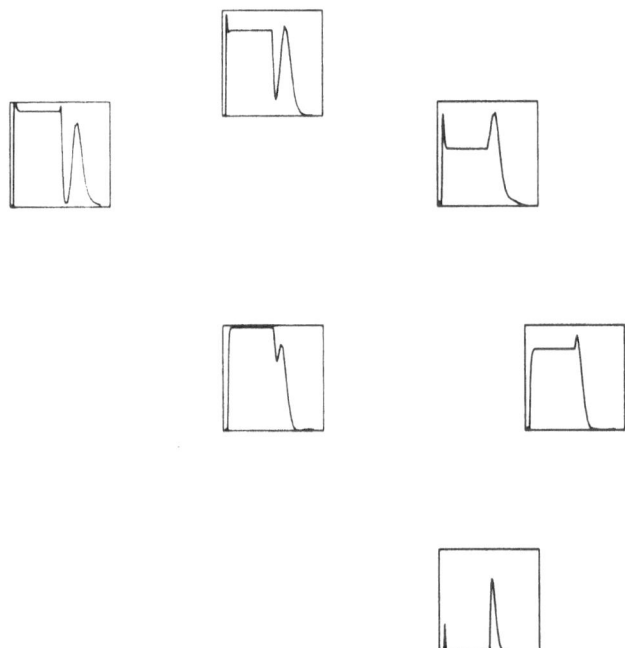

Fig.6. Similar to Fig. 5 except that directors at surface were tilted at $2^O$ rather than at $8^O$. Rise time is roughly doubled; little change otherwise. Angles of incidence are $0^O$ and $30^O$.

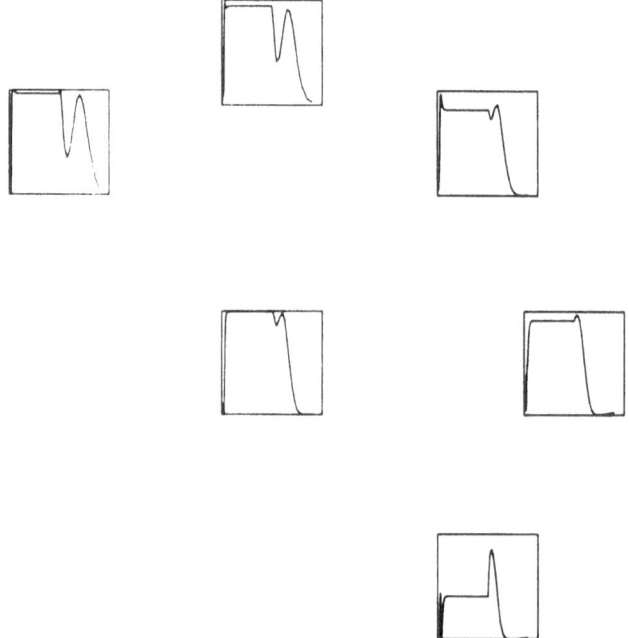

Fig. 7. Similar to Fig. 5 except that optical anisotropy is reduced to half. Optical features at $30^O$ angle of incidence closely resemble those for material of Fig. 5 at about $20^O$.

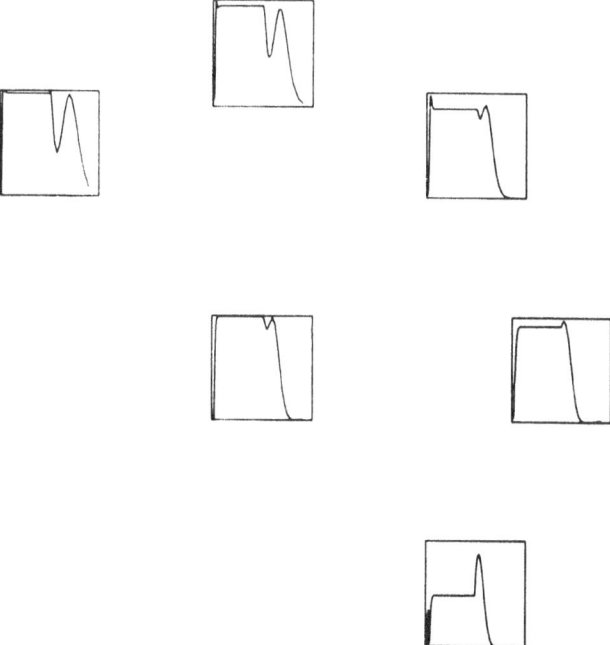

Fig. 8. Similar to Fig. 5 except that sample thickness is only 7 μm and total time is only 1/4 second. Except for faster time scale, curves closely resemble those in Fig. 7.

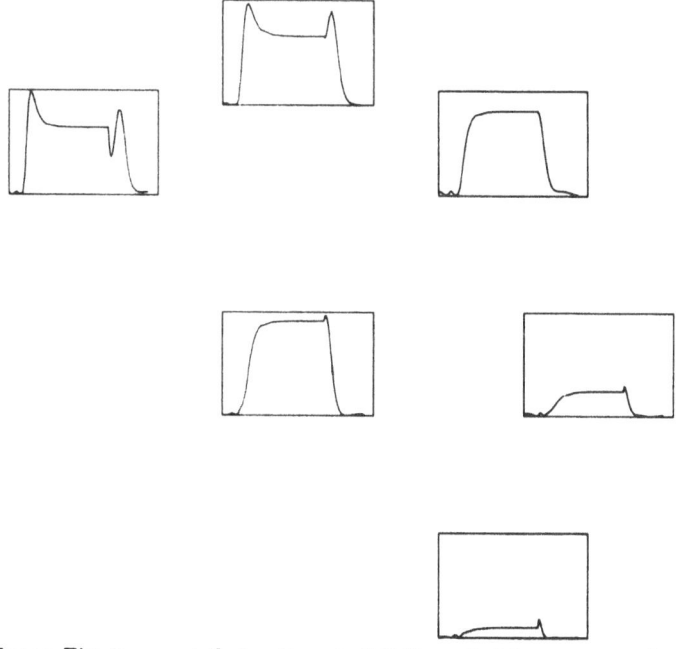

Fig. 9. Similar to Fig. 5 except that voltage is 6.7 V applied for one second rather than 17.5 V for half a second. Rise times are much longer and azimuthal asymmetry of optical features at equilibrium is larger. Contrast between "on" and "off" is generally diminished. Decay times are somewhat reduced.

There has been a considerable amount of discussion of the importance of backflow in the dynamic optical properties of twist cells. Figure 10 shows the transmittance computed with backflow halted but all other parameters the same as in Fig. 5. The small transmission peak for normally incident light that occurs shortly after the voltage is turned off is absent without backflow. In most directions, however, the fluctuations in optical transmission after the field is turned on or off are similar whether backflow is included in the computation or not. Time constants predicted without backflow are somewhat larger than when it is included.

Some of the parameters in Table 1 were obtained by adjusting parameters in curves such as those along the horizontal axis in Fig. 10 to fit observed data. The moderate differences between the curves along the horizontal axis in Fig. 10 and Fig. 5 give an indication of the error in $\gamma_1$ caused by neglecting backflow. Parameters obtained through equilibrium transmission measurements were not affected. The surface tilt was obtained by fitting both equilibrium data and dynamic data and the two values were consistent.

There is another phenomenon that is important in twist cell device design that cannot be even qualitatively accounted for if backflow is ignored. In twist cells made with liquid crystals having frequency-dependent reversible dielectric anisotropy, a reversal of tilt and twist has been observed when the directors are forced down by changing the applied field frequency so that the dielectric anisotropy becomes negative[13,1]. Without backflow, the directors never move outside the first quadrant, as they do on the right side of Fig. 4. Backflow gets them started in the wrong direction when they are being forced down[1] unless the field applied to stand them on end was quite weak.

There have been suggestions that the addition of a cholesteric dopant to the nematic liquid crystal in a twist cell might improve performance by reducing optical bounce and eliminating uncertainty in the sense of twist. Figure 11 shows the effect of adding a quarter turn of natural twist to the other parameters. The small bounce in transmittance of normally incident light largely disappears and the bounce at oblique incidence is reduced[2]. There is an appreciable, but not a large, reduction in time required for the cell transmittance to drop back from the "on" to the "off" value. A slight natural cholesteric twist seems to be advantageous in twist cell devices.

Figure 12 shows the transmittance at equilibrium configuration as a function of $V/V_c$ on a scale from zero to 4, with a vertical line at $V = V_c$. Even though the tilt is $8°$ rather than zero so that there is no true critical voltage, the optical properties change quite abruptly at $V = V_c$. The dash at the right on each graph is located at the position for $V/V_c = 4$ but actually represents the transmittance for infinite voltage; the ideal Maltese cross pattern.

The transmittance at equilibrium as a function of $V/V_c$ depends on ratios of elastic constants but not on their absolute values. Although the ratios of elastic constants for MBBA are typical of many nematic liquid crystals, the relative value of the bend constant $k_{33}$ may vary over a wide range, particularly close to a nematic-to-smectic transition temperature where it may

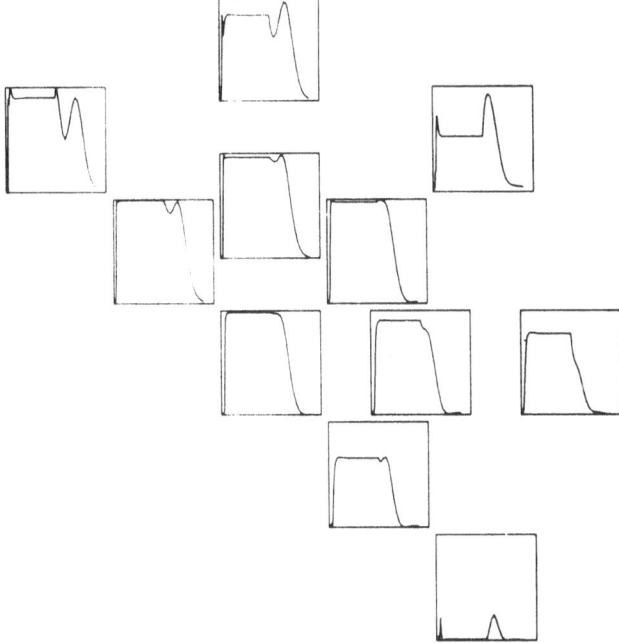

Fig. 10. Similar to Fig. 5 except that backflow has been ignored. Overall rise and decay times are lengthened by several percent, and some but not all optical bounce features disappear.

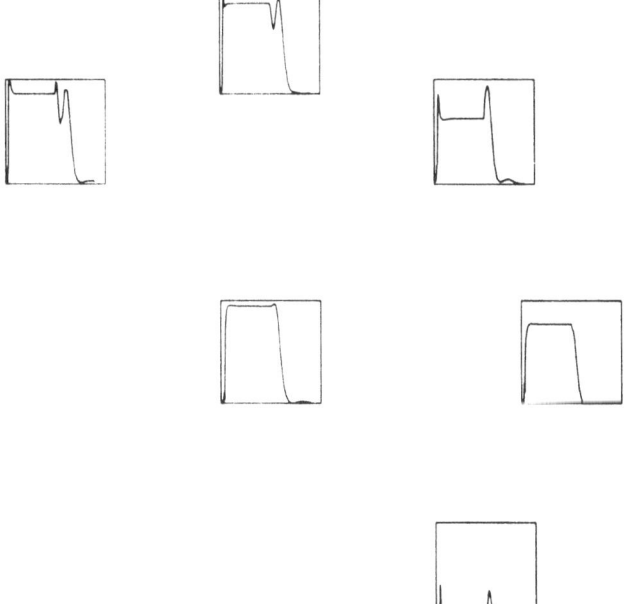

Fig. 11.  Similar to Fig. 5 but a quarter turn of strain-free cholesteric twist is assumed over the cell thickness. Curves show less backflow-associated optical bounce, but rise and decay times are shortened by backflow.

D. W. BERREMAN

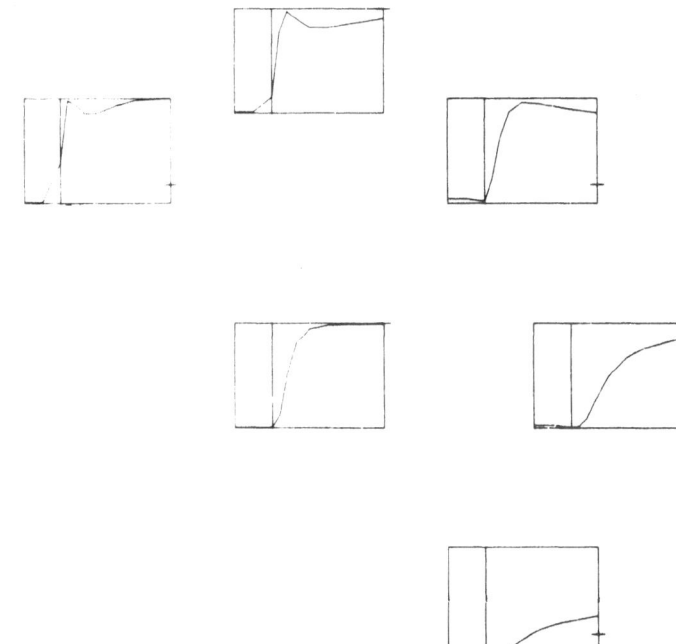

Fig. 12. Equilibrium transmittance as function of $V/V_c$. Parameters from Table 1. Abscissa length is 4 and vertical line is at $V/V_c = 1$. Right-hand dash represents the high-voltage Maltese cross limit.

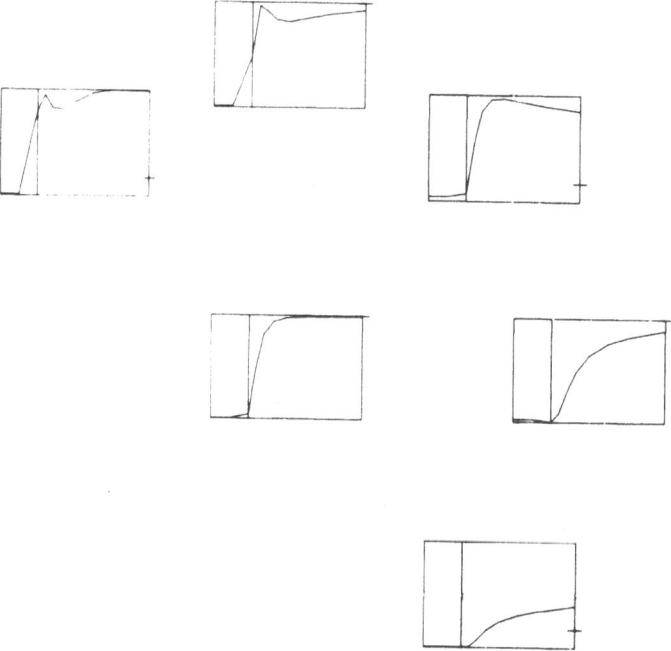

Fig. 13. Same as Fig. 12 except that $k_{33}$ is halved.

become very large. We computed transmittance curves with values of the twist constant $k_{22}$ doubled and halved and found only very minor changes. Thus only the ratio $k_{11}/k_{33}$ has a significant effect on transmittance. Figure 13 shows the effect of doubling the ratio by halving $k_{33}$, while Fig. 14 shows the effect of halving the ratio by doubling $k_{33}$. As expected, changing the ratio by altering $k_{11}$ gave almost identical curves, which are not shown.

Figure 13 shows somewhat more abrupt transitions than Fig. 12, which suggests that increasing $k_{11}/k_{33}$ might be useful in maintaining high contrast in multiplexed displays. Figure 14 shows more gradual transitions that might be useful in half-tone displays, although it is clear that viewing angles would have to be severely restricted except perhaps along the 45° azimuthal direction to show proper half-tones.

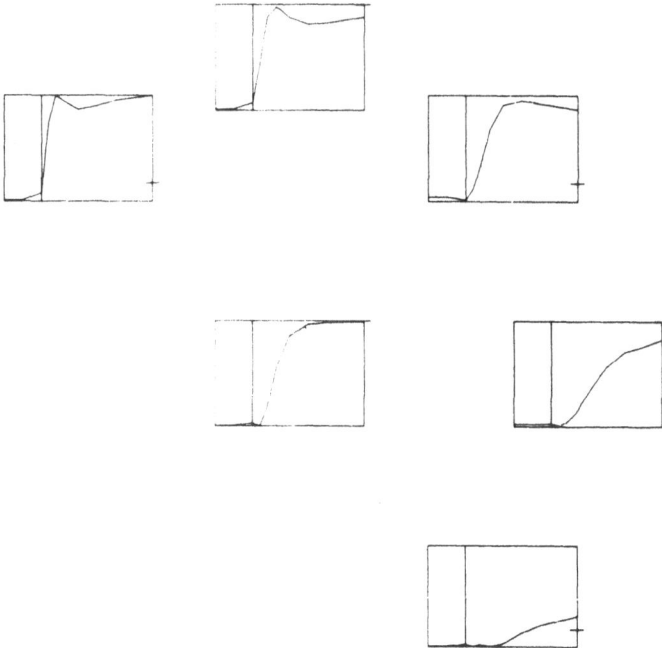

Fig. 14. Same as Fig. 12 except that $k_{33}$ is doubled.

## REFERENCES

1   C. J. Van Doorn, J. Appl. Phys. 46 (1975) 3738
2   D. W. Berreman, J. Appl. Phys. 46 (1975) 3746
3   M. Schadt and W. Helfrich, Appl. Phys. Lett. 18 (1971) 127
4   W. Helfrich, Mol. Cryst. Liq. Cryst. 21 (1973) 187
5   C. Mauguin, Bull. Soc. Fr. Mineral. 34 (1911) 71
6   G. Baur, F. Windscheid and D. W. Berreman, Appl. Phys. 8 (1975) 101
7   F. M. Leslie, Arch. Ration. Mech. Anal. 28 (1968) 265

8  J. L. Erickson, Trans. Soc. Rheol. 5 (1961) 23

9  D. O. Smith, Opt. Acta 12 (1965) 13

10  D. W. Berreman, J. Opt. Soc. Am. 63 (1973) 1374

11  D. W. Berreman, Appl. Phys. Lett. 25 (1974) 12 and 321

12  C. Gähwiller, Phys. Lett. A36 (1971) 311

13  C. J. Gerritsma, J. J. M. J. de Klerk and P. van Zantan, Solid State Comm. (to be published)

## DISCUSSION

A. G. Fischer (Universität Dortmund)

Backflow would be impeded by the end walls in a twisted nematic cell with small lateral dimensions. A "compartmentalized" display panel could be fabricated by the vacuum deposition techniques described in my paper to suppress backflow effects without a cholesteric additive.

E. Guyon (Université de Paris-Sud)

Although backflow produces some undesirable consequences in twist cells, it reduces the effective viscosity of the liquid crystal. That is why viscosities measured by ultrasonic attenuation are much lower than those measured in thin cells where wall effects impede flow. Therefore, instead of attempting to reduce backflow, it might be desirable to enhance it in order to obtain faster response. This might be done by applying a polymeric coating to relax the non-slip condition for flow at the boundaries.

# CHOLESTERIC TEXTURE AND PHASE CHANGE EFFECTS

E. P. RAYNES

Royal Radar Establishment, Malvern, Worcestershire, England

## SUMMARY

This paper describes the more useful ways of using thin layers of cholesteric liquid crystals as display devices. There are two metastable configurations of thin layers of cholesterics, the Grandjean (or planar) texture and the focal conic texture. The transformation between textures by either an electric field or a thermal process forms the basis of the texture change effect. It is also possible by applying an electric field to completely unwind the cholesteric helix to form a pseudo-nematic. This phase change effect has several unusual features which make it attractive for use in displays: the materials can be tailored to operate over a wide range of threshold voltages, the response times are shorter than in the conventional liquid crystal displays, and it is possible to multiplex 100 characters. The dynamics of the phase change are complex, and under certain conditions there are potentially useful hysteresis effects. An important technique for improving the legibility of cholesteric devices is the conversion of the scattering into absorption by using pleochroic dyes. It is possible that the use of dyes in phase change devices will produce displays with equivalent contrast but much higher brightness than twisted nematic devices. All these electrooptic effects are critically compared, where possible also with other liquid crystal devices. The present state of the art and future trends are examined.

## 1. CHOLESTERIC COMPOUNDS

Although electrooptic effects using nematic liquid crystals are the more widely studied and commercially exploited, cholesteric liquid crystals provide many interesting and useful alternative phenomena which are being intensively investigated. The helical arrangement of the director in a cholesteric liquid crystal is compared in Fig. 1 to the unidirectional arrangement within a nematic material. It is usually more convenient to illustrate the orientation in the cholesteric phase by a helix (Fig. 2).

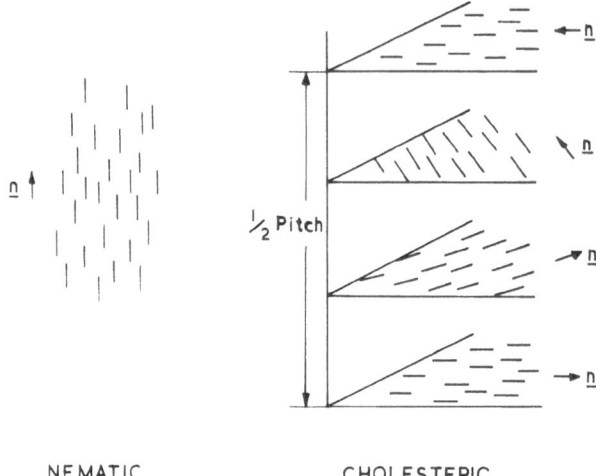

NEMATIC                              CHOLESTERIC

Fig. 1.  Diagrammatic representations of a nematic and a cholesteric liquid crystal.

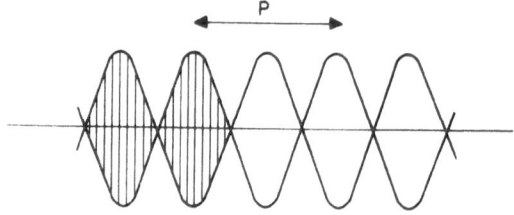

Fig. 2. The cholesteric helix and pitch length P.

Many pure cholesteric liquid crystals exist with pitch lengths $P \approx 0.1$ $\mu$m but for display applications cholesterics with longer pitches (1 - 10 $\mu$m) are required. They must also have well defined dielectric anisotropies. It is therefore common to use in displays cholesteric mixtures formed by adding small amounts of pure cholesterics (1 - 10% by weight) to a nematic liquid crystal. The cholesteric additive can even be a pseudo-cholesteric in the sense that, although it does not itself show a cholesteric liquid crystal phase, its virtual clearing point is not too far below its melting point. There are two main types of cholesterics which can be used:

1) Esters of cholesterol

2) Nematic molecules with a branched alkyl chain

$$CH_3\ CH_2\ CH(CH_3)\ CH_2\ O \text{—}\bigcirc\text{—}\bigcirc\text{—}\ CN\ ,$$

i.e. 2Me4OCB.[1] The pitch of a mixture is inversely proportional to the concentration c of the cholesteric additive:

$$1/P \propto C. \tag{1}$$

Figure 3 illustrates this for mixtures of cholesteryl nonanoate (CN) and the nematic compound 4'-n-pentyl-4-cyanobiphenyl (5CB).[2] By suitable choice of the nematic material it is possible to produce cholesteric mixtures with either positive ($\epsilon_\parallel > \epsilon_\perp$) or negative ($\epsilon_\parallel < \epsilon_\perp$) dielectric anisotropy.

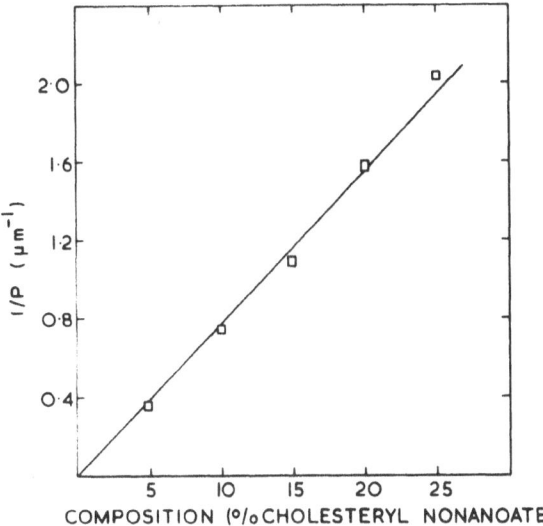

Fig. 3. Variation of pitch P with concentration of CN in 5CB at $23^\circ$ C.

There are two important effects of an electric field on a cholesteric material (Fig. 4). One is the dielectric torque which has a direction governed by the sign of the dielectric anisotropy and exists for frequencies of the applied field up to 100 kHz. The other effect is the general disruptive effect of fluid flow caused by movement of space charges when a low frequency (< 1 kHz) signal is applied. This is somewhat similar to dynamic scattering observable in certain nematic materials.

## 2. TEXTURES AND TEXTURE CHANGE EFFECTS

Figure 5 shows the two basic arrangements or textures of thin layers of cholesteric liquid crystals. In the Grandjean or planar texture the helical axis is perpendicular to the thin layer, and there is no scattering of light by the layer. The alternative arrangement is the scattering, focal conic texture

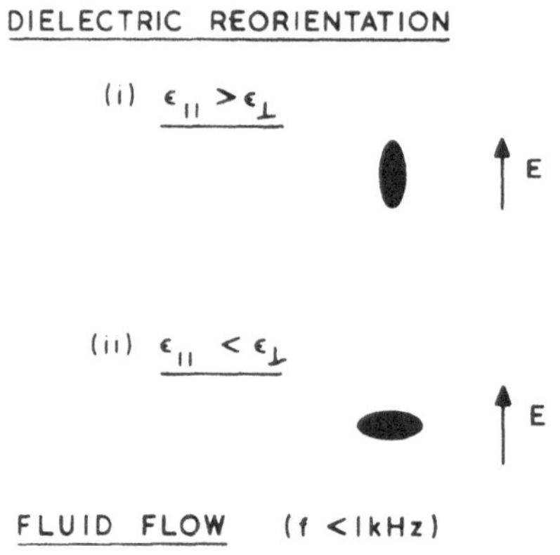

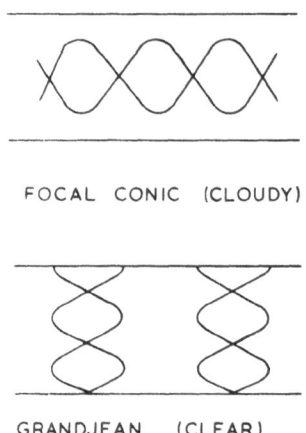

DIELECTRIC REORIENTATION

(i)  $\epsilon_{\parallel} > \epsilon_{\perp}$

↑ E

(ii)  $\epsilon_{\parallel} < \epsilon_{\perp}$

↑ E

FLUID FLOW   ( f < IkHz )

GRANDJEAN → FOCAL CONIC

Fig. 4. The two interactions of an electric field with a cholesteric.

FOCAL CONIC  (CLOUDY)

GRANDJEAN   (CLEAR)

Fig. 5. The two metastable textures of a thin cholesteric layer.

which can be very approximately described as having the helical axis parallel to the plane of the layer. In practice there are several variations of the scattering texture, some of which are well defined, but these details are not relevant here.

Although surface alignment can have some influence on the situation, both the Grandjean and focal conic textures are essentially metastable. A transition between the two textures therefore produces a change in the optical properties which remains for several hours or days. The transition was first switched electrically by Heilmeier et al.[3] who used a cholesteric mixture with $\epsilon_{\parallel} < \epsilon_{\perp}$. When a low frequency signal ($\lesssim$ 1 kHz) is applied, fluid flow converts the clear Grandjean texture into the scattering focal conic texture.

This remains until a high frequency signal ($\gtrsim 2$ kHz) is applied, producing a dielectric torque which returns the layer to the clear Grandjean texture. The effect has not found many applications in devices, partly because of the slow response and high voltages required. For example the Grandjean-to-focal conic transition requires about 30 $V_{rms}$ and takes 10 ms, and the reverse transition requires 100 $V_{rms}$ at 2 kHz to switch in 100 ms.

It is also worth noting that a thermally-induced Grandjean-focal conic transition has been used[4] in a high resolution light valve with infrared beam addressing. A resolution of 50 lines per mm over a $4 \times 4$ cm area with an addressing speed of $10^4$ picture elements per second is reported. A high frequency signal ($\gtrsim 2$ kHz) is used to erase the information and return the layer to the Grandjean texture.

### 3. CHOLESTERIC-NEMATIC PHASE CHANGE

When an electric field is applied to a cholesteric with positive dielectric anisotropy, the stable configuration is that with the helix axis normal to the field as in the scattering texture of Fig. 6. As the field is increased the dielec-

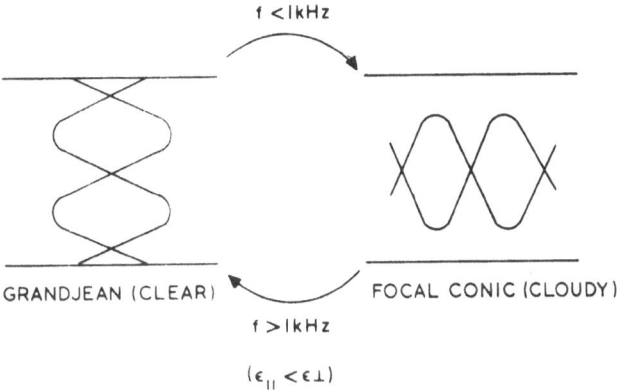

Fig. 6. The electrically induced texture change effect in a cholesteric with $\epsilon_\| < \epsilon_\perp$.

tric torque progressively unwinds the helix until the pitch becomes infinite at a critical field $E_c$ (Fig. 7). Above $E_c$ the material is essentially a homeotropically aligned nematic which does not scatter light; removal of the field restores the scattering cholesteric texture. The value of $E_c$ was first given by de Gennes[5]

$$E_c P_0 = \pi^2 \sqrt{\frac{k_{22}}{\epsilon_0(\epsilon_\| - \epsilon_\perp)}}, \tag{2}$$

where $P_0$ is the pitch in zero field. As a liquid crystal electrooptic effect, the phase change has several unusual features. The transition occurs at a threshold field, not at a threshold voltage, and the value of the threshold field can

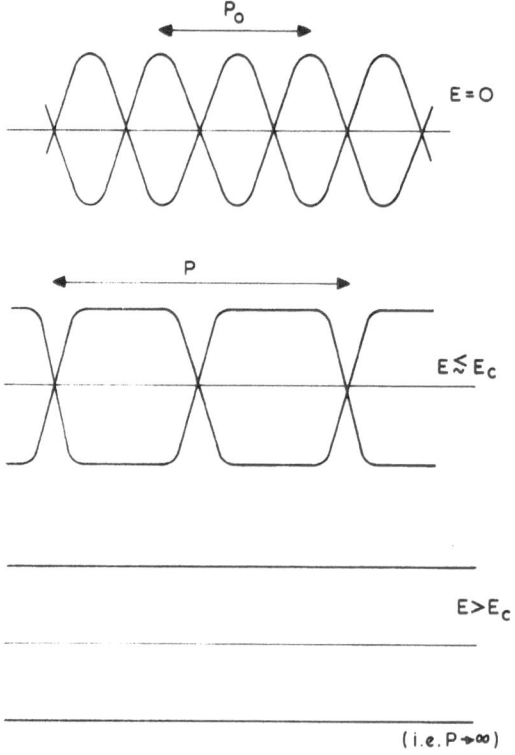

Fig. 7. The electrically induced phase change effect in a cholesteric with $\epsilon_\parallel > \epsilon_\perp$.

be varied over a wide range by altering the composition of the cholesteric mixture. From equations (1) and (2) the threshold voltage $V_c$ is given by:

$$V_c \propto c\, d, \tag{3}$$

where d is the thickness of the layer. Figure 8 shows the linear variation of $V_c$ with concentration c of CN in 5CB[2] for $12\,\mu m$ layers. Even higher threshold voltages can be obtained by including more cholesteric in the mixture or by using thicker layers. A high threshold field implies shorter response times,[6,7] and these can be as fast as 1 ms. This quick response together with the hysteresis described below makes the phase change very suitable for use in complex multiplexed displays.

## 4. DYNAMICS OF THE PHASE CHANGE

Although several authors[6-12] have studied the dynamics of the cholesteric-nematic phase change, many of the details still remain obscure. We therefore present a rather simplified picture in an attempt to explain the main features. The transition from cholessteric to nematic by the progressive unwin-

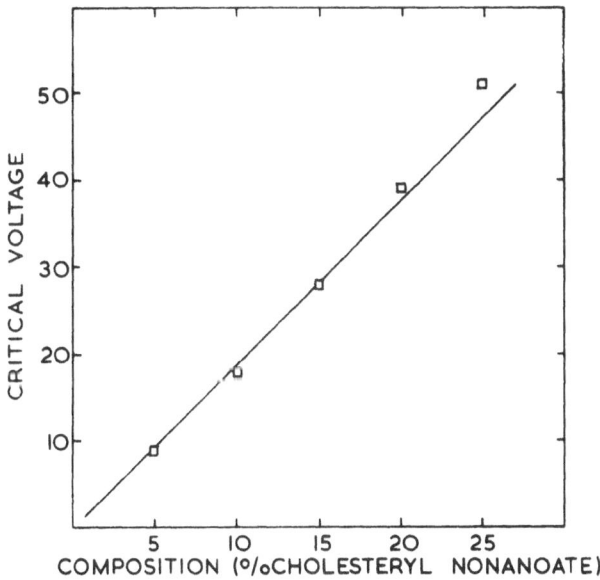

Fig. 8.  Variation of threshold voltage $V_c$ with concentration of CN in 5CB in 12 $\mu$m layers at $23^\circ$ C.

ding of the helix (Fig. 7) involves the removal of the regions of twist. These regions of twist or disclination lines can only be removed or created by a discontinuity. The phase change is therefore quite different from other liquid crystal effects which involve uniform, gradual transitions between states. As the voltage is increased, the cholesteric transforms into a nematic with residual disclinations remaining which shrink to small points and finally disappear some time after the cholesteric-nematic transition is complete. When the field is removed, nucleation and growth of disclination lines occur to produce the cholesteric phase. Residual disclinations remaining from the cholesteric-nematic transition, surface inhomogeneities and dust particles all act as nucleation points, and their density considerably affects the speed of the nematic-cholesteric transition. The sequence of pictures in Fig. 9 illustrates the growth of the cholesteric phase from a few nucleation points in a layer in which the applied field has been lowered to just below $E_c$. The nucleation of growth of disclination lines normally limits the response times to 1 ms for a cholesteric-nematic transition. Much shorter response times (50 $\mu$s) have been reported[7] but these were for incomplete transitions, and the high density of nucleation sites still present probably caused the quicker response.

Greubel has recently shown[12] that a completely different transition is possible which does not involve disclination lines, and it occurs when the nucleation process is retarded. Homeotropic boundary conditions favor the new transition and have the added advantage of providing a surface free of

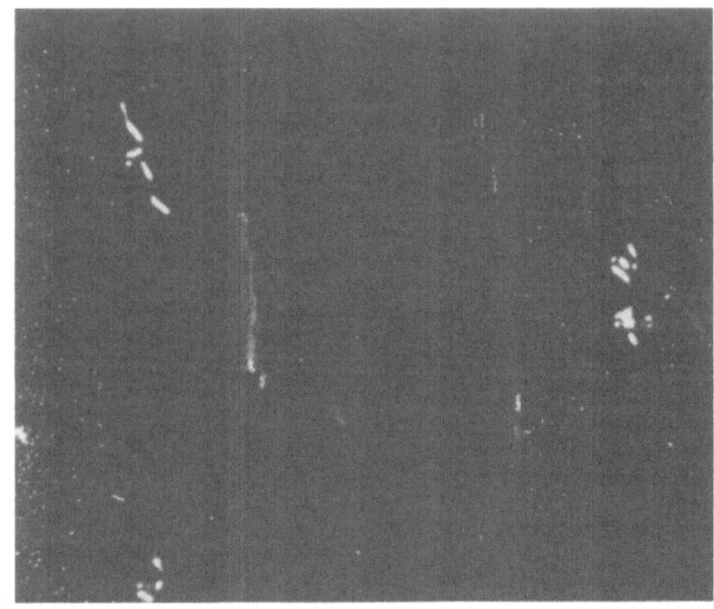

Fig. 9a.

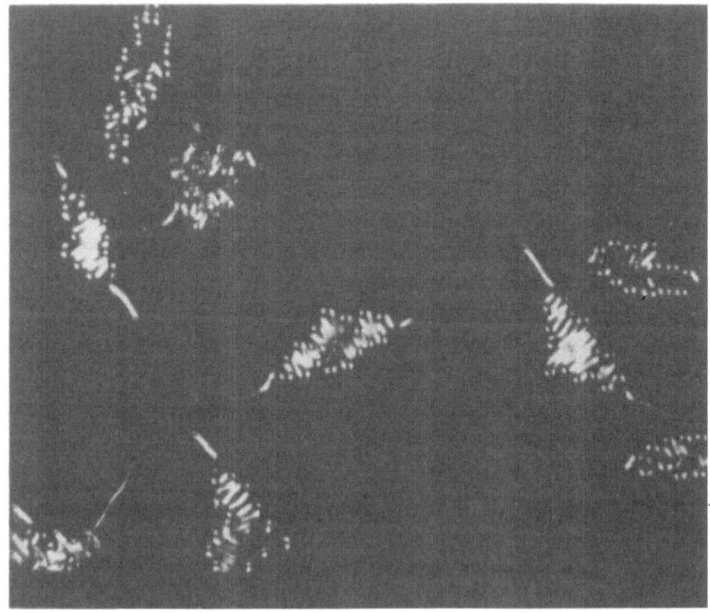

Fig. 9b.

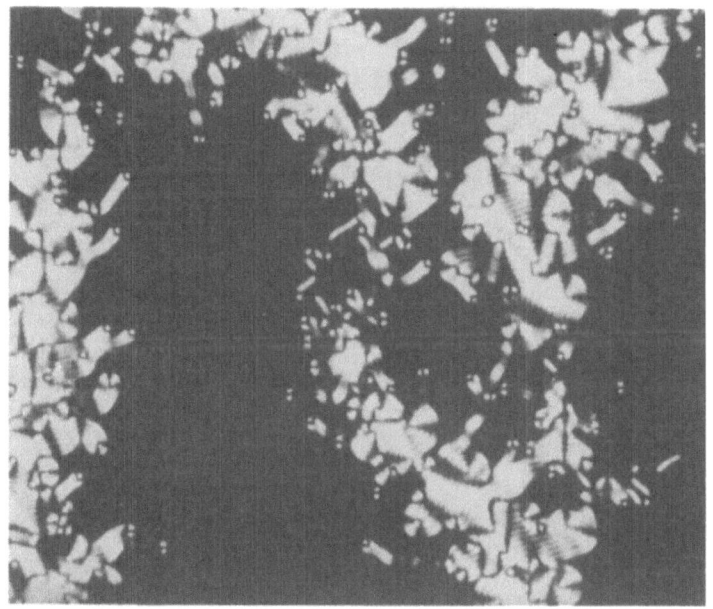

Fig. 9c.

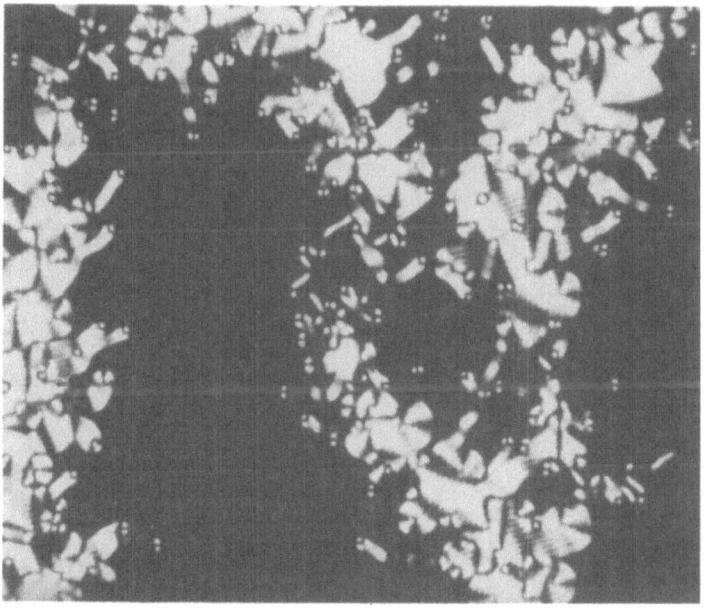

Fig. 9d.

Fig. 9e.

Fig. 9. The growth of the cholesteric phase via disclination lines from the nematic phase: (a) - (e) time increasing, crossed polarizers.

defects which could act as nucleation sites for disclination lines. Greubel's results are illustrated in Fig. 10, and show the usual cholesteric-nematic transition as the voltage is increased above 15 V. As the voltage is lowered little happens initially because of the small number of nucleation sites. However at 6 V a sudden transition occurs and the microscope shows that it does not require nucleation. Greubel pointed out that a uniform relaxation of the nematic layer is possible without nucleation by the process shown in Fig. 11. Using the continuum theory, the threshold field ($E_{c\downarrow}$) for this uniform transition from the nematic state is found to be:

$$E_{c\downarrow} = \pi \sqrt{\frac{k_{33}}{\epsilon_0(\epsilon_\parallel - \epsilon_\perp)} \left[ \left( \frac{2k_{22}}{P_o k_{33}} \right)^2 - \frac{1}{d^2} \right]}. \qquad (4)$$

Comparison with the usual phase change threshold $E_c$ from (2) gives for the ratio of the two critical fields:

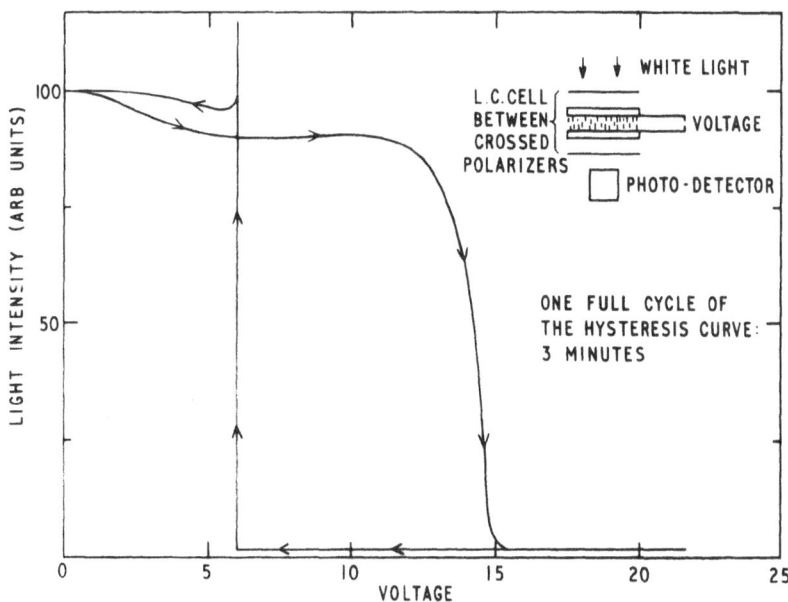

Fig. 10. Hysteresis in the phase change effect from Greubel.[12]

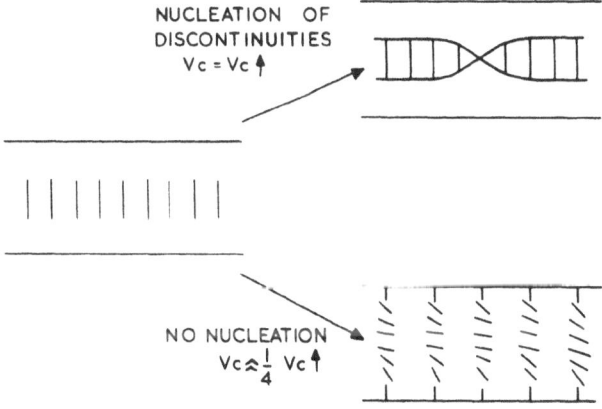

Fig. 11. The two relaxation mechanisms for the nematic-to-cholesteric transition.

$$\frac{E_c}{E_{c\downarrow}} = \pi \sqrt{\frac{k_{22}k_{33}}{4k_{22}^2 - k_{33}^2\left(\frac{P_o}{d}\right)^2}} . \tag{5}$$

Using typical values for $k_{22}$ and $k_{33}$ the value of $E_c/E_{c\downarrow}$ is calculated to be $\approx 2.5$ if $P_o < d$; this is close to the observed value in Fig. 10. There are two important consequences of this uniform transition:

1) The hysteresis shown in Fig. 10 can become very pronounced when $E_{c\downarrow}$ approaches zero. From (5) we see that $E_{c\downarrow} = 0$ if:

$$\frac{P_o}{d} \geqslant \frac{2k_{22}}{k_{33}} . \tag{6}$$

Typically this happens for $12\,\mu m$ layers if the cholesteric concentration $c < 2\%$. When this type of layer is transformed into the nematic state, the only relaxation process possible when the field is removed is the slow nucleation and growth of disclination lines. The clear nematic state therefore remains for a long time.

2) When the uniform transition with no disclination lines occurs, a scattering texture is initially formed which changes to a clear Grandjean texture after several seconds. Conversely when the nucleation dominated transition occurs the scattering texture is permanent. The layer therefore has a memory for several hours of the type of transition that had occurred from the nematic phase.

## 5. PLEOCHROIC DYES AND PHASE CHANGE

The texture and phase change effects described above use light scattering to produce a visible effect. A powerful technique for improving the contrast of phase change displays involves the addition to the liquid crystal of pleochroic dyes. These dye molecules absorb light polarized along their long axis, but transmit other polarizations (Fig. 12). The use of pleochroic dyes in liq-

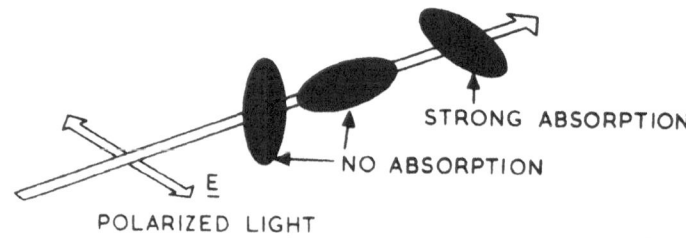

Fig. 12. Light absorption by pleochroic dyes.

uid crystals was first suggested by Heilmeier et al.,[13] but it was not until White and Taylor[14] that significant progress was made and high contrast ratios were obtained.

White and Taylor pointed out that the cholesteric-nematic phase change is the ideal effect to use with the dyes. This can be seen from Fig. 13. If the

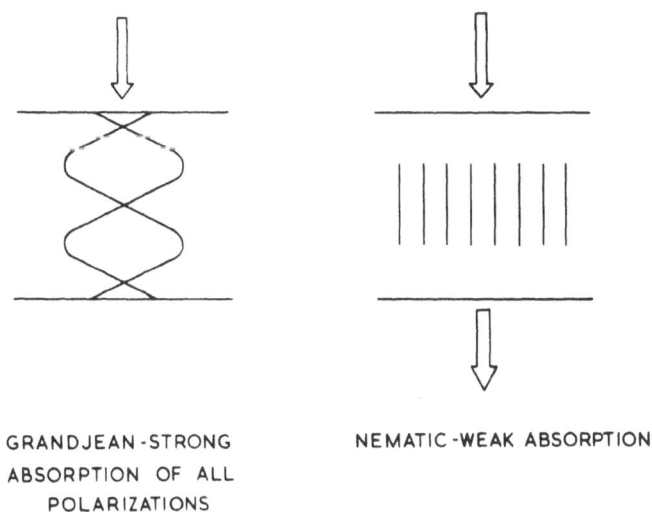

GRANDJEAN-STRONG
ABSORPTION OF ALL
POLARIZATIONS

NEMATIC-WEAK ABSORPTION

Fig. 13. Light absorption by a phase change device containing pleochroic dyes.

off-state is the Grandjean texture, all planes of polarization of the light can be absorbed by the dyes, providing the pitch of the cholesteric is not too long. In the limiting case of a very long pitch the structure behaves like a twisted nematic and "guides" the polarized light so only half of the light is absorbed. The homeotropic nematic state absorbs little of the light so it is possible to produce a high contrast, high brightness display. The focal conic state also has a high absorption, so the contrast ratio does not depend significantly upon the texture of the cholesteric phase. It would therefore seem possible to produce high brightness, high contrast displays which require no polarizers and no particular surface alignment. White and Taylor also appreciated the significance of the order parameter S of the dye, where S is given by the usual expression:

$$S = \frac{1}{2} \left( \overline{3\cos^2\theta} - 1 \right) , \tag{7}$$

where $\theta$ is the angle between the molecular axis and the liquid crystal director. A dye with molecules of a size and shape similar to liquid crystal molecules has $0.5 < S < 0.6$. This corresponds to $\theta_{rms} \approx 30^{\circ}$, and this large fluctuation produces a marked absorption in the clear nematic state.

Longer dye molecules average out some of the liquid crystal fluctuations so that they can have order parameters higher than that of the host liquid crystal. For example we have obtained an order parameter of 0.79 for a purple dye, which correponds to $\theta_{rms} \approx 20°$. It is interesting to compare this figure with the order parameter of about 0.94 for the dye used in Polaroid sheet. The optical performance of dyes in the phase change effect can be judged from the calculated contrast ratios for a Grandjean cholesteric-to-nematic transition (Fig. 14). We have chosen the absorption such that the

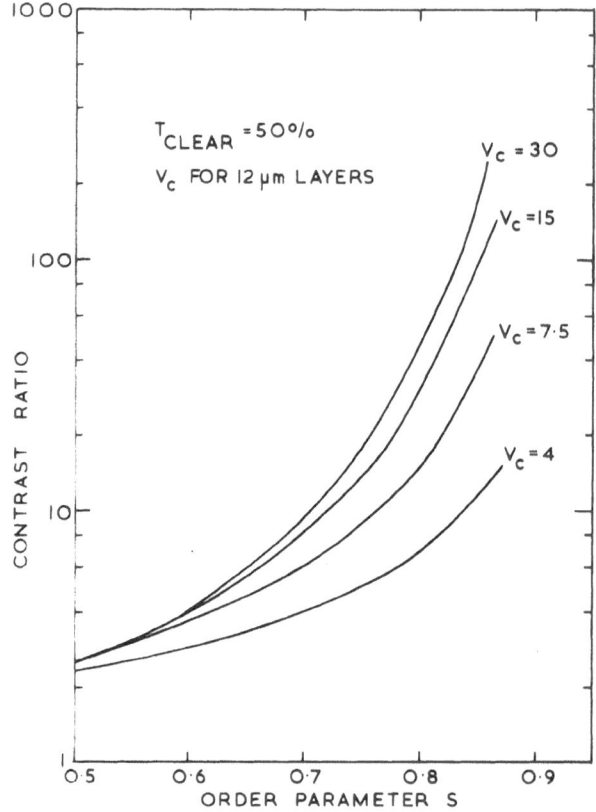

Fig. 14. Theoretical contrast ratios for phase change displays with pleochroic dyes.

clear nematic state transmits 50% of the light, and rather than quote pitch lengths, the threshold voltages for $12\,\mu$m thickness are given. Figure 14 shows that it is possible to obtain contrast ratios greater than 10 : 1 for devices with a reasonably low threshold voltage ($\approx 8$ V) if $S \gtrsim 0.75$.

The phase change effect with pleochroic dyes seems to be a strong contender for a low complexity, high contrast, high brightness display and can be used for enhanced contrast in complex multiplexed arrays. However it is worth remembering that several problems remain to be solved. It is difficult to produce a dye of any desired color and at the same time preserve a high order parameter. The solubility, photochemical and electrochemical stability of most dyes also present considerable problems which must be solved before dyes in displays can be accepted commercially.

## 6. ASSESSMENT OF THE EFFECTS

So far we have described the main features of the more useful electrooptic effects in cholesterics and we now compare them with the better known effects in nematics. Many aspects of liquid crystal display performance are equally applicable to both cholesteric and nematic displays. This is not surprising because the cholesteric mixtures used are composed mainly of nematic compounds. Therefore parameters like the temperature range, lifetime, and most of the purely constructional properties are essentially identical with those of nematic displays. There are however two important points which are particularly relevant to phase change displays. The thickness of the layer must be accurately controlled because the phase change has a threshold field, and it is usual to bias the display with a fixed voltage close to threshold during multiplexing. The presence of a dye also makes thickness variations very obvious. With selected sheet glass of 1.5 mm thickness, it is possible with care to make displays up to 8 cm × 4 cm with a thickness variation of 1 - 2 $\mu$m (i.e. 10%). Float glass is necessary if larger or flatter displays are required, but the thickness of the glass (at least 3 mm) starts to create problems with edge connectors and with parallax in reflective displays. The other point about phase change displays is that, although the material used may be cholesteric over a wide temperature range (e.g. $-10^\circ$ C to $+70^\circ$ C), the threshold field varies significantly with temperature. This restricts the temperature over which a complex display can be held sufficiently close to threshold to be multiplexed. Typically the threshold field can vary by 20% over the range $20^\circ$ C to $50^\circ$ C.

The tremendous advantage of memory offered by the cholesteric texture change effect has not been exploited to any great extent although the effect has been known for several years. The Grandjean-to-focal conic transition has parameters very similar to those of dynamic scattering. The reverse transition however requires comparatively large power ($\approx 0.1$ W cm$^{-2}$) because of the high frequency, high voltage requirement. These properties are summarized in Table 1.

The main properties of the cholesteric-nematic phase change effect are summarized in Table 2. Without dyes the optical performance is similar to that of dynamic scattering with all the visibility problems that accompany a scattering effect. The inclusion of a pleochroic dye produces an absorption device which can have a 10 : 1 contrast ratio together with a clear state transmission of 50%. By comparison, the typical clear state brightness for a reflective twisted nematic device is only 20%. However pleochroic dyes suitable for use in liquid crystals are still at an early stage of development and several problems remain to be solved before they can be regarded as a serious rival to the twisted nematic display. The response times of the phase change

TABLE 1

Properties of the Texture Change Effect

| | |
|---|---|
| Optical effect: | clear ⇋ scattering |
| Write conditions: | $V \approx 30\ V_{rms}\ (\lesssim 1\ kHz)$ |
| | Power $\approx 100\ \mu W\ cm^{-2}$ |
| | $\tau \approx 10\ ms$ |
| Erase conditions: | $V \approx 100\ V_{rms}\ (\gtrsim 2\ kHz)$ |
| | Power $\approx 10^{-1}\ W\ cm^{-2}$ |
| | $\tau \approx 100\ ms$ |

TABLE 2

Properties of the Phase Change Effect

| | | |
|---|---|---|
| Optical effect: | (no dye) | clear ⇋ scattering |
| | (with dye) | absorbing ⇋ transparent |
| | light transmission | 5% ⇋ 50% |
| Working voltage: | | $5 - 50\ V_{rms}$ |
| Response times: | | 100 - 1 ms |
| | (Considerable hysteresis and bistability effects are possible) | |
| Power consumption: | | $1 - 100\ \mu W\ cm^{-2}$ |
| Special features: | Large capability of being multiplexed. Polarizers and surface alignment not essential. | |

are sufficiently short to allow about 100 numbers to be multiplexed, which gives it a considerable advantage for use in complex displays. Although the various hysteresis and bistability effects will undoubtedly be of use in complex multiplexed displays, it is too early to be specific about the nature and extent of their use.

In summary it is obvious that cholesterics offer a number of variations and advantages over nematics, but in most cases further research is needed before the extent of these advantages becomes clear.

# REFERENCES

1   G. W. Gray, K. J. Harrison, J. A. Nash and E. P. Raynes, Electron. Lett. 9 (1973) 616
2   J. Constant and E. P. Raynes, Electron. Lett. 9 (1973) 561
3   G. H. Heilmeier and J. E. Goldmacher, Proc. IEEE 57 (1969) 34
4   H. Melchior, F. J. Kahn and D. Maydan, Appl. Phys. Lett. 21 (1972) 392
5   P. G. de Gennes, Solid State Comm. 6 (1968) 163
6   G. H. Heilmeier, L. A. Zanoni and J. E. Goldmacher, in J. F. Johnson and R. S. Porter (ed.), 'Liquid Crystals and Ordered Fluids', Plenum, New York (1970) 215
7   E. Jakeman and E. P. Raynes, Phys. Lett. 39A (1972) 69
8   J. J. Wysocki, J. Adams and D. J. Olechna, in J. F. Johnson and R. S. Porter (ed.), 'Liquid Crystals and Ordered Fluids', Plenum, New York (1970) 419
9   J. J Wysocki, Mol. Cryst. Liq. Cryst. 14 (1971) 71
10  T. Ohtsuka and M. Tsukamoto, Jap. J. Appl. Phys. 12 (1973) 22
11  R. A. Kashnow, J. E. Bigelow, H. S. Cole and C. R. Stein, Appl. Phys. Lett. 23 (1973) 290
12  W. Greubel, Appl. Phys. Lett. 25 (1974) 5
13  G. H. Heilmeier and L. A. Zanoni, Appl. Phys. Lett. 13 (1968) 91
14  D. L. White and G. N. Taylor, J. Appl. Phys. 45 (1974) 4718

# DISCUSSION

D. W. Berreman (Bell Laboratories)
When you quoted 20% brightness for a reflective twisted nematic device, did you consider depolarization by a diffuse reflector?

E. P. Raynes
No, the figure is based simply on the absorption which accompanies four passes through Polaroid HN 42 sheets.

C. J. Gerritsma (Philips)
The erasing voltage required for the texture change effect can be reduced by using a material with a more negative dielectric anisotropy. The $\alpha$-cyano stilbenes synthesized by van der Veen in our laboratory have $\Delta\epsilon = -5$ and can be erased with about 30 V, but they are not yet stable.

J. P. Hurault (LEP)
Is it really true that no surface preparation is required for phase change displays?

E. P. Raynes
Unlike the twist cell where surface alignment is essential, contrast does not depend strongly on surface alignment. It may be desirable to use a controlled surface treatment for uniform appearance, but even then the homeotropic alignment would be preferred and this is easier to obtain than the planar alignment needed for twist cells.

F. J. Kahn (Hewlett-Packard)
Since the order parameters for the dye and for the host are different, the response times might also differ. Have you found such an effect?

E. P. Raynes
We have seen no appreciable difference in response times for the phase change effect with and without dyes. The results which I report were measured without dyes.

T. J. Scheffer (Brown Boveri)
The order parameter describes thermal fluctuations of individual molecules about the director on a microscopic scale while response time is a macroscopic effect involving reorientation of the director. Response time for guest and host must be identical from first principles*. Does dye concentration affect its order parameter?

E. P. Raynes
We have not studied a wide range of concentrations. The effect doesn't seem to be important at the doping levels of 0.5 - 1% which are needed for displays, but the clearing temperature could be lowered.

J. Kirton (Royal Radar Establishment)
I might add that the achievement of adequate solubility in combination with high order parameter is a real accomplishment. The dyes used in this work were prepared by G. W. Gray, D. Coates and D. McDonnell at the University of Hull.

J. Nehring (Brown Boveri)
Can you tell us the composition of your purple dye with high order parameter?

E. P. Raynes
Red dyes with high order parameter and good stability are not uncommon, but the attempt to obtain absorption at longer wavelengths by the addition of side groups frequently lowers the order parameter and, because of the narrower absorption gap, degrades the stability. We hope to publish our solution to this problem soon.

T. J. Scheffer (Brown Boveri)
Do you think equation (4) is correct?

E. P. Raynes
Although equation (4) was obtained from Greubel's paper we have also derived it from first principles and we believe it to be completely correct*.

*) Comment subsequent to meeting.

E. Guyon (Université de Paris-Sud)

The terminology and concepts used to describe phase transitions may be generally useful in the study of non-emissive electrooptic displays phenomena which often involve a symmetry breaking transition. In particular for the nematic-cholesteric transition, the two processes described can be considered as a Landau-like transition and a transition of the nucleation type suggested by de Gennes (in T. Riste (ed.), "Fluctuations, Instabilities and Phase Transitions", Plenum (1976) Chapt. 1). An analogy can be drawn to Type II superconductors: the high field transition to the normal state is Landau-like whereas the first entry of vortices at the lower critical field can be compared to the first entry of pitch singularities in a nematic-to-cholesteric transition.

# LIQUID CRYSTAL COLOR DISPLAYS

T. J. SCHEFFER

Brown Boveri Research Center, Baden, Switzerland

## SUMMARY

Many electrooptic effects in liquid crystals can display information in color. Some of the more practical display schemes employ the guest-host effect, tunable retardation, and several different combinations of twisted nematic cells with external optically anisotropic layers. These schemes are compared on a psychophysical basis for color gamut, brightness and contrast ratio. The effects of temperature, cell thickness and viewing angle and the influence of certain material properties are also included. The orientability and solubility of dyes are considered for the guest-host effect and for the preparation of high-efficiency pleochroic polarizers for use in combination with twisted nematic cells.

## 1. PSYCHOPHYSICAL BASIS OF COLOR

Throughout this paper we will specify colors by their coordinates on the 1931 CIE chromaticity diagram[1] shown in Fig. 1. Every possible color sensation is uniquely represented by a point on this diagram which can be specified by giving the x and y chromaticity coordinates. The colors at the outer boundary of the diagram have maximum saturation, and the colors gradually become more faded toward the center of the diagram. The color hue changes on path going around the white central region. Brightness, the third type of color variation, would correspond to a third coordinate perpendicular to the x-y plane.

An unknown color can be located on the chromaticity diagram by direct measurement using a colorimeter or, alternatively, its position can be calcu-

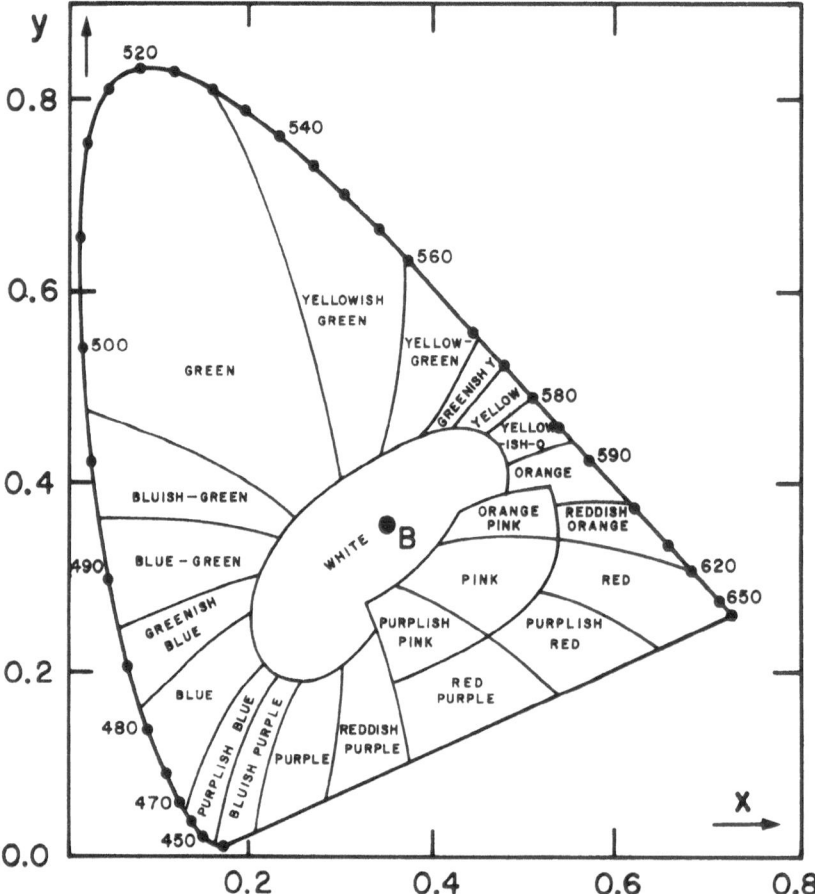

Fig. 1. 1931 CIE chromaticity diagram. The saturation of the colors gradually decreases from the outer boundaries to the center of the diagram.

lated from the spectral distribution of the radiation. We will apply this second method to determine the colors that can be obtained from displays employing the various electrooptical effects.

## 1.1. Calculation of Luminous Reflectance and Color Coordinates from the Reflection Spectrum

Since the color of an object can differ when it is viewed in daylight or in articificial light, it is important in discussing the color of passive displays to specify the spectral distribution of the particular illumination used. The CIE has defined the spectral irradiance $H(\lambda)$ of three standard sources which they denote by the letters A, B and C. Source A corresponds to incandescent illumination, source B to noon sunlight and source C to total daylight including sunlight and skylight. Unless otherwise specified, we used source B, corresponding to noon sunlight, in our calculations.

The chromaticity coordinates (x,y) of the perceived color of an object whose reflection spectrum is given by $R(\lambda)$ are obtained from the expression

$$x = \frac{1}{Q} \int H(\lambda) \ \overline{x}(\lambda) \ R(\lambda) \ d\lambda$$

and (1)

$$y = \frac{1}{Q} \int H(\lambda) \ \overline{y}(\lambda) \ R(\lambda) \ d\lambda \ ,$$

where Q is defined by

$$Q = \int H(\lambda)\overline{x}(\lambda)R(\lambda)d\lambda \ + \ \int H(\lambda)\overline{y}(\lambda)R(\lambda)d\lambda \ + \ \int H(\lambda)\overline{z}(\lambda)R(\lambda)d\lambda \ .$$

The luminous reflectance B of the object corresponds to its perceived brightness compared to an ideal 100% reflecting surface and is given by

$$B = \frac{\int H(\lambda)\overline{y}(\lambda)R(\lambda)d\lambda}{\int H(\lambda)\overline{y}(\lambda)d\lambda} \ , \tag{2}$$

where $\overline{x}(\lambda)$, $\overline{y}(\lambda)$ and $\overline{z}(\lambda)$ are the three standard tristimulus weighting functions. The integrals in (1) and (2) need to be evaluated only over the visible spectrum, which extends from 380 nm to 780 nm. We evaluated these integrals by applying Simpson's rule at the 5 nm intervals for which the source and tristimulus functions are tabulated in Ref. 1. This analysis can also be applied to calculate the chromaticity coordinates of light transmitted through a color filter. For this case the transmittance of the filter $T(\lambda)$ is substituted for $R(\lambda)$ in (1) and (2). B is then the luminous transmittance which compares the perceived brightness of the colored light that is transmitted by the filter with the brightness of the unfiltered source.

The luminous reflectance B is an important factor for a passive reflective display. The display will appear very dark if its luminous reflectance is too low in comparison with the objects around it, e.g. a watch display mounted in a silver case. This consideration is not so important for a transmissive or projection display because the surroundings generally can be made much darker than the region to be viewed so the display always appears relatively bright.

### 1.2. Maximum Luminous Reflectance for a Passive Display

A colored reflective display necessarily appears darker than a 100% reflecting white surface simply because certain portions of the spectrum of the incident illumination must be absorbed by the display in order for it to appear colored. MacAdam[2] has determined the largest luminous reflectance that is theoretically possible for any color. Figure 2 shows the maximum value of B

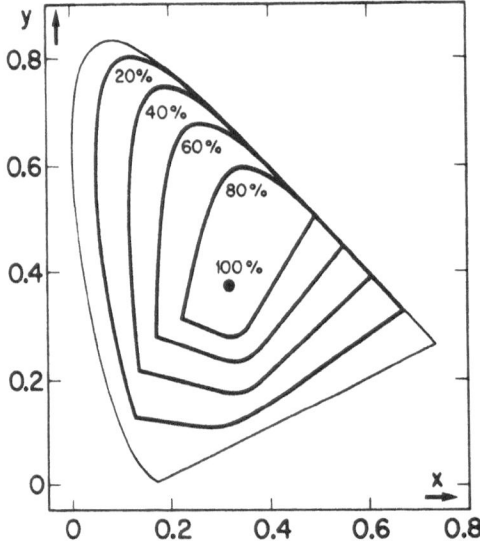

Fig. 2. Contour curves indicate maximum luminous reflectance for materials showing the indicated chromaticities when illuminated by CIE source C (daylight).

for any point on the chromaticity diagram as the height above the x-y plane indicated by the contour lines. Pure white, in the center of the diagram, has a maximum luminous reflectance of 100%, and all other colors lie below this value. It is possible for saturated oranges, yellows and yellow-greens to have large luminous reflectances, but large values of B are theoretically impossible for saturated greens, reds, blues and purples. The luminous reflectance of a passive display should be as close as possible to the theoretical limit.

MacAdam has shown that the reflectance curves corresponding to colors of maximum efficiency are either single band-pass or single band-stop curves (Fig. 3) where the reflectance is either 0 or 100%. The more the reflectance

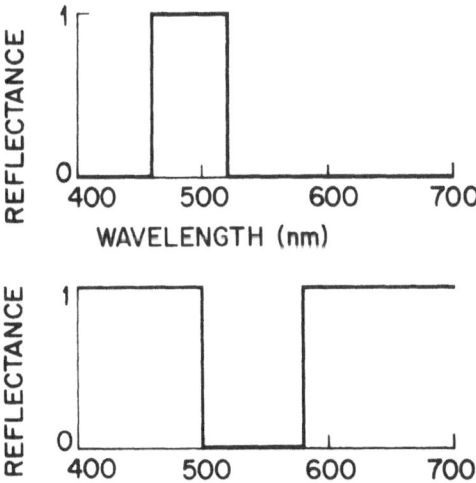

Fig. 3. Examples of band-pass (upper figure) and band-stop (lower figure) spectrophotometric curves required for colors of maximum luminous reflectance

curve of a particular color deviates from its ideal curve, the less efficient is the color. The contour lines of Fig. 2 were computed from the two types of ideal curves. Band-stop curves give the brighter colors in the part of the color diagram where the contour lines extend from the lower right part of the boundary to the second (and sharper) bend in the lower left part of the diagram. Band-pass curves give the brighter colors in the remaining upper part of the diagram.

## 2. TUNABLE RETARDATION

The birefringence of nematic liquid crystals is quite large, and the orientation of the optic axis of the nematic is easily influenced by an applied voltage. These two properties give the basis of a voltage-tunable color filter. Such effects have been investigated by a number of researchers[3,4,5,6] but there has been no objective evaluation of the colors. In this section we present a simple theoretical treatment which gives the voltage dependence of the colors as well as the effects of temperature, thickness and viewing angle.

### 2.1. Optical Path Length Difference of a Deformed Nematic Layer

The electric field induced reorientation of the optic axis in a nematic layer, known as the Freedericksz transition, is well understood[7]. Consider a nematic layer of thickness e with its optic axis uniformly parallel to the layer nor-

mal z. Application of a voltage U greater than a threshold voltage $U_o$ causes
the optic axis to deform in the layer, as shown in Fig. 4. The deformation

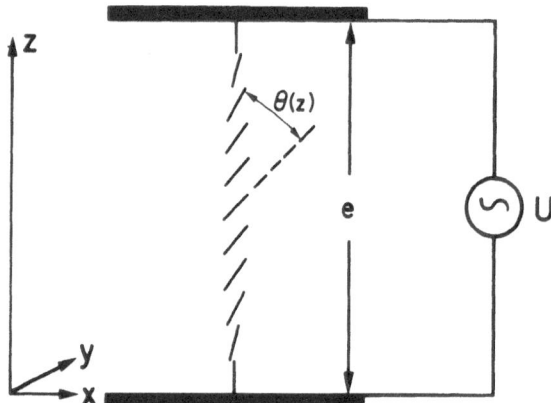

Fig. 4.   Shape of Freedericksz deformation in an initially homeotropic nematic layer

angle $\theta(z)$ is the angle between the local optic axis in the nematic layer and
the layer normal z. The threshold voltage $U_o$ is given by

$$U_o = \pi \sqrt{\frac{-k_{33}}{\epsilon_o \Delta \epsilon}} \qquad (3)$$

where $k_{33}$ is the bend elastic constant and $\Delta \epsilon < 0$ is the dielectric anisotropy. $U_o$ is generally on the order of a few volts. Solutions of $\theta(z)$ are given
for the general case of a nematic liquid crystal with arbitrary elastic constants and dielectric anisotropy[7], but we will present the simpler solutions
obtained by assuming equal bend and splay elastic constants and a small $\Delta \epsilon$.
Under these conditions the maximum deformation angle $\theta_m$ occurring in the
middle of the layer is given by the implicit expression

$$U/U_o = (2/\pi) \ F(\pi/2, q) , \qquad (4)$$

where $q = \sin\theta_m$ and F is the elliptic integral of the first kind defined by

$$F(\psi, k) = \int_0^\psi (1-k^2 \sin^2\alpha)^{-1/2} \ d\alpha .$$

The deformation angles $\theta(z)$ in other parts of the layer are computed from

$$\sin\theta(z) = q \, sn \, (\pi \, \frac{z \, U}{e \, U_o}, q) \tag{5}$$

where sn is a Jakobian elliptic function[8].

Equations (4) and (5) make it possible to calculate the optical path length difference $\Gamma$ between the ordinary and extraordinary rays passing through the deformed nematic layer. The most general expression for $\Gamma$ is given by[9]

$$\Gamma = 2 \int_0^{e/2} \left[ \frac{(a^2-b^2) \sin 2\theta(z) \cos\phi \, \sin i}{2c^2} \right.$$

$$+ \frac{1}{c} \left( 1 - a^2 \sin^2\phi \sin^2 i - \frac{a^2 b^2}{c^2} \cos^2\phi \sin^2 i \right)^{1/2}$$

$$\left. - \frac{1}{b} \left( 1 - b^2 \sin^2 i \right)^{1/2} \right] dz , \tag{6}$$

where $a = 1/n_e$, $b = 1/n_o$ and $c^2 = a^2 \sin^2\theta(z) + b^2 \cos^2\theta(z)$. $n_o$ and $n_e$ are the ordinary and extraordinary refractive indices of the nematic. The angles i and $\phi$ are the angle of incidence and the angle between the plane of incidence and the x-z plane (see Fig. 5). $\theta(z)$ gives the orientation of the optic axis in the deformed nematic layer. For normally incident light, i = 0 and (6) reduces to[10]

$$\Gamma = e \, n_o \left[ \frac{2}{\pi} \, \frac{U}{U_o} \, (1+\gamma q^2)^{1/2} \, F(\frac{\pi}{2}, k) - 1 \right] \tag{7}$$

where

$$k = q \frac{n_o}{n_e} (1+\gamma q^2)^{-1/2} \quad \text{and} \quad \gamma = \left( \frac{n_o}{n_e} \right)^2 - 1 .$$

The optical path length difference is also a function of wavelength if there is dispersion in $n_e$ and $n_o$. The phase difference $\delta$ between the two rays leaving the layer shows additional wavelength dependence:

$$\delta(\lambda) = 2\pi \, \Gamma(\lambda)/\lambda . \tag{8}$$

The transmission of a liquid crystal layer placed between crossed ideal polarizers and oriented with the x-axis (Fig. 4) at a 45° angle with the transmission axis of the first polarizer is given by

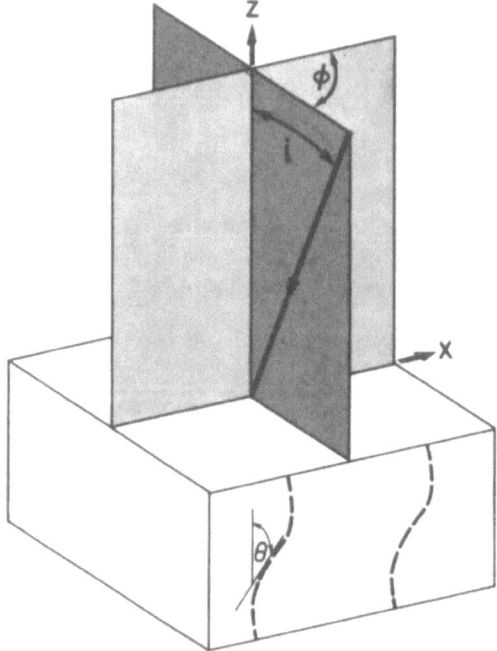

Fig. 5. Definition of angles for tunable retardation effect

$$T_\perp(\lambda) = 1/2 \sin^2 \delta(\lambda)/2 . \tag{9}$$

Rotating one of the polarizers by $90^\circ$ giving parallel polarizers yields

$$T_\parallel(\lambda) = 1/2 \cos^2 \delta(\lambda)/2$$

$$= 1/2 - T_\perp(\lambda) . \tag{10}$$

White light entering these filters emerges as colored light because $T_\perp(\lambda)$ and $T_\parallel(\lambda)$ are strong functions of wavelength. The chromaticity coordinates of these colors can be computed by substituting $T_\perp(\lambda)$ and $T_\parallel(\lambda)$ in place of $R(\lambda)$ in (1).

### 2.2. Voltage Dependence of Colors

The voltage dependence of the colors for a 10 $\mu$m thick nematic layer between crossed polarizers was calculated from the previous equations. For $n_o$ and $n_e$ we used the experimental values reported by Brunet-Germain[11] for MBBA at 25$^\circ$C, although it should be understood that the results of the calculation will be valid for MBBA only within our approximations of equal elastic constants and small $\Delta\epsilon$. The four parameters of Sellmeir's dispersion equation[12] (including a near UV and far IR absorption term) were adjusted to give the best fit to Brunet-Germain's measurements which were reported at five discrete wavelengths.

We used the following procedure to compute the voltage dependence of the interference colors. First, for a given $U/U_o$ (4) was employed to obtain the corresponding value of q. These values were substituted into (7) together with the $n_e$ and $n_o$ obtained from Sellmeir's equation to compute the optical path length difference $\Gamma$. This last step was repeated at 5 nm wavelength intervals to compute $\Gamma$ over the entire visible spectrum. Equations (8) and (9) were then used to obtain the transmission curve between crossed polarizers. The results for two reduced voltages $U/U_o$ = 1.5 and 1.8 are shown in Fig. 6.

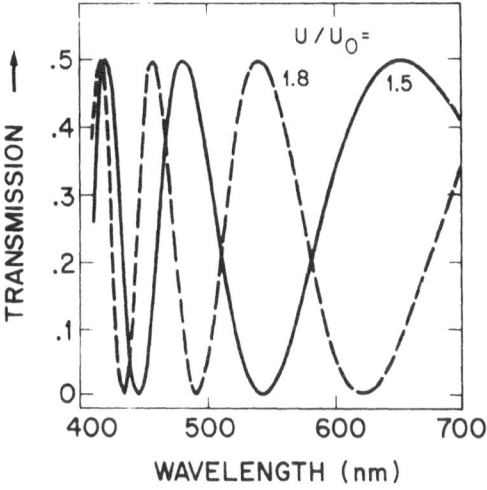

Fig. 6. Calculated transmission spectra of 10 $\mu$m thick MBBA layer between crossed polarizers for $U/U_o$ = 1.5 (solid curve) and $U/U_o$ = 1.8 (dashed curve)

The curves are of an oscillatory nature which is characteristic of simple interference phenomena. The chromaticity coordinates corresponding to these two reduced voltages, and many other voltages as well, were calculated from (1) and are plotted as a continuous curve in Fig. 7.

As the voltage is increased above the threshold value, the interference colors describe a curve on the color diagram that spirals inward toward the white point. The order of the interference colors, beginning at one, increases by one each time the curve crosses the straight line connecting the white point to the middle of the lower boundary of the diagram. The color corresponding to $U/U_o$ = 1.1 is first-order yellow. For $U/U_o$ = 1.2, 1.3, 1.5 the colors are second-order blue, green and red. $U/U_o$ = 1.8 corresponds to third-order green.

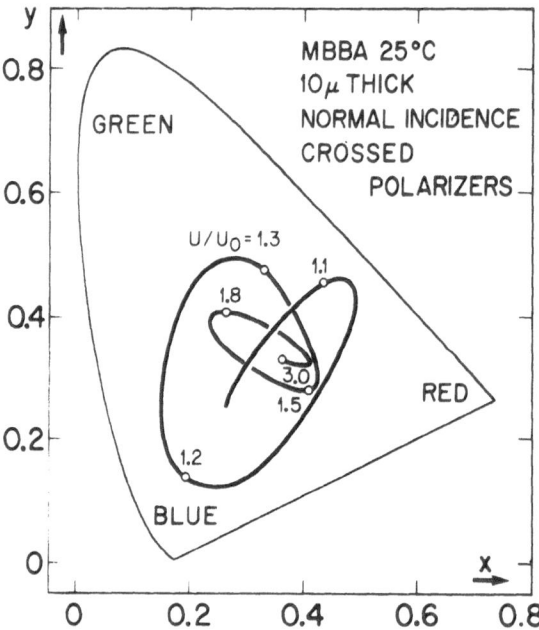

Fig. 7. Calculated voltage dependence of transmission colors for tunable retardation effect in MBBA.

## 2.3. Effect of the Nematic Material on Colors

A second calculation was carried out using refractive index data for butoxy-benzoic acid[13] rather than for MBBA in order to determine the effect that a different liquid crystalline material would have on the obtainable colors. The color ranges for the two materials are given in Fig. 8 for comparison. Noticeable differences between the two curves are seen, especially for the higher order colors. The color ranges differ primarily because the refractive indices of the two nematics show different amounts of dispersion. The dispersion in MBBA is quite large owing to a strong absorption band in the near UV which makes the colors strongly non-Newtonian. Similar behavior would be expected for other Schiff base nematics as well as from azo and azoxy compounds. The dispersion in butoxybenzoic acid is much less than for MBBA and the colors are closer to being Newtonian. Nematics belonging to the ester and biphenyl families would show a behavior similar to the butoxybenzoic acid.

## 2.4. Temperature Dependence

A disadvantage of the tunable retardation effect is that the colors vary strongly with temperature. This variation results from the large temperature dependence of the elastic constants, the dielectric constants and the refractive indices which are all unavoidable characteristics of the nematic phase.

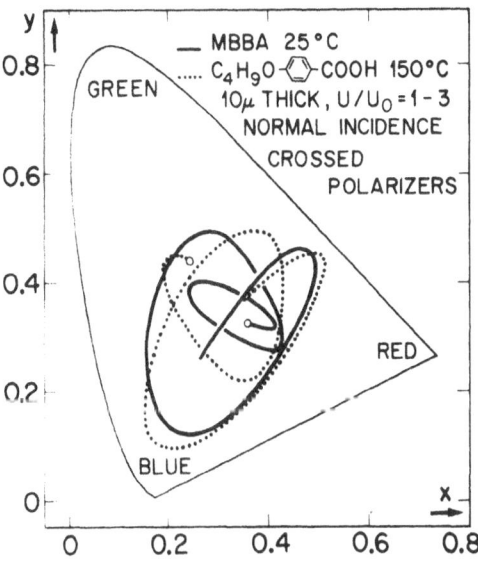

Fig. 8. Calculated range of transmitted tunable retardation colors for MBBA (solid curve) and butoxybenzoic acid (dotted curve). Differences are caused by different amounts of dispersion in the refractive indices

Consider the specific example of MBBA where both $k_{33}$[14] and $\Delta\epsilon$[15] are known as a function of temperature (Fig. 9). The threshold voltage, com-

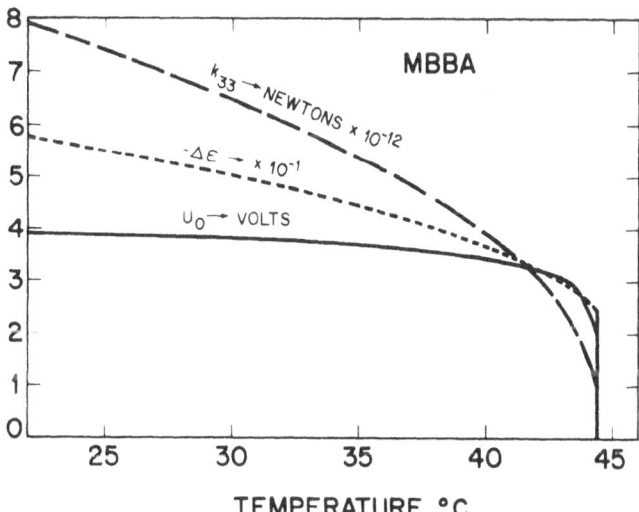

Fig. 9. Measured values of bend elastic constant $k_{33}$ (dashed curve) and dielectric anisotropy $\Delta\epsilon$ (dotted curve) for MBBA as a function of temperature. The calculated threshold voltage $U_o$ is a decreasing function with temperature

puted from (3), decreases at higher temperatures causing a stronger deformation for the same applied voltage. This would have the effect of increasing the optical path length difference $\Gamma$ if it were not for the even larger decrease of the birefringence at higher temperatures (Fig. 10).

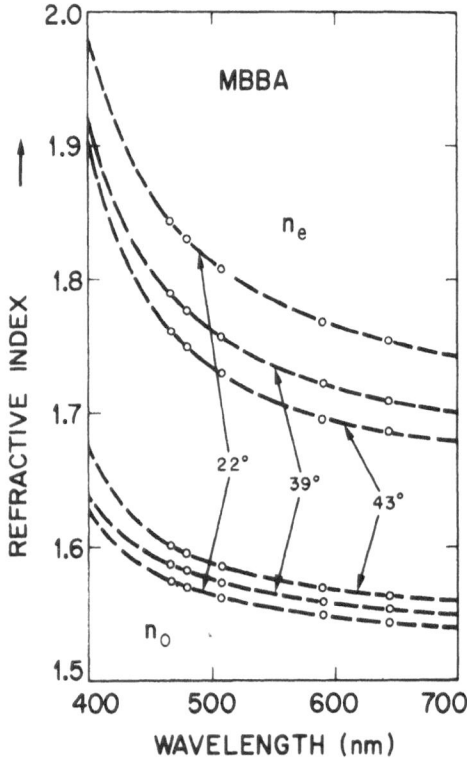

Fig. 10. Refractive index curves for MBBA showing strong dispersion and temperature dependence. The open circles are Brunet-Germain's[11] experimental points and the dashed curves have been calculated from Sellmeir's dispersion equation

The chromaticity coordinates corresponding to the transmitted color when 7 volts are applied across a 10 $\mu$m thick MBBA layer were calculated at the nine temperatures for which refractive index data for MBBA are available[11]. These colors describe a curve which is shown on the diagram of Fig. 11. The third-order green color at room temperature passes through blue at higher temperatures to become a second-order red color near the isotropic transition temperature of MBBA.

## 2.5. Thickness and Angle Dependence

The tunable retardation colors are very sensitive to thickness variations of the nematic layer. The colors shown on the diagram of Fig. 12, computed

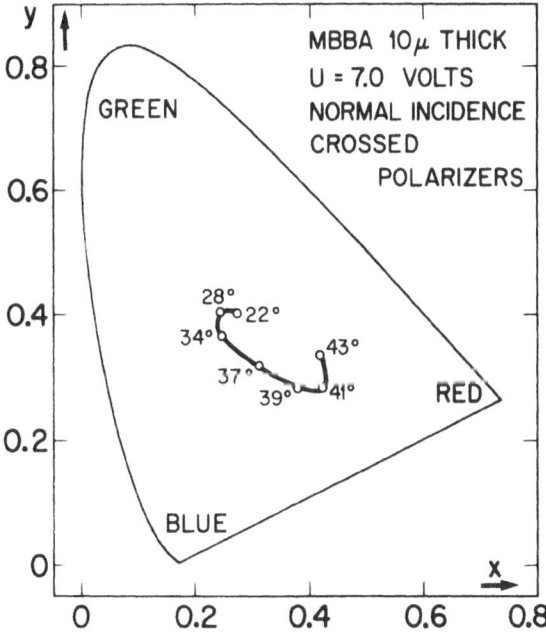

Fig. 11. Temperature dependence of transmitted tunable retardation colors

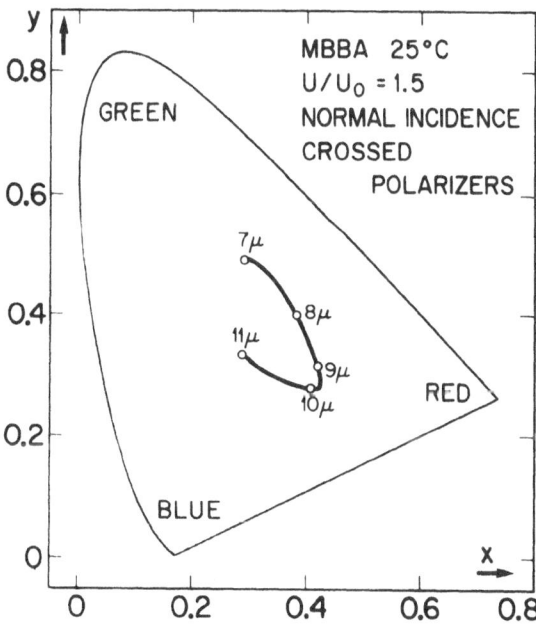

Fig. 12.  Dependence of transmitted tunable retardation colors on nematic layer thickness

for MBBA with $U/U_o = 1.5$, increase through nearly one complete order as the cell thickness is increased from 7 $\mu$m to 11 $\mu$m. A cell thickness tolerance within $\pm 0.25$ $\mu$m over the full area of the display would be required for good color uniformity.

The tunable retardation colors are also very strongly dependent on viewing angle. We computed the optical path length difference of the deformed nematic layer as a function of viewing angle from (6), evaluated by numerical integration. Figure 13 shows the range of colors obtained by varying the

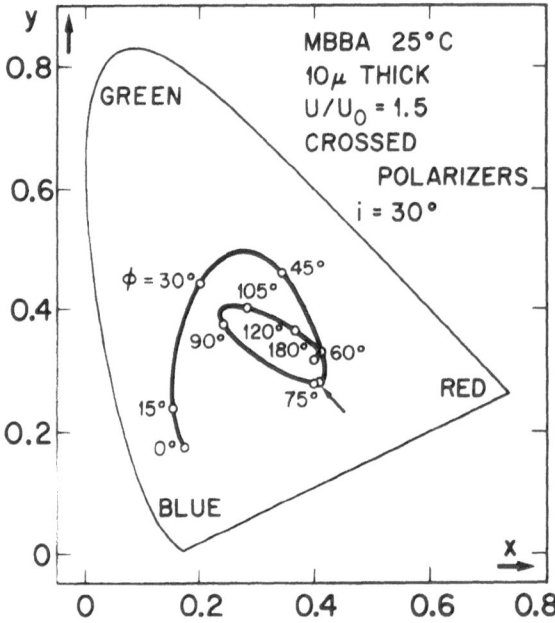

Fig. 13. Variation of tunable retardation colors with azimuthal angle $\phi$ for a constant 30° angle of incidence. Azimuthal angles $\phi$ and 360°-$\phi$ give the same color. Arrow indicates color at normal incidence.

azimuthal angle for a 30° incident light beam. The colors span nearly two full orders, one order lying below the second-order red color occurring for normally incident light (arrow in Fig. 13) and one order above it.

## 2.6. Discussion and Conclusion

We have examined the range of colors that can be obtained with the tunable retardation effect on the objective basis of the CIE chromaticity diagram. Colors of any given hue can be obtained by applying the appropriate voltage to the nematic layer, but many of the colors are of low saturation. The thickness and angle dependence of the colors limit the application of this effect to projection displays which permit the use of small electrooptic cells of thick optically flat glass and where the conditions of cell illumination can be precisely defined by the design of the optical system. Special precautions are required to keep the nematic layer at a constant temperature, because of the strong temperature dependence of the colors.

There are several variations to the conventional tunable retardation effect. The geometry just described with the initially homeotropic state between crossed polarizers has the advantage that the field-off state is dark and the display acts as a light valve as well as a tunable color filter. The difficulty with this geometry, however, is that the optic axis in the nematic layer is not restricted solely to the x-z plane (Fig. 4) but can assume different azimuthal angles in different parts of the nematic layer, giving the transmitted colors a mottled appearance. This angle degeneracy can be prevented by specially treating the surfaces of the display cell to give the optic axis a slight uniform tilt bias in the field-off state[3]. The bias must be kept small or else the light valve character of the display will be lost.

Another variation of the tunable retardation effect employs a nematic with a positive dielectric anisotropy and starts with an initial alignment that is nearly parallel to the cell surfaces. Application of a voltage to the cell reorients the optic axis of the layer to make it perpendicular to the boundaries. This scheme avoids the degeneracy problem of the initially homeotropic geometry, but the display can never be made completely dark because of the residual retardation of the transition layers near the cell boundaries where the optic axis is still parallel to the surface. Furthermore, this scheme has a larger temperature dependence because the deformation and the birefringence both act to decrease the optical path length difference at higher temperatures.

## 3. TWISTED NEMATIC CELL WITH EXTERNAL BIREFRINGENT LAYERS

In the tunable retardation scheme the birefringent layer is also the active nematic electrooptical cell. In this section we describe color switching displays that employ passive birefringent layers which are separate from the liquid crystal electrooptical cell. This type of display can only switch between two discrete colors, but the colors show no temperature or thickness dependence.

### 3.1. Transmission Displays

Consider the transmission display system shown in Fig. 14 which consists of a 90° twisted nematic cell[16] between parallel polarizing layers with a birefringent layer interposed at 45° between the twisted nematic cell and the final polarizer. In the field-off state of the cell, the direction of polarization of the light from the first polarizer is rotated by 90°. The transmission spectrum of the system is given by $T_\perp(\lambda)$ of (9). In the field-on state the transmission is $T_\parallel(\lambda)$ given by (10). The optical path length difference is now a fixed property of the birefringent layer rather than a variable property of the electrooptical cell. The methods described in Section 2.1 can be used to compute the range of colors possible for a birefringent material having no

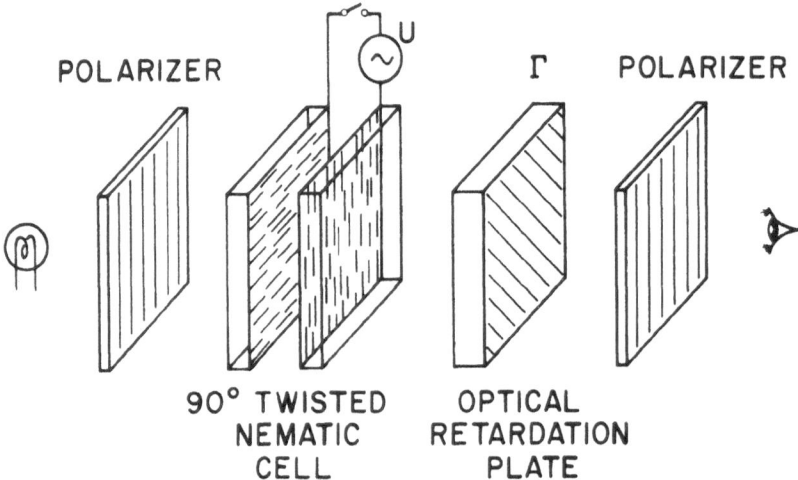

POLARIZER                    Γ        POLARIZER

90° TWISTED          OPTICAL
NEMATIC           RETARDATION
CELL                 PLATE

Fig. 14. Two-color transmission display using a 90° twisted nematic cell and a passive birefringent sheet.

dispersion in its refractive indices. The results are given in the color diagrams of Fig. 15 for the field-off state and Fig. 16 for the field-on state. These diagrams appear similar to Fig. 7 for the tunable retardation effect except that now the numbers indicated along the spiral curve refer to the fixed retardation Γ of the external birefringent layer (in units of 100 nm) rather than to the reduced voltage applied to the nematic layer. The differences between the color range of Fig. 15 and those of Fig. 8 are a result of the different amounts of dispersion in the refractive indices of the birefringent materials. The colors indicated in Figs. 15 and 16 follow the Newtonian scale since we have assumed Γ to be a constant independent of wavelength. Figures 15 and 16 form a complementary system of colors; that is the two colors corresponding to a given Γ add together to give white light.

Experimental transmission curves measured with the scheme in Fig. 14 are given in Fig. 17 for the yellow/blue complementary pair of colors obtained with Γ = 650 nm[17]. Birefringent sheets obtained by stretching polymer films are most suitable for this type of display. Such sheets are commercially available in the standard values of Γ = 140, 200, 280 and 560 nm. A large number of other values can be obtained by appropriate additive and subtractive combinations of these standard sheets, the Γ values of the two sheets adding when their slow axes are parallel and subtracting when their slow axes are at right angles.

The colors of the display no longer depend upon the thickness of the electrooptic cell, since they are determined only by the retardation Γ of the external birefringent layer, and the uniformity of the commercially available birefringent sheets is more than adequate to ensure the absence of color variation over the display. Transmissive displays large enough to be placed in front of a cathode ray tube have proved feasible using this technique[18]. Furthermore, the colors are not temperature dependent. The angle dependence

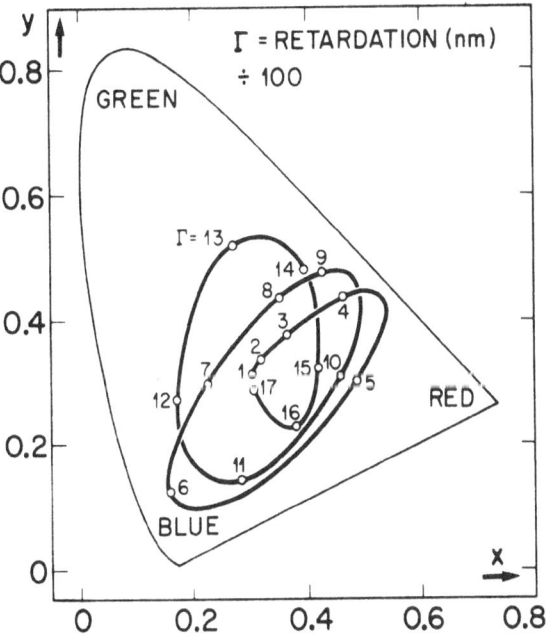

Fig. 15.  Transmission colors of the **field-off** state of the display in Fig. 14 as a function of the optical path difference Γ of the passive birefringent sheet.

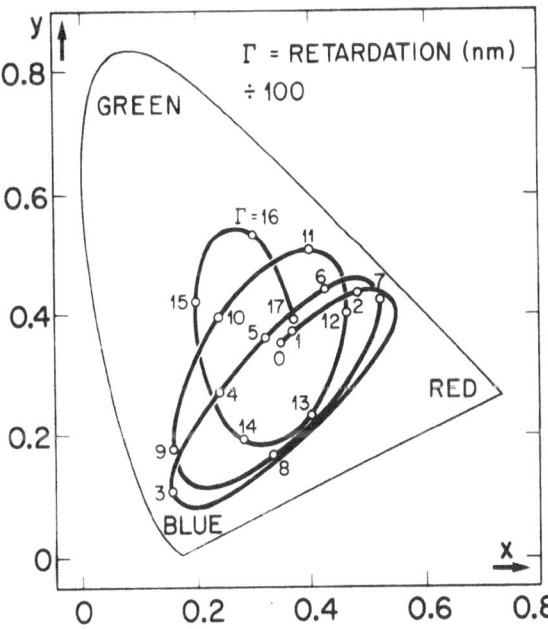

Fig. 16.  Transmission colors of the **field-on** state of the display in Fig. 14 as a function of the optical path length difference Γ of the passive birefringent sheet.

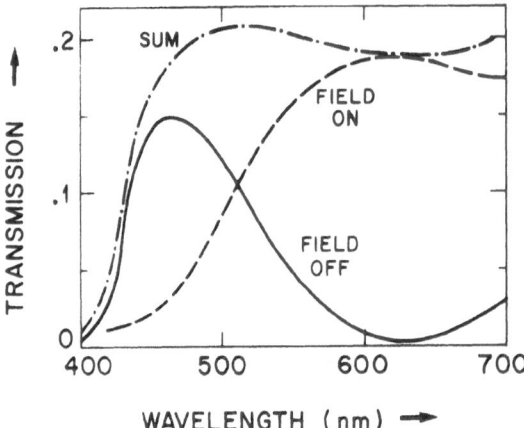

Fig. 17. Experimental transmission curves for the display of Fig. 14 with $\Gamma = 650$ nm. The transmitted curves are complementary since the sum curve (upper dashed line) is nearly a constant. Low transmission near 400 nm is caused by absorption in the liquid crystal.

of this display type is considerably less than for tunable retardation displays because the optic axis in the external birefringent layer lies in the plane of the sheet rather than being directed out of this plane as it is in the nematic layer. Consider the example of a birefringent layer consisting of a stretched polyvinyl alcohol film ($n_e$ = 1.560 and $n_o$ = 1.526[19]) with an optical path length difference of $\Gamma$ = 1000 nm when viewed at normal incidence. According to Fig. 15 a second-order red color is obtained. For a fixed angle of incidence of 30° the minimum value of $\Gamma$ is 836 nm, which occurs at an azimuthal angle of 0° and corresponds to a yellow-green color. The maximum value, occurring at an azimuthal angle of 90°, is $\Gamma$ = 1057 nm which corresponds to a third-order red-purple. The total range of colors therefore spans about half of one order instead of the two orders covered by the tunable retardation effect under similar conditions (Fig. 13).

The relative brightness of the colors is surprisingly high considering that the oscillatory shapes of the transmission curves are quite different from the band-pass and band-stop curves which yield maximum luminous transmittance (Fig. 3). The efficiency of the birefringence colors transmitted between crossed polarizers as a function of $\Gamma$ was calculated by comparing the luminous transmittance obtained from (9) and (2) with the maximum luminous transmittance for the same color obtained from the curves in Fig. 2. The results are given in Fig. 18. The efficiency cannot exceed 50% because half of the incident light is absorbed by the polarizers which we have assumed to be ideal.

We have so far considered only displays employing single birefringent sheets, and the colors obtained under these circumstances are shown in Fig. 15 and 16. This is just the simplest case of a much more general class of birefringent color filters consisting of two or more separate birefringent layers

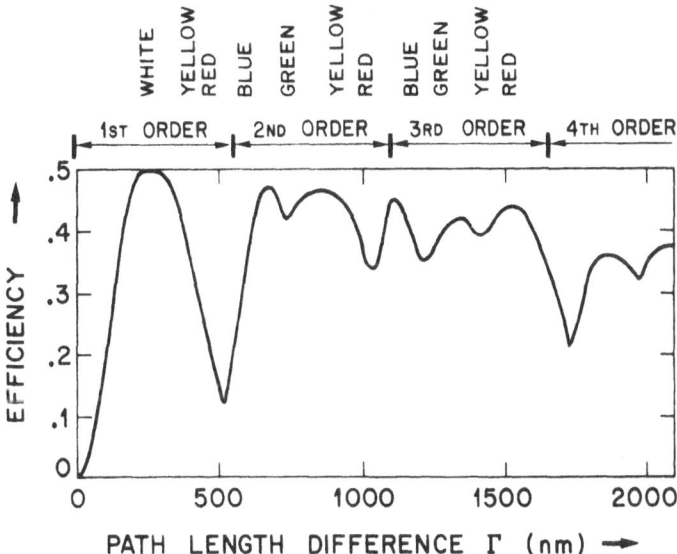

Fig. 18. Luminous efficiency of birefringence colors transmitted between crossed ideal polarizers calculated for C illumination. Maximum efficiency is 0.5 because ideal polarizers absorb half of the light.

oriented at special angles. Ammann[20] reviews procedures for synthesizing networks from multiple birefringent layers whose transmission spectra can be arbitarily specified. With these networks it would be possible, at least in principle, to obtain other, more saturated colors than those shown in Fig. 15 and 16.

### 3.2. Reflection Displays

The insensitivity of the colors to temperature and to cell thickness variations as well as the relative insensitivity of the colors to viewing angle makes it worthwhile to consider reflective displays employing this effect. A simple reflective display can be made by placing a metallic reflector behind the final polarizer of Fig. 14. The reflectivity of the device as a function of wavelength is given by

$$R_{off}(\lambda) = 1/2 \sin^4(\pi\Gamma/\lambda) \qquad (11)$$

and

$$R_{on}(\lambda) = 1/2 \cos^4(\pi\Gamma/\lambda) \qquad (12)$$

instead of (9) and (10). The colors that can be obtained in the field-off state are given in Fig. 19. Note that for a given $\Gamma$ the hues of the colors are practically the same as for the transmission case of Fig. 15, but they are much

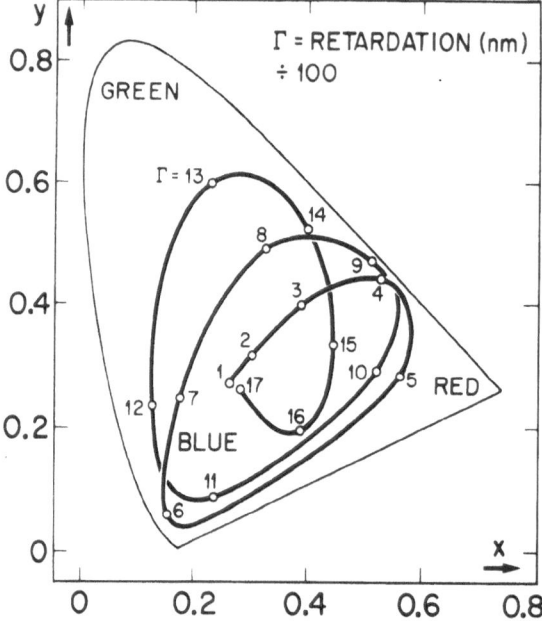

Fig. 19. Computed colors possible in the field-off state of a reflective mode display using an external birefringent sheet. Colors are more saturated than for transmission display of Fig. 15.

more saturated. The efficiency curve for these colors is very similar to Fig. 18. Unfortunately, the luminous reflectance of practical displays employing this effect is unacceptably low. The colors appear dark because of excess light absorption in the nonideal polarizing layers. These losses are accentuated in a reflective system because the light has to pass four times through a polarizer before it reaches the observer.

Shanks describes a reflective display that uses only a single polarizer[21]. The basic design is illustrated in Fig. 20 and consists of a polarizer, a 45° twisted nematic electrooptical cell, a birefringent sheet and a reflector. The optic axis of the birefringent layer is parallel to the orientation direction of the nematic liquid crystal at the adjacent side of the twist cell. In the field-off state the birefringent plate has no effect since the light is polarized parallel to its optic axis and the display appears white, but in the field-on state the E-vector of the polarized light makes a 45° angle with the optic axis of the birefringent layer and the same colors are generated as would be produced if a birefringent layer of retardation $2\Gamma$ were placed between parallel polarizers in transmission. The reflection curves for this case are given by

$$R_{off}(\lambda) = 1/2 \qquad (13)$$

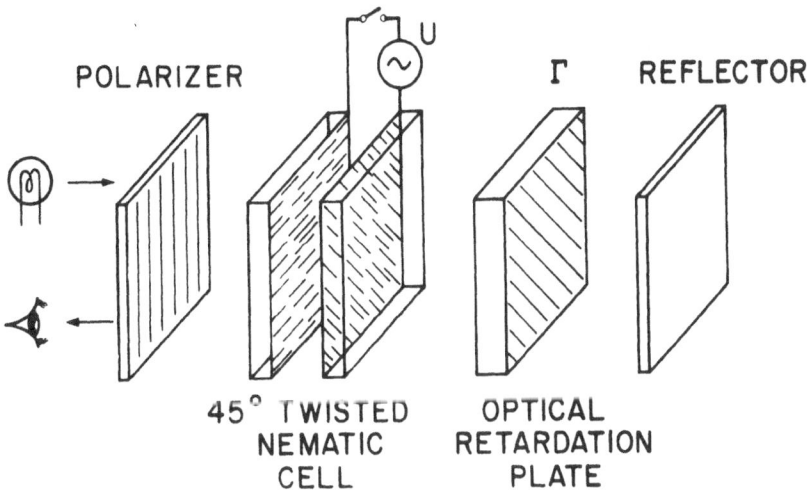

POLARIZER                                $\Gamma$      REFLECTOR

45° TWISTED          OPTICAL
NEMATIC          RETARDATION
CELL            PLATE

Fig. 20.  Reflective display scheme using a 45° twisted nematic cell and a single polarizer.

and

$$R_{on}(\lambda) = 1/2 \cos^2(2\pi\Gamma/\lambda) . \qquad (14)$$

Rotating the birefringent layer in Fig. 20 by 45° interchanges $R_{off}$ and $R_{on}$ in the expressions (13) and (14). Note that a black/white display is possible if the birefringent layer is an achromatic quarter-wave plate.

Display schemes using a 45° twist cell generally are not practical because the cell has a residual optical retardation in its field-on state caused by the transition layers near the cell boundaries where the optic axis of the nematic remains nearly parallel to the surface. This residual retardation causes no problem in a 90° twist cell because the E-vector of the light is either parallel or perpendicular to the optic axis of the transition layer. In a 45° twist cell, on the other hand, the optic axis of the second transition layers makes a 45° angle with the E-vector of the polarized light and the effects of its optical retardation must be considered in the performance of the display. In Fig. 20, for example, the optical retardation of the transition layer will either add to or subtract from the retardation of the external birefringent sheet, and this will impose a cell thickness and voltage dependence to the colors. The residual optical path length difference of a 45° twist cell for $U \gg U_o$ is given by

$$\Gamma \simeq \frac{(n_e-n_o)e}{\pi} \frac{U_o}{U} \qquad (15)$$

This amounts to about a quarter wavelength retardation for a typical 10 $\mu$m thick cell with $n_e$-$n_o$ = 0.2 and an applied voltage of 5 times the threshold value.

## 4. NEMATIC GUEST—HOST EFFECT

The earliest liquid crystal display scheme using pleochroic dyes was de-scribed by Heilmeier and Zanoni[22] in 1968, and later in more detail by Heilmeier, Castellano and Zanoni[23]. Pleochroic dye molecules have the pro-perty that the amount of light that they absorb depends on their orientation to the E-vector of the incident radiation. In this effect the pleochroic dye (guest) molecules are dissolved in a nematic liquid crystal (host) which acts as a matrix to orient the dye molecules. Application of an electric field to the nematic liquid crystal layer reorients the embedded dye molecules to-gether with the nematic molecules and causes a change in the light absorbed by the layer. The pleochroic dye molecules used in this effect have rigid elongated structures similar to nematic liquid crystal molecules.

Consider the transmissive display scheme illustrated in **Fig. 21** which consists of a linear polarizer in front of a parallel oriented layer of nematic/

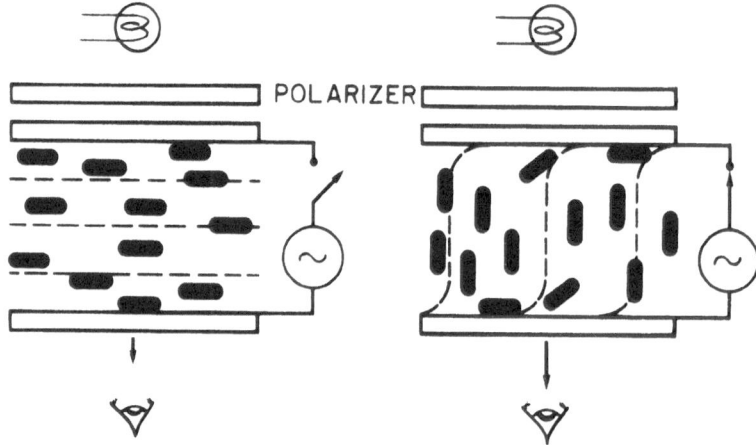

Fig. 21. Transmission display using nematic guest-host effect. Black shapes symbol-ize oriented pleochroic dye molecules.

dye mixture sandwiched between two transparent electrodes. The lozenge-shaped forms in the figure represent schematically the oriented pleochroic dye molecules. In the field-off state, the E-vector of the polarized light is parallel to the long axes of the dye molecules and the absorption of light is maximized. In the field-on state the nematic layer reorients the dye mole-cules and the absorption of light is minimized.

In Fig. 22 are shown the transmission spectra of a 17 $\mu$m thick parallel-oriented layer consisting of 1% Indophenol Blue dissolved in the three com-ponent nematic host comprising equal amounts of p-butoxy-, p-hexyloxy and p-octanoyloxybenzylidene-p'-aminobenzonitrile[16]. The curves were measured in polarized light for both the field-on and field-off states and have

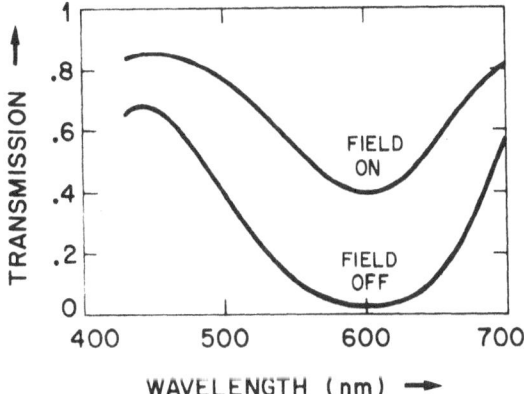

Fig. 22. Polarized transmission curves for a 1% mixture of Indophenol Blue in a nematic host.

been corrected for surface reflection losses. Note that a considerable amount of the polarized light is still absorbed by the dye in the field-on state, causing the display to retain a bluish color and resulting in a poor contrast between the on and off states. This poor contrast has two primary causes: failure of the transition axis of the dye molecule to coincide with its long axis, and incomplete ordering of the dye molecule by the nematic host. The first cause can be overcome, but the second cause, incomplete ordering, is fundamental to the nematic phase and can never be completely eliminated.

The black shapes in Fig. 21 represent only the average orientation of the dye molecules in the nematic host. Actually both dye and nematic liquid crystal molecules are tumbling anisotropically about this direction of average orientation. The degree of orientation of the long axis of a dye molecule or a liquid crystal molecule can be expressed[24] in terms of an order parameter S where

$$S = <3 \cos^2\alpha - 1>/2 \qquad (16)$$

averaged over time. $\alpha$ is the angle between the average direction and the long molecular axis at a given instant in time. A rough idea of the order parameter for nematics is obtained from the Maier-Saupe theory[24] which predicts S = 0.7 at room temperature for a nematic substance whose isotropic transition occurs at 65° C. The order parameter of the dissolved dye is not necessarily the same as that of the nematic host; it can even be higher.

The order parameter of the transition axis of the dye, the quantity of interest for a guest-host display, is measured from the intensity of the polarized optical absorption band corresponding to that particular transition[25]. The transmission spectrum of a parallel oriented guest-host layer is measured in linear polarized light with the E-vector parallel to the optic axis of the sample, giving $T_\parallel(\lambda)$, and with the E-vector oriented perpendicular to the optic axis, giving $T_\perp(\lambda)$. S is obtained from[26]

$$S = \frac{\log T_\parallel - \log T_\perp}{\log T_\parallel + 2\log T_\perp} \tag{17}$$

which can be evaluated at any wavelength where there is significant absorption by the pleochroic dye. Equation (17) is valid and the computed S value is independent of wavelength as long as the dye is a pure compound with only one optical transition. Equation (17) is generally evaluated at the wavelength where most light is absorbed by the dye. The chemical structures of Indophenol Blue and Sudan Black B are shown in Fig. 23. These structures

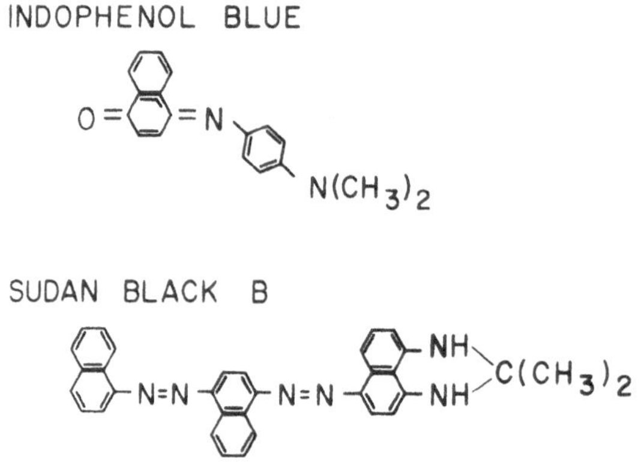

INDOPHENOL BLUE

SUDAN BLACK B

Fig. 23. Typical pleochroic dyes for the nematic guest-host effect.

illustrate the characteristics of guest-host dyes, which are long, rigid, extended aromatic molecular systems. The dyes must not contain ionic substituents because these would decrease their solubility in the nematic host, increase the power consumption of the display and promote electrochemical reactions. The larger order parameter of Sudan Black B (0.63)[26] compared with Indophenol Blue (0.55) illustrates the general rule that longer dye molecules tend to have larger order parameters in the same host. Unfortunately, the longer dye molecules are also less soluble and so it is necessary to compromise between order parameter and solubility.

## 5. GUEST–HOST INTERACTION COMBINED WITH THE CHOLESTERIC PHASE CHANGE EFFECT

White and Taylor[26] have recently described a major improvement over the nematic guest-host effect which requires no polarizing sheets or surface orientation and yet can give brighter displays with higher contrast. The basic effect is illustrated in Fig. 24 for the reflective mode. The new idea is the

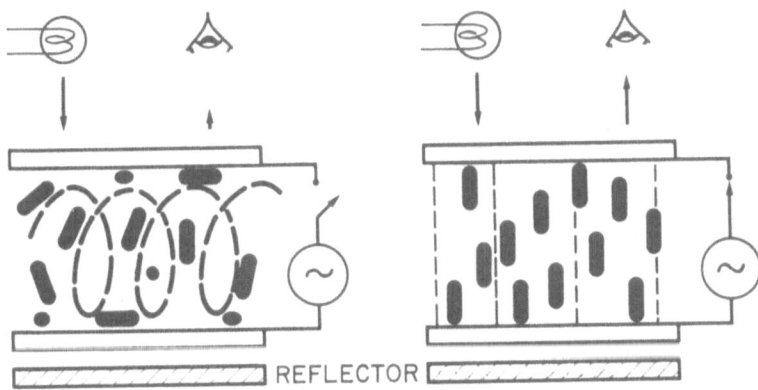

Fig. 24. Reflective display using cholesteric guest-host effect.

addition of a small amount of chiralic dopant to the guest-host nematic mixture to generate a chiral nematic mesophase which absorbs light of all polarizations equally well. In the field-off state this cholesteric structure adopts either the focal conic texture with the helicoidal axis more or less randomly oriented in the layer (shown in Fig. 24) or the planar texture where the helicoidal axis is uniformly perpendicular to the layer. In practice both textures are generally found in the same display, the planar texture occurring in regions of the display where there are never any fields applied and the focal conic texture occurring in the active display region. In the field-off state the dye molecules absorb light regardless of its polarization, for either texture. In the field-on state the chiral nematic structure gives way to a uniform homeotropic configuration (Fig. 24) which absorbs much less light since the long axes of the dye molecules are perpendicular to the E-vector of the light.

We can best compare the optical performance of the nematic and the cholesteric guest-host displays by considering the brightness and contrast ratio characteristics of the two displays. The reflectivity of a nematic guest-host display, assuming an ideal polarizer and reflector, can be written as

$$R_{off} = 1/2 \exp \left[ -2(2S+1)\alpha_o D \right] \qquad (18)$$

and

$$R_{on} = 1/2 \exp \left[ -2(1-S)\alpha_o D \right] \qquad (19)$$

where $R_{off}$ and $R_{on}$ are the reflectances in the field-off and field-on states, S is the order parameter of the transition axis of the dye, $\alpha_o$ is the attenuation constant of the same concentration of the dye in an isotropic host and D is the cell thickness. Equations (18) and (19) may be combined to eliminate $\alpha_o$ and D yielding

$$R_{on} = 1/2 \, C^{(S-1)/3S} , \tag{20}$$

where C is the contrast ratio of the display defined by $C = R_{on}/R_{off}$. Equation (20) describes a family of brightness versus contrast curves for the nematic guest-host scheme. The shape of the curves depends only upon the order parameter of the dye; the dye concentration and cell thickness determine the operating point on the curve. A similar analysis for the cholesteric guest-host effect yields

$$R_{off} = \exp[-2\alpha_o D] \tag{21}$$

for the field-off state in the focal-conic texture and

$$R_{on} = \exp[-2(1-S)a_o D] \tag{22}$$

for the field-on state. Eliminating $\alpha_o$ and D from these equations yields

$$R_{on} = C^{(S-1)/S} . \tag{23}$$

The curves corresponding to (20) and (23) are drawn in Fig. 25 for the two cases S = 0.5 and S = 0.7. The brightness curves for the nematic guest-

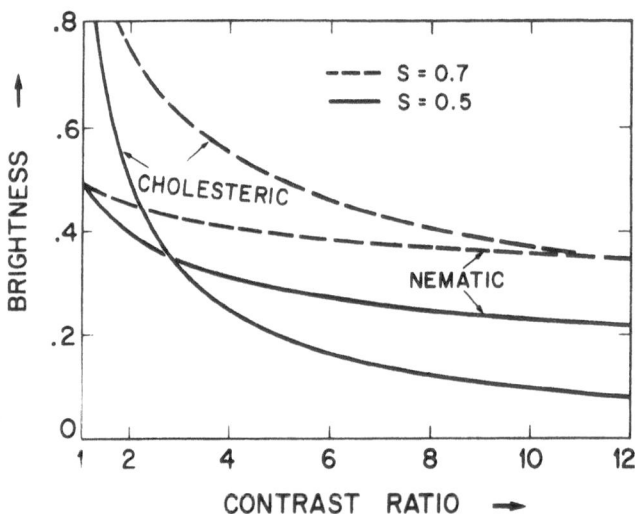

Fig. 25. Brightness versus contrast for nematic and cholesteric guest-host displays.

host effect all start at 0.5 since the ideal polarizer absorbs half of the light. The curves for the cholesteric guest-host effect start at unity but eventually cross the corresponding nematic curves since, for a given S, the exponent is

three times larger in (23) than in (20). The two solid curves for S = 0.5 intersect at C = 2.83. For a given contrast ratio C < 2.83 the cholesteric scheme
will give the brighter display; the nematic scheme will give the brighter display if C > 2.83. The contrast ratio for either scheme may be arbitrarily chosen by appropriately adjusting the cell thickness and dye concentration. For
a dye with S = 0.7 (dotted lines in Fig. 25), the two curves intersect much
farther out at C = 11.31. For a practical display employing a neutral dye
with S = 0.7 and having a contrast ratio comparable to a newspaper (i.e. C =
5-6), the cholesteric scheme will give a brighter looking display. The brightness difference in practice will be even larger than indicated in Fig. 25,which
assumes an ideal polarizer.

The advantage of the cholesteric guest-host effect is that an improved
optical performance is accompanied by a simplified cell technology, the display no longer requiring a polarizer or surface orientation procedures. This
effect needs a higher operating voltage than the conventional guest-host
scheme does, but this is accompanied by a steep optical response curve and
hysteresis[27], which are both useful for multiplexing.

## 6. DISPLAYS EMPLOYING COLOR SELECTIVE POLARIZERS[17,28]

Let us now consider a different display type where oriented pleochroic dyes,
similar to those described earlier, are located in external sheets rather than
being mixed into the liquid crystal itself. The reflective display scheme in
Fig. 26 can switch two unrelated colors. The first two elements are selective

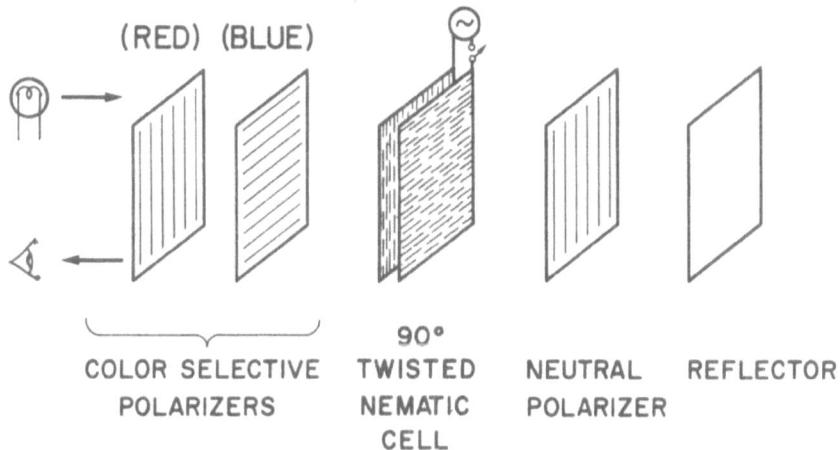

Fig. 26: Two color reflective display using two different selective polarizers oriented
at right angles.

polarizers, a red polarizer containing a vertically oriented red pleochroic dye,
and a blue polarizer containing horizontally oriented blue pleochroic dye
molecules. Next follow a 90° twisted nematic cell, an ordinary neutral po-

larizer oriented to absorb light with E-vectors in the vertical direction, and a metallic reflector.

To understand how this display works it is helpful to consider the unpolarized display illumination as consisting of two orthogonally polarized components, one with vertical E-vector and one with horizontal E-vector. When the twist cell is turned on, certain wavelength bands of the vertically polarized component of the incoming light are absorbed in the red polarizer and the rest of the light is completely absorbed in the neutral polarizer. The only light that can reach the observer comes from those spectral regions of the horizontally polarized component that are not absorbed in the blue polarizer. In the field-off state the horizontally polarized component is completely absorbed, but only certain spectral regions of the vertically polarized component are absorbed in the red polarizer. A display with segmented electrodes would therefore show blue characters against a red background. Rotating either the neutral polarizer or the two selective polarizers by 90° would interchange the colors.

The positions of the two selective polarizers can be interchanged without effect. In principle, the two selective polarizers could even be interchanged with the neutral polarizer, but in practice the neutral polarizer should not be the first element in the display because this configuration gives objectionable black edges around the switched-on elements when the display is viewed at an angle. This effect results from the small separation between the electrodes and the reflector which permits the E-vector of obliquely incident light to be rotated by 90° as it goes in through an off-region and to come out through an on-region without a second rotation. The reflected beam is thus completely absorbed by the front polarizer and the parts of the display near the borders of the activated electrode segments appear black. This effect does not occur when the neutral polarizer is next to the reflector or when the display is operated in the transmission mode without a reflector.

We have prepared selective polarizers of various colors by employing the standard technique[19] of staining stretched polyvinyl alcohol film (PVA) with water-soluble pleochroic dyes. In our procedure a 60 mm wide, 70 mm long and 50 $\mu$m thick PVA strip is mounted in the sliding metal frame shown in Fig. 27a. The frame is then submerged in water to soften the PVA. The strip is slowly stretched to about 5 times its original length and firmly fixed (Fig. 27b) because the PVA is quite elastic in this state. The stretched PVA is then dipped into a 0.1% aqueous solution of the dye where it is stained in approximately two minutes to the strength desired for reflection mode displays. Polarizers designed for transmission mode displays are left in the solution for a longer time. After rinsing and drying, the PVA strip is quite rigid and can be removed from the holder.

Fig. 28 shows the transmission spectra of a stretched PVA sheet dyed with Solantine Red 8BL. The two curves were measured in linearly polarized light and have been corrected for loss due to surface reflections. In the upper curve for $T_\perp$ the E-vector of the light is perpendicular to the stretching direction, and very little light is absorbed. In the lower curve for $T_\parallel$ the pleochroic dye molecules absorb strongly, resulting in a deep red transmitted

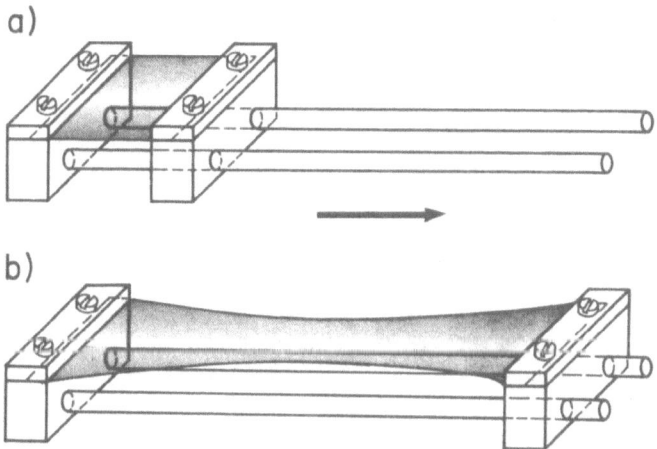

Fig. 27. Polyvinyl alcohol sheets in a) unstretched and b) stretched states.

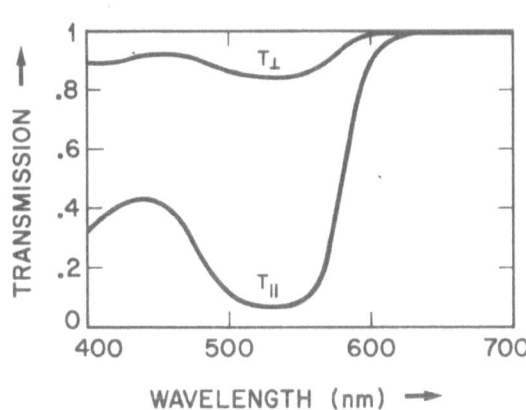

Fig. 28. Polarized transmission curves of a selective polarizer prepared with Solantine Red 8BL.

color. The small absorption for $T_\perp$ indicates that the dye molecules are not perfectly oriented in the stretched PVA. The apparent order parameter for Solantine Red 8BL in this polarizer is computed from (17) to be 0.84, a value that has yet to be achieved in a liquid crystal guest-host system.

It is interesting to see how the apparent order paremeter of the dye depends upon the extent to which the PVA has been stretched. The results for Solantine Red 8BL are shown in Fig. 29, where it is seen that the dye becomes better and better oriented the more the PVA is stretched. We determined the stretch ratio of the PVA by drawing a circle on the unstretched PVA and then taking the ratio of the major and minor axes of the ellipse into which the circle deformed when the PVA was stretched. This ratio is larger than the linear elongation ratio of the PVA strip because the width of

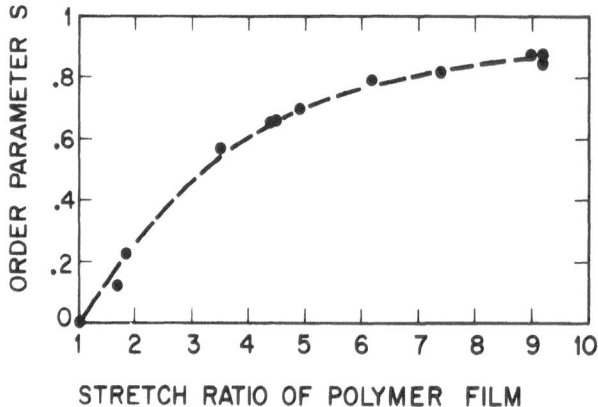

Fig. 29. Order parameter of Solantine Red 8BL as a function of the amount that the polyvinyl alcohol sheet has been stretched.

the PVA sheet decreases in the stretching process. We did not make samples with stretch ratios larger than 10 because the films ruptured when stretched beyond this point.

The order parameters of different pleochroic dyes used in selective polarizers show considerable variation. Examples of some of these dyes are given in Fig. 30. Their structures are similar to the guest-host dyes in Fig. 23, except that the selective polarizer dyes have the additional highly ionic

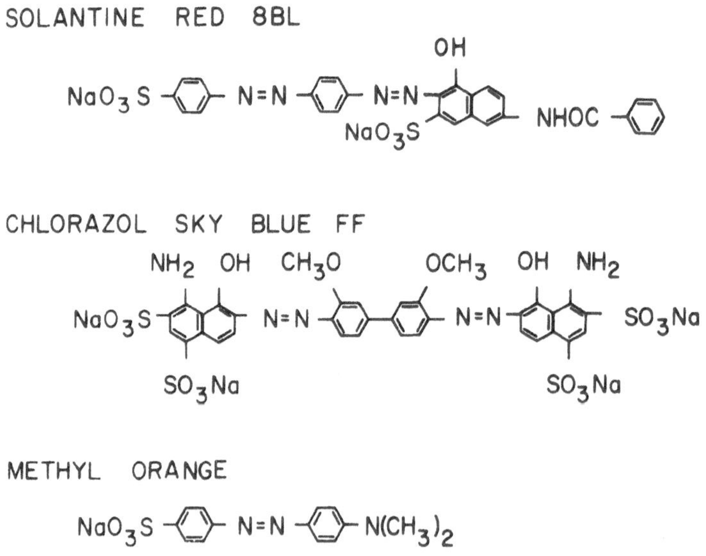

SOLANTINE RED 8BL

CHLORAZOL SKY BLUE FF

METHYL ORANGE

Fig. 30. Typical pleochroic dyes used for preparing selective polarizers.

sulfonate substitutents (-$SO_3$Na) which promote their solubility in water. As a general rule longer dye molecules tend to orient better than do shorter molecules. We investigated some very long triazo and tetraazo dyes but found that they are unable to stain the PVA to the required strength even though they were quite soluble in water.

Ordinary neutral sheet-type polarizers are also made by staining stretched PVA, but in this case the PVA is stained with inorganic iodine instead of with organic dyes. The iodine forms a complex with the PVA, building up chains of iodine atoms which are much longer than any organic dye molecule. The apparent order parameter can exceed 0.97[29].

This effect is particularly suited for reflective displays, since the colors do not depend upon temperature or the thickness of the nematic layer. Furthermore, the colors are not angle dependent, but the viewing angle restrictions of the twisted nematic cell still apply. This effect can give bright reflective displays because only one sheet-type neutral polarizer is required and, as Fig. 28 shows, selective polarizers are completely transparent in certain wavelength regions.

## 7. CHOLESTERIC LAYER AS ACTIVE DISPLAY ELEMENT

Cholesteric liquid crystals are of interest for displays because of their very attractive iridescent reflection colors. The optical properties of cholesteric liquid crystals are well understood; the colors result from the interference of light in the periodic internal structure of the mesophase. Cholesteric liquid crystals have been extensively studied in an attempt to develop a practical display where the hue of the iridescent reflection colors could be controlled with an applied voltage. The results of these investigations have been rather disappointing and the outlook is not very promising, for the very fundamental reason that the electrical and optical effects are incompatible. The cholesteric liquid crystal must be in its planar texture to exhibit the iridescent reflection colors, but it must assume the focal conic texture in order to undergo field-induced pitch dilation which is necessary to change the reflection color (assuming that a longitudinal electric field is applied as would be the case in a display with transparent electrodes). A longitudinal field applied to the planar texture produces the grid pattern deformation[30] which is a precursor of the transformation to the scattering focal conic texture and eventual pitch dilation. Only an in-plane (transverse) electric field has the proper direction to unwind the helix while maintaining the planar texture, but this is not a practical geometry.

## 8. CHOLESTERIC LAYER AS PASSIVE DISPLAY ELEMENT

Let us therefore consider display applications where the planar cholesteric layer is employed only as a passive element, with no voltage applied to it. One novel application where a cholesteric layer is employed as a colored polarizing reflector[31] is shown in Fig. 31. The upper diagram shows the various polarization states of the light that propagates through the display in the

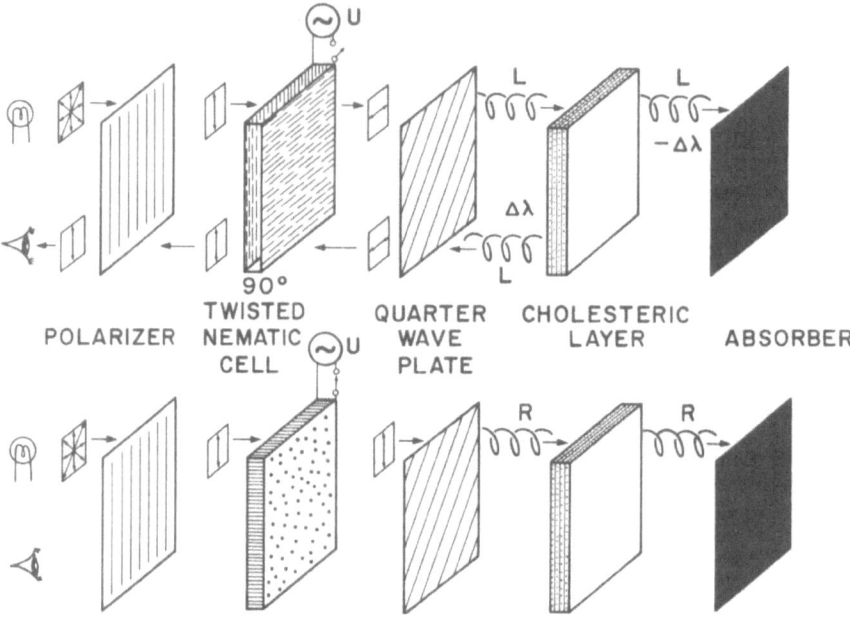

Fig. 31. Twisted nematic display employing an external cholesteric layer as a colored polarizing reflector.

field-off state. Linear polarized light leaving the front polarizer is rotated 90° by the twisted nematic element and converted to left-circular polarized light by the quarter-wave plate behind it. Light outside the reflection band of the cholesteric layer (-Δλ) is transmitted to the black layer where it is absorbed, and light within the reflection band of the cholesteric layer (Δλ) is totally reflected as left-circular polarized light by the assumed left-handed cholesteric structure. The quarter-wave plate transforms this reflected light into linear polarized light which is rotated 90° by the twist cell and transmitted through the front polarizer, giving the display the characteristic iridescent color of the cholesteric layer. In the field-on state (lower diagram of Fig. 31) the quarter-wave plate generates right-circular polarized light which is completely transmitted by the left-handed cholesteric structure and then absorbed in the black layer. This display configuration with segmented electrodes shows black characters against an iridescent background. Rotating either the polarizer or quarter-wave plate by 90° would give the converse effect.

This display effect is unusually bright because only one sheet type polarizer is employed, and the quarter-wave plate and cholesteric layer act as a nearly ideal colored polarizing reflector[31]. At present this type of display is impractical for large scale production because it requires two separate liquid layers. This situation can change, however, as soon as other technological advances make it possible to polymerize planar-textured cholesteric liquid crystals into thin sheets.

## REFERENCES

1 see, for example, D. B. Judd and G. Wyszecki, 'Color in Business, Science and Industry' 2nd ed., Wiley, New York (1963)

2 D. L. MacAdam, J. Opt. Soc. Am. 25 (1935) 249 and 361

3 F. J. Kahn, Appl. Phys. Lett. 20 (1972) 199

4 M. F. Schiekel and K. Fahrenschon, Appl. Phys. Lett. 19 (1971) 391

5 G. Assouline, M. Hareng and E. Leiba, Electron. Lett. 7 (1971) 699

6 R. A. Soref and M. J. Rafuse, J. Appl. Phys. 43 (1972) 2029

7 H. Gruler, T. J. Scheffer and G. Meier, Z. Naturforsch. 27a (1972) 966

8 P. F. Byrd and M. D. Friedman, 'Handbook of Elliptic Integrals for Engineers and Physicists' Springer, Berlin (1954) 8

9 M. Francon in S. Flügge (ed.) 'Handbuch der Physik XXIV', Springer, Berlin (1956) 442

10 J. Nehring, A. R. Kmetz and T. J. Scheffer, J. Appl. Phys. February 1976 (to be published)

11 M. Brunet-Germain, C. R. Acad. Sc. Paris 271B (1970) 1075

12 M. Born and E. Wolf, 'Principles of Optics' 3rd ed., Pergamon, Oxford (1965) 96

13 Landolt-Börnstein, 6th ed., Vol. II, Part 8, Springer, Berlin (1962) 4-555

14 I. Haller, J. Chem. Phys. 57 (1972) 1400

15 Measurements from this laboratory

16 M. Schadt and W. Helfrich, Appl. Phys. Lett. 18 (1971) 127

17 T. J. Scheffer, J. Appl. Phys. 44 (1973) 4799

18 E. P. Raynes and I. A. Shanks, Electron. Lett. 10 (1974) 114

19 E. H. Land and C. D. West, in J. Alexander (ed.), 'Colloid Chemistry' Vol. 6, Reinhold, New York (1946) 172

20 E. O. Amman, in E. Wolf (ed.), 'Progress in Optics' IX, North Holland, Amsterdam (1971) Chapter IV

21 I. A. Shanks, Electron. Lett. 10 (1974) 90

22 G. H. Heilmeier and L. A. Zanoni, Appl. Phys. Lett. 13 (1968) 91

23 G. H. Heilmeier, J. A. Castellano and L. A. Zanoni, Mol. Cryst. Liq. Cryst. 8 (1969) 293

24 A. Saupe, Angew. Chem. Internat. Edit. 7 (1968) 97

25 A. Saupe and W. Maier, Z. Naturforsch. 16a (1961) 816

26 D. L. White and G. N. Taylor, J. Appl. Phys. 45 (1974) 4718

27 W. Greubel, Appl. Phys. Lett. 25 (1974) 5

28 S. Kobayashi and F. Takeuchi, Proc. S.I.D. 14/4 (1973) 115

29 E. H. Land, J. Opt. Soc. Am. 41 (1951) 957

30 W. Helfrich, J. Chem. Phys. 55 (1971) 839

31 T. J. Scheffer, J. Phys. D: Appl. Phys. 8 (1975) 1441

## DISCUSSION

M. F. Schiekel (AEG-Telefunken)
Have you considered the effect of the numerical aperture of a tunable retardation projection display on its color gamut and saturation?

T. J. Scheffer
My calculations were done for collimated light at the specified angles. With practical projection optics, light passing through a point in the liquid crystal cell is collected over a cone of angles determined by the numerical aperture of the system. The color of the corresponding spot focussed on the projection screen is therefore a mixture of the colors calculated for these various angles. Consequently a larger numerical aperture yields less saturated colors with a somewhat altered voltage dependence.

I. F. Chang (IBM)
Would you comment on the degradation problems associated with dyes in liquid crystal display devices?

T. J. Scheffer
The dyes used in practical displays must be thermally, photochemically and electrochemically stable. White and Taylor report the 5-nitro-2-aminothiazole dyes to be thermally and photochemically stable but they noted some degradation under ac operation after three months. An additional complication with dye mixtures occurs upon filling thin cells, where a chromatographic separation of the dye components is observed owing to differences in the adsorption of the dyes on the inner glass surfaces.

F. J. Kahn (Hewlett-Packard)
Of the color effects you describe, which has the best angular field of view?

T. J. Scheffer
All displays with active elements based on refraction effects (e.g. twisted nematic, tunable retardation) have restricted fields of view; bulk absorption in guest-host displays depends only weakly on viewing angle.

SHORT COMMUNICATION

# SPECTRUM OF VOLTAGE CONTROLLABLE COLOR FORMATION

## WITH NEMATIC LIQUID CRYSTALS

H. MADA and S. KOBAYASHI

Tokyo University of Agriculture and Technology, Koganei, Japan

A number of reports on color display devices with nematic liquid crystals have been published[1-6]. In a previous paper we have reported on the optimum conditions for voltage controllable color formation (VCCF) with a panel operated in the tunable birefringence mode in which a nematic liquid crystal with a positive dielectric anisotropy ($\Delta\epsilon > 0$) was utilized. We also reported on its spectral width[6], where the effect of absorption anisotropy was neglected. This communication describes the whole transmission spectrum by taking absorption anisotropy into account.

The optimum condition for VCCF in a parallel oriented cell with molecules of $\Delta\epsilon > 0$ is achieved by placing the cell at $45°$ between crossed polarizers. In this condition, the transmitted intensity $I_T$ of monochromatic light through the cell is given by

$$I_T = [\frac{1}{2}(T_e \text{-} T_o)]^2 + T_e T_o \sin^2(\frac{\pi d \Delta n}{\lambda}) \tag{1}$$

where $T_e$ and $T_o$ stand for the normalized amplitudes of the transmitted extraordinary and ordinary rays, respectively; and where $\Delta n$ is the voltage-dependent effective birefringence. $T_e$ and $T_o$ were measured by setting parallel polarizers either parallel or perpendicular to the optic axis of the cell. Then $I_T = T_e^2$ for the extraordinary ray in the first case, and $I_T = T_o^2$ for the ordinary ray in the second case. $T_e$, $T_o$, and $\Delta n$ are wavelength-dependent; $\Delta n$ and thus $T_e$ are voltage-dependent, giving rise to VCCF.

In Figure 1, measured data for the spectral transmission of the extraordinary ray $T_e^2$ and ordinary ray $T_o^2$ at zero applied field are shown. The liquid crystal material used was a 4:1 mixture of MBBA and BBCA (butoxy-benzylidene-cyanoaniline). The quantities $[(T_e \text{-} T_o)/2]^2$ and $T_e T_o$ are also shown in Fig. 1 by dashed lines. These quantities, which correspond to the first term and the factor of the second term in (1), give the envelopes of the VCCF spectrum. In the case where there is no absorption anisotropy, $T_e = T_o$ and the first term in (1) vanishes. Then the VCCF cell can extinguish all wavelengths equally well. For the anisotropic absorption of shorter wavelengths shown in Fig. 1, this term exhibits a peak corresponding to a reduc-

H. MADA and S. KOBAYASHI

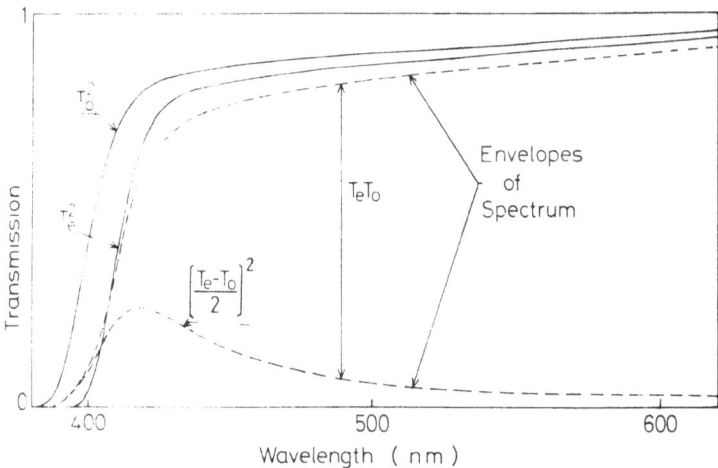

Fig. 1. The spectra of $T_e^2$, $T_o^2$, $[(T_e \cdot T_o)/2]^2$, and $T_e T_o$.

tion in the capability to extinguish short wavelengths. This transmission background degrades the color purity of VCCF.

In Fig. 2, a comparison between measured and calculated VCCF spectra is shown. The solid curve is the normalized transmission measured by applying 4 $V_{rms}$ (5 kHz) to the panel, and the dashed line is calculated from (1) using the theoretical variation of the effective birefringence with wavelength

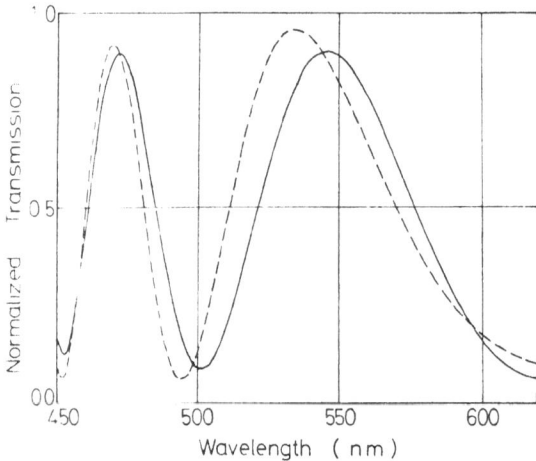

Fig. 2. Experimental (solid) and calculated (dashed) spectra of voltage controllable color formation for 4:1 mixture of MBBA:BBCA at 4 $V_{rms}$, 5 kHz, 25° C. Cell thickness was about 16 $\mu$m.

and applied voltage obtained from an independent study[7]. Qualitative agreement is recognized. We attribute the discrepancy to a very small error in the assumed $\Delta n$, which has a strong temperature dependence. The position of the peaks in the spectrum $\lambda_m$ is very sensitive to changes in $\Delta n$, e.g. $\delta\lambda_m = (3\text{-}15) \times 10^3 \, \delta\Delta n(nm)$.

We conclude that the variation observed in the envelope of the VCCF spectrum results from anisotropic absorption.

## REFERENCES

1   G. Assouline, M. Hareng and E. Leiba, Electron. Lett. 7 (1971) 699
2   S. Kobayashi, F. Takeuchi and T. Shimomura, Proc. 5th Conf. Solid State Devices, August 1973, Tokyo, Supplement to J. Jap. Soc. Appl. Phys. 43 (1973) 131; S. Kobayashi and T. Shimomura, Proc. Internat. Conf. on Liq. Cryst., December 1973, Bangalore (in press)
3   T. J. Scheffer, J. Appl. Phys. 44 (1973) 4799
4   S. Sato and M. Wada, IEEE Trans. Electron Devices ED-21 (1974) 171
5   G. H. Gooch and H. A. Tarry, Electron. Lett. 10 (1974) 2
6   H. Mada and S. Kobayashi, Rev. Phys. Appl. 10 (1975) 147
7   H. Mada and S. Kobayashi, Mol. Cryst. Liq. Cryst. (in press)

# CHEMICAL COMPOSITION AND DISPLAY PERFORMANCE

D. DEMUS

Martin-Luther-Universität Halle, German Dem. Rep.

## SUMMARY

Low melting liquid crystalline substances of different chemical structure are reviewed. The relations between melting and clearing point and the molecular structure are discussed. For practical purposes it is necessary to lower the melting points by the use of mixtures. The assumptions for calculation of eutectic melting diagrams are given. By comparison of calculated with experimental results some problems, especially the occurrence of mixed solid crystals and unstable solid and liquid crystalline states, are demonstrated.

The dielectric properties of liquid crystals are discussed with respect to their molecular structure. In particular, nematics with low frequency relaxation regions are considered. The dynamic properties of liquid crystals depend on the viscosities. The rheological behavior of liquid crystals and its temperature dependence is discussed briefly. The paper concludes with remarks on the double refraction of liquid crystals and their decomposition under the influence of heat, electric fields, radiation and chemical reagents.

## 1. INTRODUCTION

Recognition of the useful electrooptical properties of liquid crystals stimulated great efforts in the synthesis of new mesomorphic substances. In 1960 about 1500 liquid crystalline substances were known and now — 15 years later — the number exceeds 6000. Despite this high number of compounds no ideal substance is known, the properties of which fulfill all desirable requirements. By the use of mixtures it is possible to adapt some properties of the substance to a special purpose, but at present mixtures also exhibit restrictions with respect to ideal substances.

It is the purpose of this paper first to give a short survey on the chemistry of practically useful liquid crystals. Then some problems arising from the use of mixtures are discussed. The connections between molecular structure and properties of liquid crystals — dielectric, rheological, optical and some chemical properties — are reviewed. For the preparation of this paper results of our own group as well as the literature have been used. However, it is not intended to give a complete review of the literature. Only some topics are surveyed which lie within our own field of work and are assumed to be also of interest to those who are concerned with the electrooptics of liquid crystals.

## 2. LOW MELTING LIQUID CRYSTALLINE COMPOUNDS

In recent years several reviews on the chemistry of liquid crystals with respect to their practical use have been published[31-36]. A compilation of all liquid crystalline substances known up to 1972 can be found in the book of Demus et al.[37]. The following tables of selected examples give a survey on the chemistry of low melting substances.

A general remark concerning the melting temperatures of the substances is necessary. Because a generally valid melting theory is lacking we are forced to use empirical rules for the prediction of melting points. For obtaining low melting temperatures the size of the aromatic part of a molecule has to be restricted. The following structures yield the greatest success in obtaining low melting liquid crystals:

In some cases heterocyclic analogs of these groups have been efficient. Branches in the molecule, especially in the third formula may lead to a remarkable decrease of the melting temperatures. $R^1$ and $R^2$ usually are aliphatic substituents. Table 1 shows the most important groups used as substituents.

Table 2 shows selected examples for structure I. The stilbenes (No. 1 and 2) have relatively low clearing temperatures. They can isomerize upon exposure to radiation. The tolanes (No. 4) are known to be sensitive to UV radiation[33]. The esters (No. 5) generally possess lower clearing points than examples 6-8. They have been synthesized with a large number of different

TABLE 1

Most Important Substituent Groups

| Substituent | Remarks |
| --- | --- |
| $C_nH_{2n+1}-$ | Low melting and clearing temperatures |
| $C_nH_{2n+1}O-$ | |
| $C_mH_{2m+1}O-(CH_2)_n-O-$ | Melting temperature often lowered more than clearing temperature |
| $C_nH_{2n+1}-\overset{\parallel}{\underset{O}{C}}-$ | Strong tendency to smectic phases |
| $C_nH_{2n+1}-\overset{\parallel}{\underset{O}{C}}-O-$ | Relatively high melting and clearing temperatures |
| $C_nH_{2n+1}-O-\overset{\parallel}{\underset{O}{C}}-O-$ | Melting temperature often lowered more than clearing temperature |
| $C_nH_{2n+1}-S-$ | Tendency to smectic phases |
| $C_nH_{2n+1}-O-\overset{\parallel}{\underset{O}{C}}-CH=CH-$ | Cis-trans-isomerisation due to radiation |
| $C_nH_{2n+1}-O-\overset{\parallel}{\underset{O}{C}}-\overset{\mid}{\underset{CH_3}{C}}=CH-$ | |
| $Cl-$ , $Br-$ | Remarkably strong dipole moments |
| $N \equiv C-$ | Strong dipole moments, leading to increase in melting temperatures |
| $N \equiv C-CH_2-$ | |
| $N \equiv C-CH_2-CH_2-$ | |
| $N \equiv C-CH=CH-$ | |
| $O_2N-$ | |

TABLE 2

Low Melting Compounds of the Type $R^1$ —⟨O⟩— M —⟨O⟩— $R^2$

| No. | Example | melt (°C) | clear (°C) | Ref. |
|---|---|---|---|---|
| 1 | $C_2H_5O$ —⟨O⟩— $\underset{\underset{CH_3}{\mid}}{C}$ = CH —⟨O⟩— $C_4H_9$ | 60 | 60 | 1 |
| 2 | $C_4H_9$ —⟨O⟩— $\underset{\underset{Cl}{\mid}}{C}$ = CH —⟨O⟩— $OC_2H_5$ | 29 | 58 | 1 |
| 3 | $C_2H_5O$ —⟨O⟩— $\underset{\underset{CN}{\mid}}{C}$ = CH —⟨O⟩— $OC_6H_{13}$ | 54 | 80.5 | 7 |
| 4 | $C_7H_{15}$ —⟨O⟩— C ≡ C —⟨O⟩— $OCH_3$ | 39 | 54 | 2 |
| 5 | $C_5H_{11}O$ —⟨O⟩— $\underset{\underset{O}{\parallel}}{C}$ — O —⟨O⟩— $C_5H_{11}$ | 39 | 55 | 3 |
| 6 | $CH_3O$ —⟨O⟩— CH = N —⟨O⟩— $C_4H_9$ | 22 | 47 | 4 |
| 7 | $C_5H_{11}$ —⟨O⟩— N = N —⟨O⟩— $OCH_3$ | 39 | 65 | 5 |
| 8 | $C_5H_{11}$ —⟨O⟩— $\underset{\underset{O}{\downarrow}}{N}$ = N —⟨O⟩— $C_5H_{11}$ | 22 | 65 | 6 |

| | | | |
|---|---|---|---|
| No. 1 - 3 | stilbene derivatives | No. 6 | azomethines |
| No. 4 | tolane derivatives | No. 7 | azo derivatives |
| No. 5 | esters | No. 8 | azoxy derivatives |

substituents. They are colorless and represent the most stable class of substances listed in Table 1. The azomethines (No. 6), known in some examples for a long time, are the most widely treated class of liquid crystalline compounds. They are known with nearly all available substituents and mostly possess very low melting points with sufficiently high clearing temperatures. Unfortunately these nearly colorless substances are unstable with respect to moisture, different chemicals, UV radiation and dc electric fields. The azo and azoxy derivatives (No. 7 and 8) are much more stable, but they exhibit a strong brown or yellow color, respectively.

The examples listed in Table 3 are analogous to the substances in Table 2, but they are branched by an additional ring substituent. Therefore it is not surprising that in examples 1-3 the clearing points are significantly lowered with respect to the unbranched compounds. The OH-substituted substances No. 4-6, however, exhibit unexpectedly high clearing points, probably due to intramolecular hydrogen bondings which enhance the longitudinal polarizability of the molecules.

While two-ring compounds with two chain atoms in the middle group M have been synthesized to a great extent and nearly all available groups have been used, the chemistry of substances with four chain atoms in M is still in the beginning. The peroxides No. 1 in Table 4 exhibit a strong tendency to decompose. Also the colorless oximebenzoates No. 2 are not very stable. The ketoximebenzoates, due to their low melting, sufficiently high clearing points and good stability seem to be a substance class of high value. Only a few examples of the azines, and nearly nothing about their properties are known. Substituted phenyl cinnamates No. 5, some examples of which have been known for a long time, are now of high interest because of the extremely high positive dielectric anisotropy found with particular substituents.

Table 5 contains liquid crystalline compounds with 3 benzene rings in the molecule. The substituted bis-benzoyloxybenzenes (No. 1-6) are colorless, very stable substances with properties variable according to the nature of the specific substituents. The substituted phenyl benzoyloxybenzoates No. 7 are relatively high melting, but because of their very high clearing points they are potentially useful as mixture components. The analogous branched derivatives No. 8 and 9 show slight depressions of the melting temperatures and a strong lowering of the clearing points. All may be assumed to be of high chemical stability.

Table 6 shows examples for substances with miscellaneous structures. There are the colorless very stable biphenyl derivatives (No. 1 and 2) and the biphenyl analogous phenylpyrimidines (No. 3). The class of azomethines is represented by the low melting substituted benzylidenaminopyridines (No. 4) which probably tend to decompose like the isocyclic azomethines. The slightly yellow cyclohexanones (No. 5), although relatively high melting, may be of value for dynamic scattering cells because of their extremely high contrasts. The class of cyclohexyl carboxylates (No. 6) exhibits high stability, low melting points, sufficiently high clearing temperatures and extremely low viscosities. These colorless substances seem to be a remarkable advance towards nematic compounds of high electrooptical quality.

D. DEMUS

TABLE 3

Low Melting Compounds of the Type  R$^1$ —⬡—M—⬡—R$^2$
with R$^3$ substituent

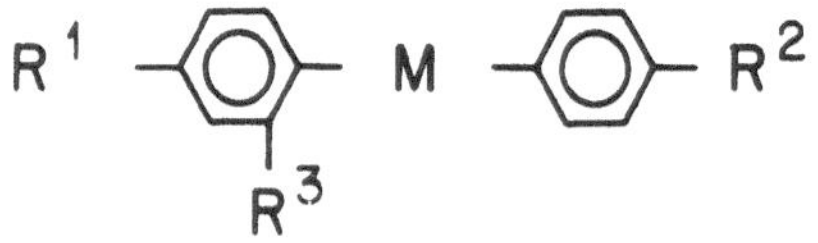

| No. | Example | melt (°C) | clear (°C) | Ref. |
|---|---|---|---|---|
| 1 | 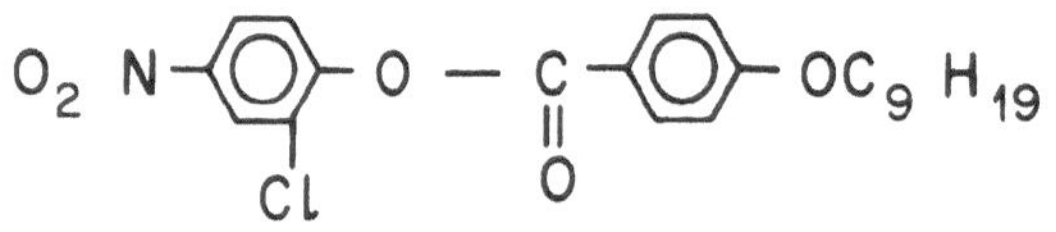 $O_2N$—⬡(Cl)—O—C(=O)—⬡—$OC_9H_{19}$ | 47.5 | 18 | 8 |
| 2 | 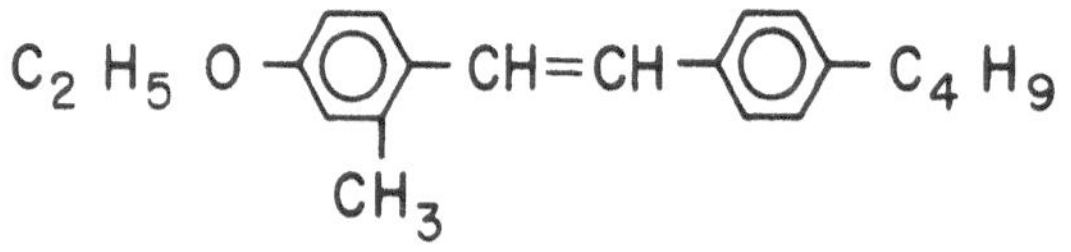 $C_2H_5O$—⬡($CH_3$)—CH=CH—⬡—$C_4H_9$ | 49 | 44 | 9 |
| 3 | 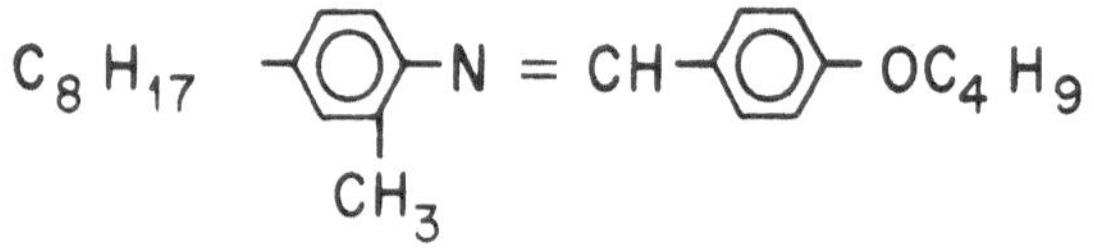 $C_8H_{17}$—⬡($CH_3$)—N=CH—⬡—$OC_4H_9$ | 50.3 | 67.2 | 10 |
| 4 | 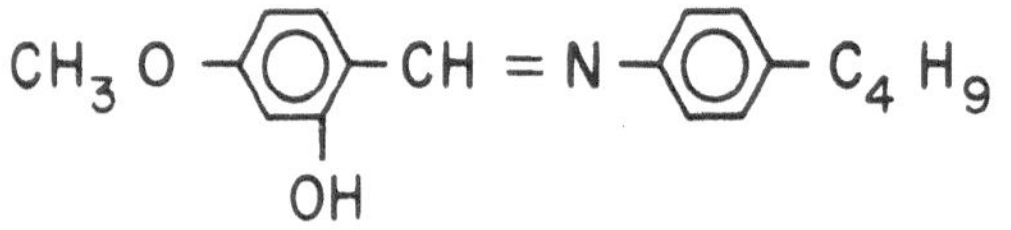 $CH_3O$—⬡(OH)—CH=N—⬡—$C_4H_9$ | 44 | 64.5 | 11 |
| 5 | 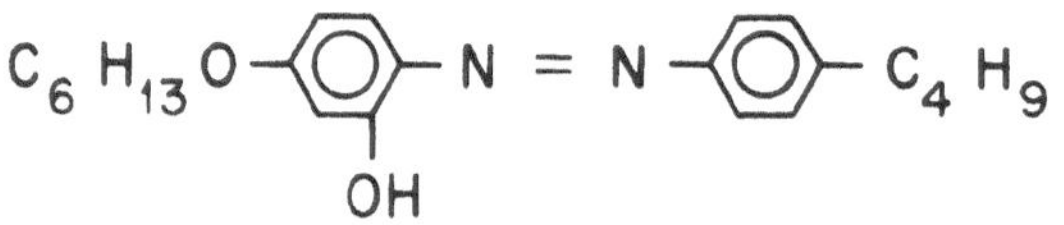 $C_6H_{13}O$—⬡(OH)—N=N—⬡—$C_4H_9$ | 17 | 82 | 12 |
| 6 | 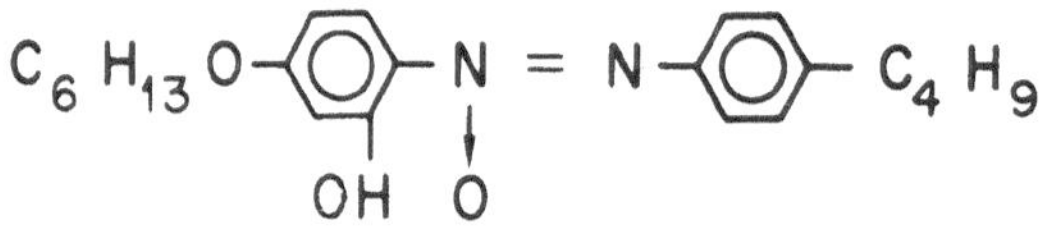 $C_6H_{13}O$—⬡(OH)—N(→O)=N—⬡—$C_4H_9$ | 57 | 93 | 12 |

No. 1  esters            No. 4  azomethines
No. 2  stilbenes         No. 5  azo derivatives
No. 3  azomethines       No. 6  azoxy derivatives

TABLE 4

Low Melting Compounds of the Type $R^1$ —⬡— M —⬡— $R^2$

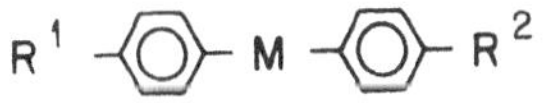

(M contains 4 chain atoms)

| No. | Example | melt (°C) | clear (°C) | Ref. |
|---|---|---|---|---|
| 1 | 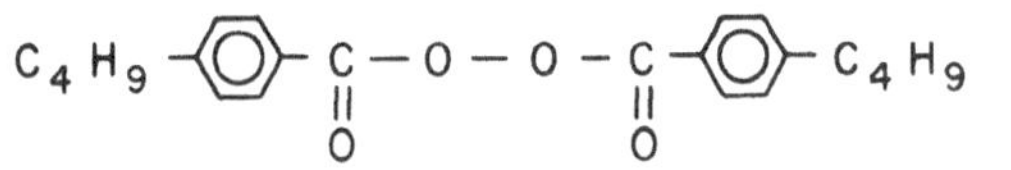 $C_4H_9$ —⬡— C—O—O—C —⬡— $C_4H_9$ (‖O, ‖O) | 25 | 55 | 13 |
| 2 | 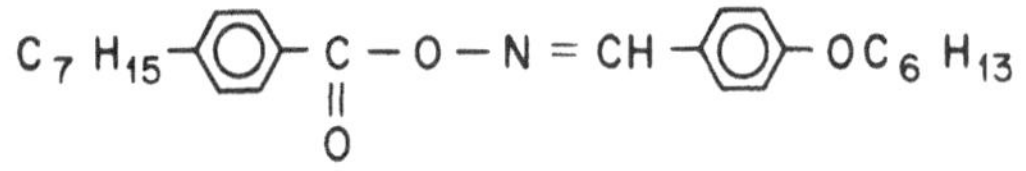 $C_7H_{15}$ —⬡— C—O—N=CH —⬡— $OC_6H_{13}$ (‖O) | 65 | 98 | 3 |
| 3 | 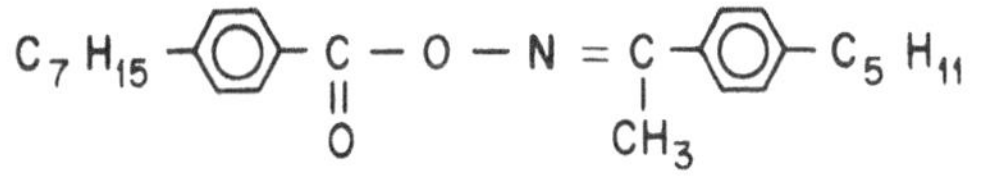 $C_7H_{15}$ —⬡— C—O—N=C —⬡— $C_5H_{11}$ (‖O, CH$_3$) | 30 | 69 | 14 |
| 4 | 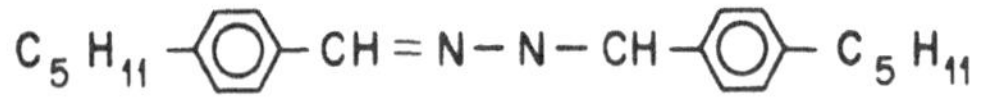 $C_5H_{11}$ —⬡— CH=N—N—CH —⬡— $C_5H_{11}$ | 61 | 102 | 15 |
| 5 | 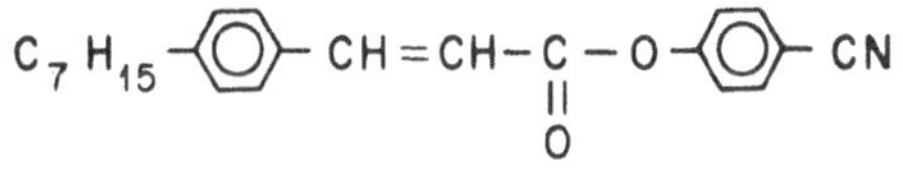 $C_7H_{15}$ —⬡— CH=CH—C—O —⬡— CN (‖O) | 57.8 | 109.8 | 16 |

No. 1  peroxides                    No. 4  azines
No. 2  oximebenzoates               No. 5  cinnamates
No. 3  ketoximebenzoates

D. DEMUS

TABLE 5

Low Melting Compounds of the Type $R^1 -\!\!\bigcirc\!\!- M^1 -\!\!\underset{R^3}{\bigcirc}\!\!- M^2 -\!\!\bigcirc\!\!- R^2$

| No. | Example | melt (°C) | clear (°C) | Ref. |
|---|---|---|---|---|
| 1 | $C_5H_{11}-\bigcirc-\underset{O}{\overset{\parallel}{C}}-O-\underset{Cl}{\bigcirc}-O-\underset{O}{\overset{\parallel}{C}}-\bigcirc-C_5H_{11}$ | 79 | 145 | 17 |
| 2 | $C_5H_{11}-\bigcirc-\underset{O}{\overset{\parallel}{C}}-O-\underset{Br}{\bigcirc}-O-\underset{O}{\overset{\parallel}{C}}-\bigcirc-C_5H_{11}$ | 76 | 134 | 17 |
| 3 | $C_8H_{17}O-\bigcirc-\underset{O}{\overset{\parallel}{C}}-O-\underset{CH_3}{\bigcirc}-O-\underset{O}{\overset{\parallel}{C}}-\bigcirc-OC_8H_{17}$ | 72.4 | 156 | 18 |
| 4 | $C_6H_{13}-\bigcirc-\underset{O}{\overset{\parallel}{C}}-O-\underset{C_2H_5}{\bigcirc}-O-\underset{O}{\overset{\parallel}{C}}-\bigcirc-C_6H_{13}$ | 44 | 80 | 19 |
| 5 | $C_6H_{13}-\bigcirc-\underset{O}{\overset{\parallel}{C}}-O-\underset{\underset{CN}{\overset{|}{CH_2}}}{\bigcirc}-O-\underset{O}{\overset{\parallel}{C}}-\bigcirc-C_6H_{13}$ | 73.5 | 79.5 | 20 |
| 6 | $C_6H_{13}O-\bigcirc-\underset{O}{\overset{\parallel}{C}}-O-\underset{CN}{\bigcirc}-O-\underset{O}{\overset{\parallel}{C}}-\bigcirc-OC_6H_{13}$ | 87 | 163 | 20 |
| 7 | $C_6H_{13}-\bigcirc-\underset{O}{\overset{\parallel}{C}}-O-\bigcirc-\underset{O}{\overset{\parallel}{C}}-O-\bigcirc-OC_6H_{13}$ | 89 | 177 | 22 |
| 8 | $\underset{Cl}{Cl-\bigcirc}-\underset{O}{\overset{\parallel}{C}}-O-\underset{Cl}{\bigcirc}-\underset{O}{\overset{\parallel}{C}}-O-\bigcirc-OC_5H_{11}$ | 71.5 | 116 | 23 |
| 9 | $C_5H_{11}-\bigcirc-\underset{O}{\overset{\parallel}{C}}-O-\underset{Cl}{\bigcirc}-\underset{O}{\overset{\parallel}{C}}-O-\bigcirc-C_5H_{11}$ | 39 | 122 | 23 |

No. 1 - 9  substituted bis-benzoyloxy-benzenes

TABLE 6

Substances with Different Structures

| No. | Example | melt ($^{\circ}$C) | clear ($^{\circ}$C) | Ref. |
|---|---|---|---|---|
| 1 | $C_6H_{13}$ —⟨O⟩—⟨O⟩— CN | 13.5 | 27 | 25,26 |
| 2 | $C_7H_{15}O$ —⟨O⟩—⟨O⟩— $NO_2$ | 36.5 | 38.5 | 25,26 |
| 3 | $C_6H_{13}O$ —⟨O⟩—⟨N,N pyrimidine⟩— $C_6H_{13}$ | 30 | 60.5 | 27 |
| 4 | $CH_3O$ —⟨O⟩— CH = N —⟨N⟩— $C_8H_{17}$ | 29.5 | 38.2 | 29 |
| 5 | $C_9H_{19}O$ —⟨O⟩— CH = (cyclohexanone) = CH —⟨O⟩— $OCH_3$ | 82 | 132 | 28 |
| 6 | $C_5H_{11}$ —⟨H⟩— C(=O) — O —⟨O⟩— $OC_5H_{11}$ | 28 | 70 | 30 |

No. 1 - 2    biphenyl compounds  
No. 3       phenylpyrimidines  
No. 4       benzylidenaminopyridines  

No. 5    cyclohexanone derivatives  
No. 6    substituted phenyl cyclo-  
         hexylcarboxylates

Surveying all substance classes which are listed in Tables 2 - 6 it may be stated that the two-ring compounds with a two-chain atom middle group are the most widely treated nematic substances. Most of the possible variants with different substituents are known and sursprising advances are scarcely to be expected. A comparable situation exists in the branched substances of Table 3 in which, however, the OH-branched compounds seem to play a special role. The chemistry of the two-ring compounds with a four chain atom middle group obviously is still in the beginning and the possible structural variants are far from being exhausted. Unbranched three-ring compounds (Table 5) in most cases exhibit too high melting points, and we also find this disadvantage in most branched three-ring compounds. Nevertheless, substances of this kind may be valuable as mixture components.

The most interesting hints for future work in the synthesis of low melting nematic materials may be gained from the examples in Table 6. The nematic biphenyls with their strong dipole moments may already be considered as a sensation; all other simple biphenyl derivatives are smectic (see for instance Demus[21,37]). With the phenyl pyrimidines we possess the first low melting biphenyl analogous heterocyclic compounds. By the use of other heterocyclic systems there appears to be a broad field of possibilities for the syntheses of new substance groups. Similar chances for future work may be predicted for the two-ring heterocyclic compounds with middle groups (see Table 6, example No. 4). At present there may be some difficulties and restrictions in the economical preparation of multifunctional heterocyclic synthesis materials, but by experience such restrictions shrink to the same extent to which the interest in the products grows.

In addition to the examples of Tables 2-6 it must be mentioned that low-melting optically active compounds are known, for instance in the class of substituted phenylbenzoates[38], azomethines[24,39] and biphenyls[26]. Optically active substances are of considerable interest as mixture components for the stabilization of twisted nematic layers[26]. The comprehensive class of cholesteric substances derived from steroids[37], despite its importance for storage mixtures and materials for the various electrooptic effects occurring in cholesterics[40-44], are beyond the scope of this short review.

## 3. LIQUID CRYSTALLINE MIXTURES

As can be seen in the tables presented above, substances are now available with stable liquid crystalline phases at ambient temperature. But for technical purposes it is necessary to maintain the nematic state down to temperatures of -5 to -20° C. Temperatures of +50 to +60° C may be considered at the upper limit. Obviously no pure compound is now known which meets the desired lower limit, and the chances of reaching the goal in the future are poor. Moreover some of the lowest melting compounds possess too low clearing points. By the use of mixtures, the temperature range of the liquid crystalline state can be extended considerably: the melting points of mixtures, especially eutectic ones, usually are lower than the melting points of the pure components; clearing points usually are elevated by the addition of components with higher clearing points, the clearing point of mixtures often being a nearly linear function of the concentration.

The melting behavior of mixtures and the calculation of melting temperatures for mixtures may be complicated by the occurrence of different solid crystalline modifications. Figure 1 shows the stability relations of a

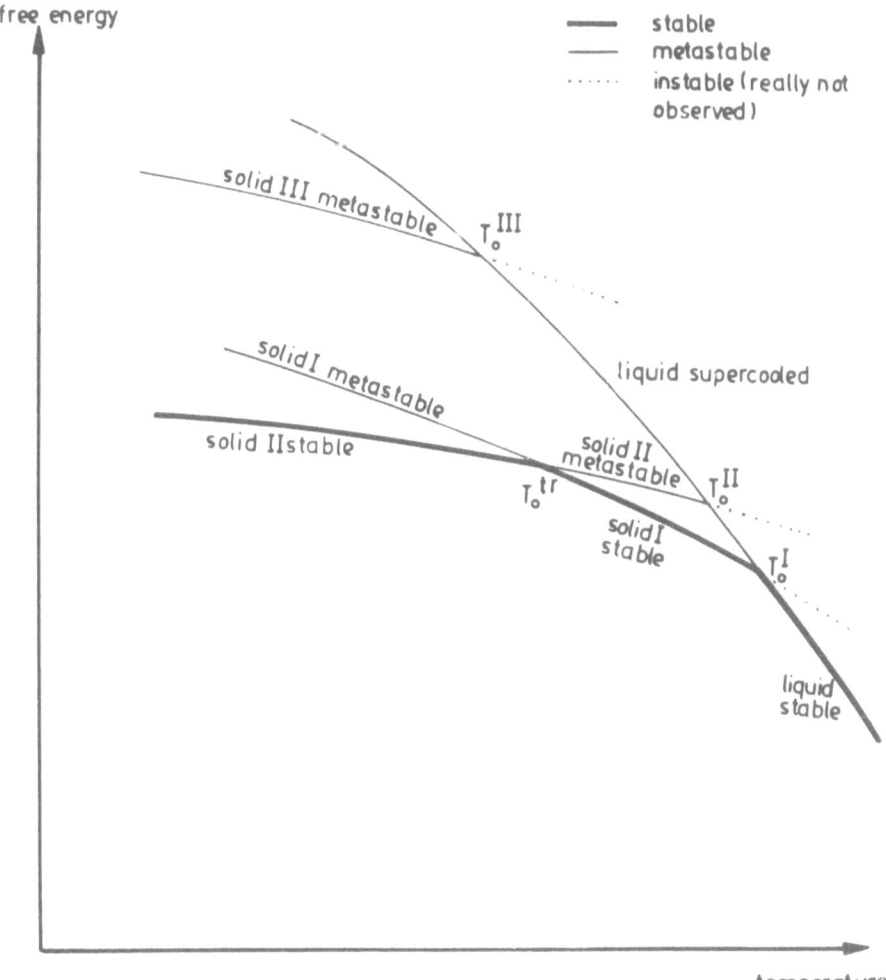

Fig. 1. Free energy versus temperature for a polymorphic substance.

substance with three solid modifications: the highest melting modification solid I melts at $T_o^I$; the transition to solid II may occur at $T_o^{tr}$ but, lacking nucleation, solid I can be supercooled into its metastable region. Alternatively, solid II can be heated up to its melting temperature $T_o^{II}$, if the nucleation of solid I at $T_o^{tr}$ is prevented. Therefore solid I and solid II each may occur in a stable and a metastable region. Often, by supercooling of the melt or crystallization from solution, substances exhibit solid modifications which are metastable in the whole temperature region. For instance, solid III in Fig. 1 may be heated up to its melting point $T_o{}^{III}$. It melts and the melt may be metastable or, depending on the conditions, crystallize to solid II or solid I. As has been observed in many cases, a spontaneous direct transition from solid III to solid I or II can also occur. Which transitions really take place depends mainly on nucleation. Compounds with several solid modifications are not exceptions, but rather are the rule. For instance, we have found substances with 4 solid modifications and 3 different melting points. If there are several solid modifications in the pure components, the melting behaviour of mixtures can become very complicated.

Melting temperatures of mixtures with ideal eutectic diagrams of state for the multicomponent systems can be calculated by the equation of Schröder-van Laar[47-50]:

$$T_i = \frac{\Delta H_{oi}}{\dfrac{\Delta H_{oi}}{T_{oi}} - R \ln x_i} \tag{1}$$

$\Delta H_{oi}$    molar heat of fusion for the component
$T_{oi}$    melting point of the pure component i [°K]
$R$    gas constant
$x_i$    mole fraction of the component i.

If the component i exhibits a transition in the solid state (as for instance solid II in Fig. 1) at the transition temperature $T_{oi}^{tr}$, (1) is valid only in the region

$$T_{oi}^{tr} < T < T_{oi} .$$

In the region $T < T_{oi}^{tr}$, it has to be replaced by

$$T_i = \frac{\Delta H_{oi} + \Delta H_{oi}^{tr}}{\dfrac{\Delta H_{oi}}{T_{oi}} + \dfrac{\Delta H_{oi}^{tr}}{T_{oi}^{tr}} - R \ln x_i} \tag{2}$$

$\Delta H_{oi}^{tr}$    molar heat of transition.

Concerning Fig. 2, (1) is valid for the curve $T_{oi}^I$-A, and also for A-C if the transition to solid II does not occur. If this transition occurs at $T_{oi}^{tr}$, (2) is to be used for the calculation of curve A-B. If only solid III is present, we may use (1) for the curve $T_{oi}^{III}$-D ($\Delta H_{oi}$ in this case means the heat of fusion of solid III). Evidently the existence of three solid modifications in one component makes three different eutectic points possible. Indeed we have found a

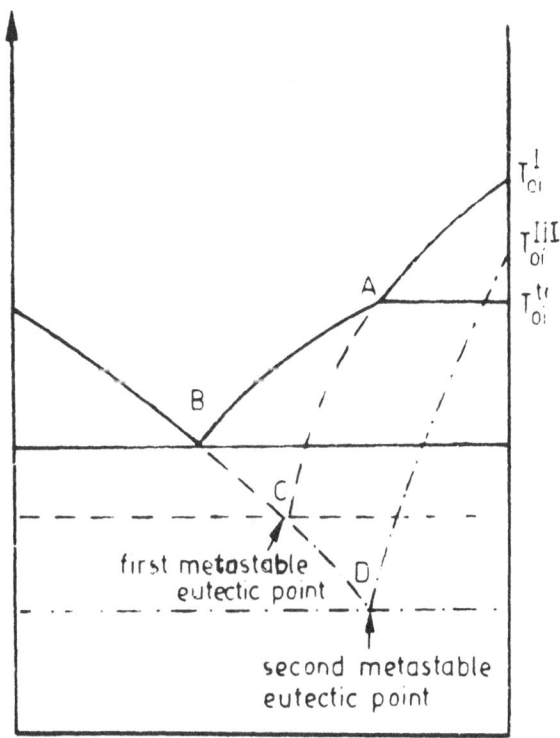

Fig. 2. Diagram of state for a binary system, including metastable states.

number of binary systems with several eutectics, and some systems of this kind have already been reported in the literature[53,54,45].

Being aware of the fact that each component can exhibit even more than two solid modifications, and that many practically used mixtures consist of more than two components, it is clear that very complicated relations may occur. If the respective system has not been investigated very carefully, a metastable eutectic point may readily be mistaken for the stable one. Smith[53] has reinvestigated the binary system Azoxyphenetol/Azoxyanisol with consideration of the unstable second solid modification of the latter (Fig. 3). By use of (1) he was able to calculate the stable as well as the metastable eutectic point. Obviously in these experiments either the stable or the metastable solid modification of azoxyanisole occurred, but the transition solid I → solid II was not found.

There are some binary mixtures with unexpectedly low melting temperatures in the literature. Being aware of the possibility that metastable eutectics might have occurred we have reinvestigated some of these mixtures. Figure 4 shows a diagram of two azomethine derivatives. Pohl and Steinsträsser[55] found eutectic melting at −14° C and 33 mole % for the first component; our own experimental value is +37° C and 35 mole %. Since we have

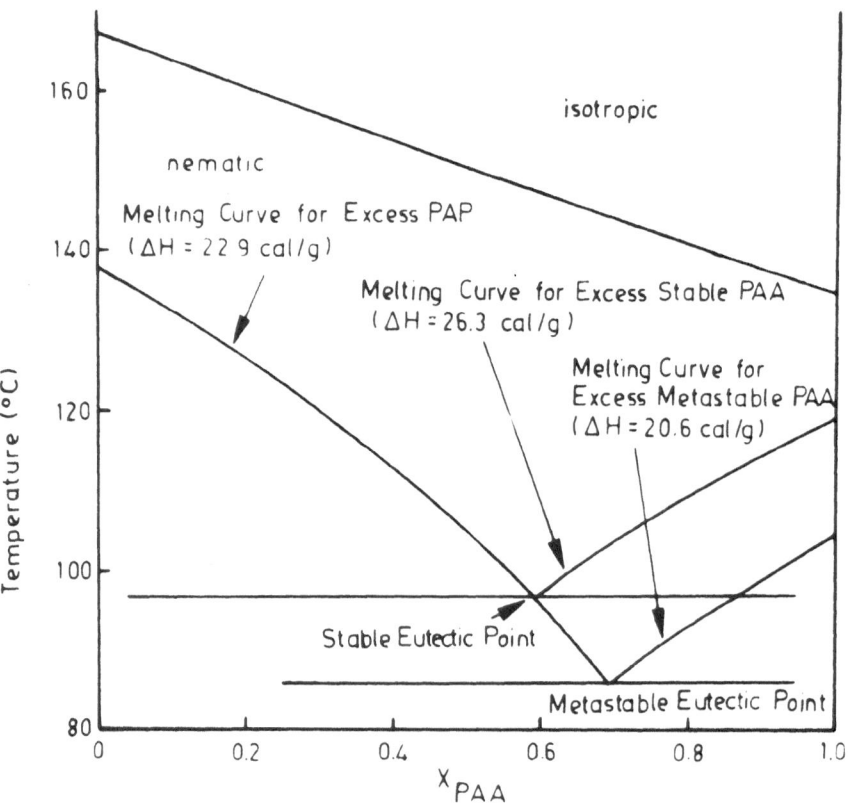

Fig. 3. Diagram of state of Azoxyphenetol/Azoxyanisol after Smith.[53]

been able to detect two solid modifications of the second component, we
have calculated the melting curves for both (see Fig. 4). The agreement between experiment and calculation is only good for the melting curve of the
first, but not for the second component. This may be due to partial occurrence of mixed crystals or to simultaneous occurrence of both solid modifications of the second component in our experiments. The eutectic point at
−14° C points to additional metastable solid modifications of one or both
components. As Sorkin and Denny[111] point out, because of the chemical
reactivity of Schiff's bases Pohl and Steinsträsser did not present binary systems, but due to the formation of additional compounds most of the systems correspond to multicomponent systems which are not well reproducible.

Also in other binary mixtures we found different melting behavior depending on the conditions. Usually, cooling of a small sample of the substance yields low-melting crystals, whereas cooling of larger quantities leads

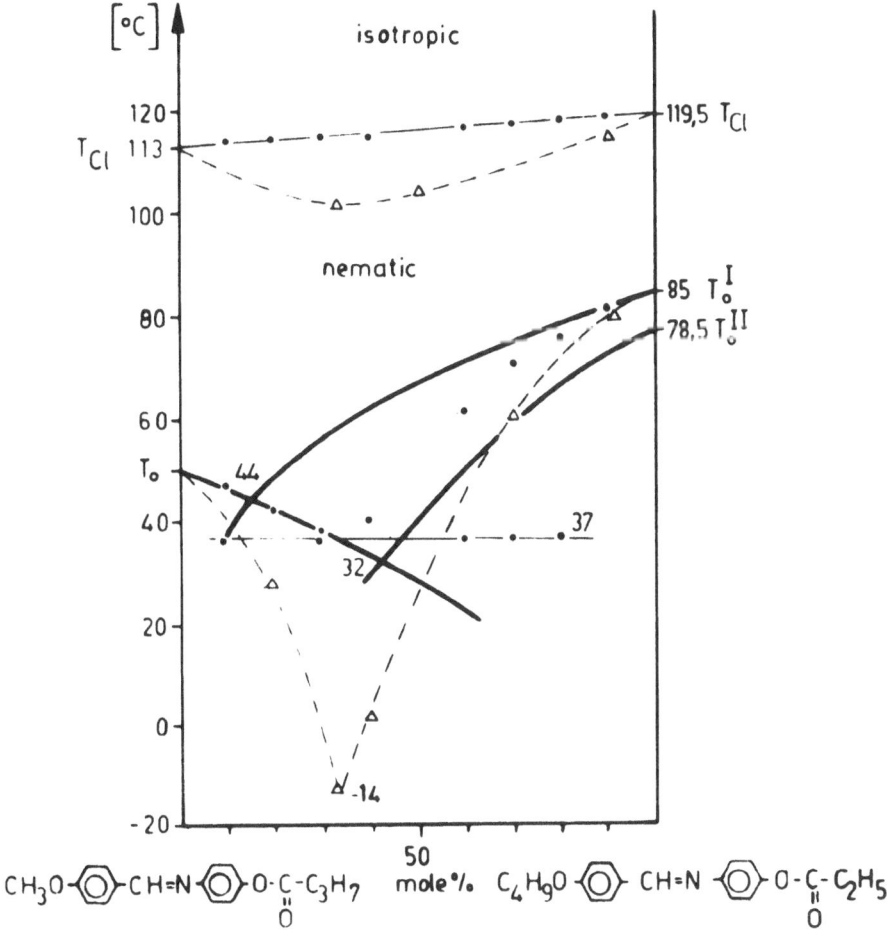

Fig. 4. Diagram of state for the system 4-Butanoyloxy-N-[4-methoxy-benzylidene]-aniline/4-propionyloxy-N-[4-butyloxybenzylidene]-aniline after Pohl and Steinsträsser[55] (experimental values: △, dotted curves) and Demus et al.[56] (experimental values: •, solid curves calculated with $\Delta H_o^I$ = 6850 cal/mole and $\Delta H_o^{II}$ = 3580 cal/mole from equation (1)).

to higher melting crystals, which may be considered as the stable ones. Multicomponent mixtures mostly show a strong tendency to supercooling. Therefore it often requires precise work to find the stable eutectic mixtures.

Several systems have been published in which the calculation of eutectic mixtures agrees well with experimental results[49-53,59] but the agreement is not good in all cases. Besides the occurrence of metastable solid modifications, the influence of which in principle can be considered in the calculation, there are restrictions to the validity of (1) and (2). The necessary assumption for their validity is the occurrence of ideal eutectic systems, i.e. mixed crystals are not allowed. Even if limited miscibility exists in the solid state, there are significant differences between theory and experiment. As Demus et al.[49] point out, miscibility in the solid state is to be expected especially for substances with similar molecular structure, such as neighboring members of a homologous series. If full miscibility occurs in the solid state, no melting point depression should be observed. The melting behavior of systems with miscibility in the solid state also may be calculated[50].

In simple cases the clearing curve of a binary system is the straight line joining the clearing points of the two components. Also in multicomponent systems the clearing point can be a linear function of the composition. Therefore it is possible by adding components with high clearing points to elevate the clearing temperature of mixtures. For instance, Gray et al.[26,52] successfully increased the clearing points of mixtures of cyanobiphenyls by adding cyanoterphenyl derivatives. If the mixture does not show ideal behaviour in the liquid crystalline state, more or less pronounced deviations for the linear function of the clearing curve may occur[50]. In a number of practical cases we have found definite minima in the clearing curves.

## 4. DIELECTRIC PROPERTIES

The dielectric properties are very important for the behavior of liquid crystals in electric fields. Therefore in the last years many experimental and theoretical investigations of liquid crystals have been undertaken. According to Maier and Meier's extension[69] of the Onsager theory, the dielectric anisotropy of nematic liquid crystals is expressed by:

$$\frac{\Delta \epsilon}{4\pi} = N \, h \, F \cdot S \left\{ \Delta \alpha - F \, \frac{\mu^2}{2kT} \, (1 - 3 \cos^2 \beta) \right\} \qquad (3)$$

$$\Delta \epsilon = \epsilon_{\parallel} - \epsilon_{\perp}$$

$$N = \frac{N_L \cdot \rho}{M}$$

$N_L$      Loschmidt's number
$\rho$        density
M        molar mass
S        degree of order
$\Delta \alpha = \alpha_{\parallel} - \alpha_{\perp}$ (anisotropy of polarizability)
h, F    factors of the Onsager theory corresponding to the inner field
$\mu$        permanent electric dipole moment
k        Boltzmann constant
T        temperature [$^\circ$ K]
$\beta$        angle between permanent electric dipole moment and molecular long axis.

Equation (3) relates dielectric properties to molecular structure. Molecules without a permanent dipole moment according to (3) exhibit a positive dielectric anisotropy, usually with low values[70], depending, of course, on the anisotropy of polarizability of the molecule. For practical purposes, substances with large positive or negative dielectric anisotropies are needed, which can only be obtained from molecules with the proper permanent dipole moments. Therefore in the discussion of dielectric properties, the dipole moments and their angle with the long molecular axis play the most important role.

Of special interest are substances with large positive dielectric anisotropy. In order to obtain maximum values of $\Delta\epsilon$, chemical groups such as $-NO_2$ or $-CN$ which introduce partial moments of about 4 D into the molecule are needed as p-substituents. Because of its chemical reactivity and its broader form which lowers the clearing points, the nitro group is less advantageous than the cyano group. In Table 7 some examples for p-CN substituted compounds are listed. In example No. 1, all three longitudinal components of the partial moments occuring in the molecule point in the same direction. Therefore the dielectric anisotropy is significantly higher than with No. 2, where the longitudinal component of the moment of the middle group is directed against the two others. Also p-CN compounds with other middle groups exhibit high positive values for $\Delta\epsilon$ (see examples 3-6 in Table 7). The middle groups according to their partial moments and polarizabilities influence the dielectric anisotropy[64,62] and in some cases may change its sign, but all p-CN substituted compounds can be expected to exhibit strong positive dielectric anisotropy. Strong partial moments in p-position and large anisotropies of the polarizability caused by extended aromatic central parts in the molecules yield high positive values of the dielectric anisotropy. But as has been pointed out, compounds with large aromatic parts will exhibit high melting points and therefore are not useful for most practical purposes. The reported $\Delta\epsilon$ values of about +20 (see in particular Titov[16], van Meter[61], Klingbiel[62], Kresse[72]) seem to be near the maximum possible in substances useful for electrooptic purposes. It must be emphasized, however, that because of temperature dependence at lower temperatures in these substances, even higher $\Delta\epsilon$ values are possible; for instance example No. 4 at 97° C shows $\Delta\epsilon$ = +35, the highest positive value reported to date.

As example No. 7 shows, the influence of several small but equally oriented partial moments may result in significant positive dielectric anisotropies. For No. 9, an example of the large number of substances with two p-alkoxy groups, the negative $\Delta\epsilon$ is due mainly to the effect of the lateral partial moment of the alkoxy groups. This effect can be observed in substances with different middle groups, and one or even two alkoxy groups may be replaced by alkanoyloxy groups[62,72].

Substances with $\Delta\epsilon < 0$ are used for the tunable retardation effect[40-44] and especially for the dynamic scattering effect. Although in the latter case $\Delta\epsilon$ = -0.2 to -0.5 is usually sufficient, there is a special interest in substances with strong negative anisotropy. By means of such substances as components

## TABLE 7

### Dielectric Properties*

| No. | Structure | $\epsilon_{\parallel}$ | $\epsilon_{\perp}$ | $\Delta\epsilon$ | Ref. |
|---|---|---|---|---|---|
| 1 | $C_4H_9O$—◯—$CH = N$—◯—$CN$ | | | +14 | 59 |
| 2 | $C_4H_9O$—◯—$N = CH$—◯—$CN$ | | | +8.8 | 59 |
| 3 | $C_6H_{13}$—◯—$\underset{O}{\overset{\parallel}{C}} - O$—◯—$CN$ | 28.0 | 11.0 | +17.0 | 16 |
| 4 | $C_4H_9$—◯—$CH = CH - \underset{O}{\overset{\parallel}{C}} - O$—◯—$CN$ | 23.2 | 6.9 | +16.3 | 16 |
| 5 | $C_7H_{15}$—◯—$\underset{O}{\overset{\parallel}{C}} - O$—◯$\underset{Cl}{}$—$\underset{O}{\overset{\parallel}{C}} - O$—◯—$CN$ | | | +19.5 | 61 |
| 6 | $C_5H_{11}$—◯—◯—$CN$ | 17 | 6 | +11 | 66 |
| 7 | $C_4H_9$—◯—$\underset{O}{\overset{\parallel}{C}} - O$—◯—$\underset{O}{\overset{\parallel}{C}} - O$—◯—$C_4H_9$ | 7.45 | 4.6 | +2.85 | 67 |
| 8 | $C_5H_{11}$—◯—$N = N$—◯—$C_5H_{11}$ | 2.8 | 2.5 | +0.3 | 70 |
| 9 | $C_5H_{11}O$—◯—$\underset{O}{\overset{\parallel}{C}} - O$—◯—$OC_8H_{17}$ | 4.41 | 4.75 | −0.34 | 65 |
| 10 | $C_4H_9O$—◯—$\underset{CN}{\overset{\mid}{C}} = N$—◯—$C_5H_{11}$ ** | | | +0.6 | 60 |
| 11 | $C_2H_5O$—◯—$\underset{CN}{\overset{\mid}{C}} = CH$—◯—$OC_6H_{13}$ | 6.3 | 11.8 | −5.5 | 7 |
| 12 | $C_4H_9O$—◯—$CH = \underset{O}{\overset{\parallel}{\bigcirc}} = CH$—◯—$OC_4H_9$ | 5.35 | 7.12 | −1.77 | 68 |

*) The $\Delta\epsilon$ values correspond to different temperatures and partly to different reduced temperatures. Therefore the absolute values are only qualitatively comparable.

**) Equimolar mixture with the corresponding ethoxy compound.

it is possible to give mixtures the desired dielectric properties. To obtain a large negative $\Delta\epsilon$, it seems reasonable to introduce into the molecules a lateral CN-group. As No. 10 (Table 7) shows, the expected effect does not always take place. In this substance, for instance, the moment of CN-group indicated by the resulting low value of $\Delta\epsilon$ must have an angle $\beta \approx 60°$, which is near the magic angle at which the longitudinal and lateral components of the moment cancel in their effect on $\Delta\epsilon$. Number 11 is an example of substances with the largest known negative $\Delta\epsilon$. Number 12 has been measured at higher temperatures, but in mixtures at lower temperatures it exhibits nearly the same strong tendency to negative $\Delta\epsilon$ as No. 11.

Steinsträsser and Pohl[71] found in some mixtures a linear dependence of $\Delta\epsilon$ on the concentration of an added component up to a certain limit. We have also investigated the concentration dependence of the dielectric properties of mixtures[72]. As Table 8 shows, the dielectric anisotropies of the mixture components add according to their mole fractions $x_i$. In mixtures of substances with very different chemical structure, however, deviations from this linearity are to be expected.

TABLE 8

Dielectric Properties of Some Esters and their Mixtures at
$45°$ C in the Low Frequency Region ($\nu < 1$ kHz)

$$R^1 - \langle O \rangle - \underset{\underset{O}{\|}}{C} - O - \langle O \rangle - R^2$$

| $R^1$ | $R^2$ | $\Delta\epsilon_i$ | $x_i$ | $x_i \Delta\epsilon_i$ |
|---|---|---|---|---|
| $C_5H_{11}O-$ | $-OC_8H_{17}$ | $-0.76$ | $0.27$ | $-0.21$ |
| $C_6H_{13}O-$ | $-OC_7H_{15}$ | $-0.53$ | $0.12$ | $-0.06$ |
| $CH_3O-$ | $-OC_6H_{13}$ | $-0.66*$ | $0.24$ | $-0.16$ |
| $C_6H_{13}-$ | $-OC_4H_9$ | $-0.21$ | $0.36$ | $-0.07$ |

calculated value:   $\Delta\epsilon = \Sigma x_i \Delta\epsilon_i = -0.50$

measured value:   $\Delta\epsilon = -0.50$

*) extrapolated to $45°$ C.

The dielectric constants $\epsilon_\|$ and $\epsilon_\perp$ suffer a relaxation step in the region of $10^{11}$ Hz which is due to a complex relaxation process. For molecules which are not symmetrical with respect to rotation about a transverse axis, Maier and Saupe[73] predicted an additional dispersion step for $\epsilon_\|$ in the MHz region. The prediction was verified experimentally with the di-n-alkyloxy-azoxy-benzenes[74-76]. The mechanism of this relaxation has been interpreted by Maier and Meier[69] as a rotation of the molecule about a transverse axis. The relaxation time $\tau_R$ is temperature dependent according to the equation of Eyring and Kauzmann:

$$\frac{d(\ln \tau_R)}{d(\frac{1}{T})} = -\frac{\Delta E^*}{RT^2} \tag{4}$$

Figure 5 shows the logarithmic plot of $\tau_R$ versus reciprocal temperature for two phenylpyrimidine derivatives. From the linear part of the curves we have calculated the activation energies $\Delta E^* = 16$ kcal/mole. The nonlinear region at higher temperatures may be interpreted as a temperature dependence of the retardation factor caused by the familiar strong temperature dependence of the order parameter in the region below the clearing point.

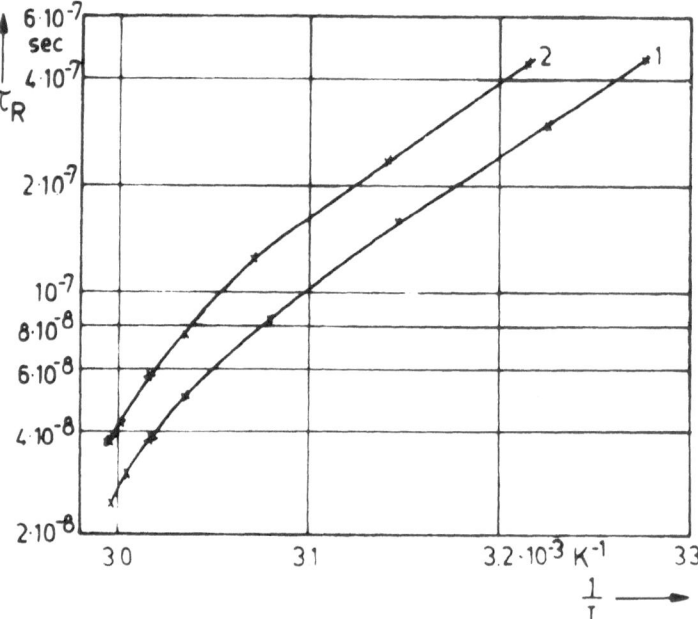

Fig. 5. Relaxation times for two substituted phenylpyrimidines[77]:
1  5-n-hexyl-2[4-n-hexyloxyphenyl]-pyrimidine
2  5-n-hexyl-2[4-n-nonyloxyphenyl]-pyrimidine

Considering the strong temperature dependence of $\tau_R$, it is not sur-prising that advances in the synthesis of low melting liquid crystalline compounds, have produced substances with relaxation times of the order $10^{-3}$ to $10^{-5}$ s[78-81]. The increase of $\tau_R$ is due not only to low temperatures but also to the enlargement of the length of the molecule, or to increased viscosities caused by the molecular structure. Substances with positive $\Delta\epsilon$ at low frequencies can change the sign of $\Delta \epsilon$ near the relaxation frequency. Consequently there is an isotropy frequency $f_o$ for which $\Delta \epsilon = 0$. Figure 6 shows the temperature dependence of the isotropy frequency of two phenyl-pyrimidines. This isotropy frequency is of considerable practical interest. At lower and higher frequencies $\Delta \epsilon$ has respectively positive and negative values. Therefore the liquid crystal can be dielectrically switched between two orientations by changing the frequency of the applied voltage. This has the advantage that the thickness-dependent natural turn-off time can be replaced by a dielectrically driven turn-off. Furthermore the threshold and slope of the voltage dependence of transmission can be influenced. The effect is useful in nematic devices with homogeneous as well as twisted alignments[80, 82,83,110].

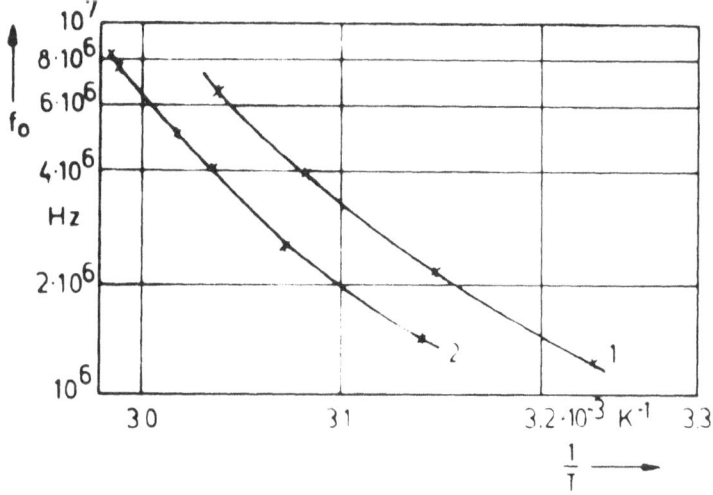

Fig. 6. Temperature dependence of the isotropy frequencies for two substituted phenylpyrimidines[77].

1   5-n-hexyl-2[4-n-hexylolxyphenyl]-pyrimidine
2   5-n-hexyl-2[4-n-nonyloxyphenyl]-pyrimidine.

## 5. VISCOSITY OF NEMATIC LIQUID CRYSTALS

The viscosity of nematic liquid crystals exhibits anisotropic behavior. According to the continuum theory there are five coefficients with the dimension of a viscosity[84]. The anisotropy of the viscosity has been determined only for a few substances, which does not allow one to establish relationships between viscosity behavior and molecular structure. In the literature, however, a number of viscosity values can be found which correspond to the minimum viscosity $\eta_1$, or are not clearly defined. We have measured substances of various chemical natures by a microcapillary method which yields values near $\eta_1$[114]. These viscosity values were used to infer the following relations between viscosity and structure. Despite the fact that not all of these values are quantitatively comparable, this intention seems to be permissible. The viscosity coefficients which are of interest for the electrooptics of liquid crystals are the molecular or microscopic ones rather than the macroscopically defined coefficients measured. Although the macroscopic and microscopic viscosities need not have the same values, it is however to be expected that there exists a relation between them. This allows one to compare the macroscopic values for different substances and to draw conclusions on their microscopic rheological behaviour.

Viscosities of liquids show a strong temperature dependence, which may be expressed by the equation

$$\log \eta = \log \eta_0 + \frac{\Delta E_{visc}}{2.3 \, RT} \tag{5}$$

$\eta_0$      integration constant
$\Delta E_{visc}$      activation energy of viscosity

The activation energies of liquid crystals lie in the region of 4 to 10 kcal/ mole. The corresponding values for the isotropic state are nearly the same. In general, the relation between viscosity and molecular structure seems to be similar for liquid crystals and isotropic liquids. Figure 7 displays the viscosities of three members of a homologous series. As might be expected, the viscosities increase with increasing length of the alkyl chain. Figure 8 allows estimation of the influence of different p-substituents on the rheological properties. Replacement of a substituent by a more polar group increases the viscosities: this can be seen by comparison of alkyl (No. 1 in Fig. 8) with alkoxy (No. 2), alkoxy (No. 4) with alkyloxycarbonyloxy (No. 5) and methoxy (No. 3) with cyano (No. 2). In Fig. 9 we compare substances with different middle groups. Obviously the cyclohexylcarboxylate (No. 1) exhibits the lowest viscosity, but the benzoate (No. 2) and the azomethine (No. 3) seem to possess values of the same order if the different substituents are taken into consideration.

Since measurements on pure substances of different chemical natures are rare, the viscosities of different mixtures at room temperature are listed in Table 9. The rules derived from the examples in Fig. 7-9 are confirmed. The viscosity of mixtures generally is not a simple function of the viscosities of the components.

## 6. ELECTRICAL CONDUCTIVITY

The electrical conductivity of a liquid crystal depends strongly on its purity. Liquid crystals are available with electrical conductivities in the region $10^{-7}$ to $10^{-13}$ $(\Omega \, cm)^{-1}$. The conductivity, even in the best purified substances, is due to impurities present in the substance or arising from chemical or electrochemical processes.

As Gaspard et al.[97] point out, the charge carriers in highly purified nematic liquid crystals are due to the dissociation of weak electrolytes. This

TABLE 9

Viscosities of Substances at $20^{\circ}$ C

| Structure | | | $\eta$(cSt) | Ref. |
|---|---|---|---|---|
| $CH_3O$—⟨O⟩—N=N—⟨O⟩—$C_4H_9$ (with O below N=N) | | "Merck 4" | 30 | 102 |
| $C_4H_9O$—⟨O⟩—CH=N—⟨O⟩—CN | 33,3% | | | |
| $C_6H_{13}O$—⟨O⟩—CH=N—⟨O⟩—CN | 33,3% | | 170 | 84 |
| $C_7H_{15}C$—O—⟨O⟩—CH=N—⟨O⟩—CN (C with O below) | 33,3% | | | |
| $CH_3O$—⟨O⟩—CH=N—⟨O⟩—$C_4H_9$ | | | 30 | 84 |
| $CH_3O$—⟨O⟩—N=N—⟨O⟩—$C_4H_9$ (with O below) | 60% | | | |
| $CH_3O$—⟨O⟩—N=N—⟨O⟩—$C_2H_5$ (with O below) | 30% | "Merck 7A" | 43 | 102 |
| $C_6H_{13}O$—⟨O⟩—C—O—⟨O⟩—C—O—⟨O⟩—$C_4H_9$ (C=O groups, CN on last ring) | 10% | | | |
| $CH_3O$—⟨O⟩—C—O—⟨O⟩—$C_5H_{11}$ (C=O) | 57% | | | |
| $C_6H_{13}O$—⟨O⟩—C—O—⟨O⟩—$C_5H_{11}$ (C=O) | 29% | "Merck 9A" | 95 | 102 |
| $C_6H_{13}O$—⟨O⟩—C—O—⟨O⟩—C—O—⟨O⟩—$C_4H_9$ (C=O groups, CN on last ring) | 14% | | | |
| $CH_3O$—⟨O⟩—C—O—⟨O⟩—$OC_6H_{13}$ (C=O) | 17,8% | | | |
| $C_5H_{11}O$—⟨O⟩—C—O—⟨O⟩—$OC_8H_{17}$ (C=O) | 31,5% | | | |
| $C_6H_{13}O$—⟨O⟩—C—O—⟨O⟩—$OC_7H_{15}$ (C=O) | 21,9% | | 100 | 87 |
| $C_6H_{13}$—⟨O⟩—C—O—⟨O⟩—$OC_4H_9$ (C=O) | 28,8% | | | |
| $C_5H_{11}$—⟨H⟩—C—O—⟨O⟩—CN (C=O) | | | 30* | 87 |

*) extrapolated

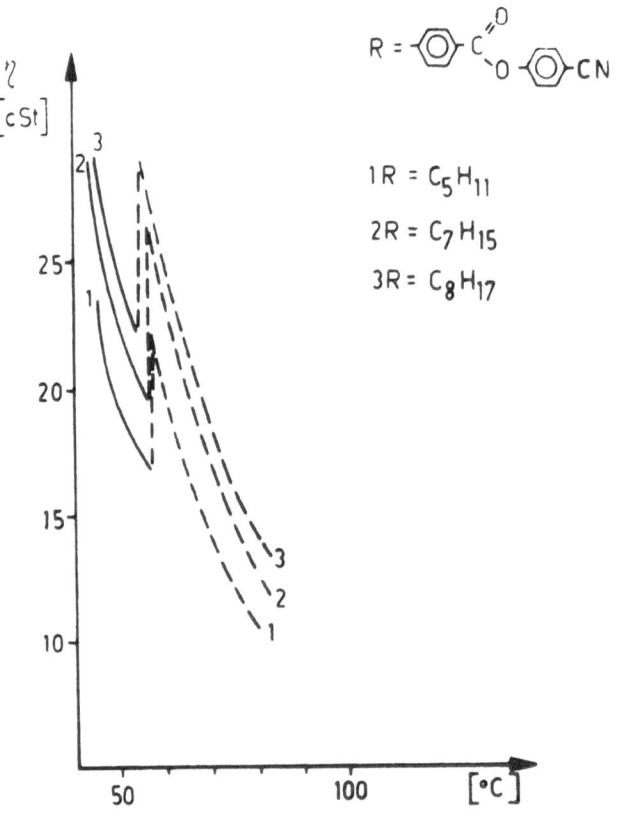

Fig. 7. Temperature dependence of the viscosities of 4-cyanophenyl 4-n-alkyl-benzoates. Dashed curve: isotropic state; solid curve: nematic state.

dissociation is a relatively slow process. Therefore the current/voltage characteristic has an initial ohmic region which saturates when the current is limited by the low dissociation velocity of the electrolyte (Fig. 10). The weak slope in this saturation region and the strong increase at still higher voltages are due to injection phenomena at the electrodes which lead to formation of additional charge carriers. This formation of new charge carriers may be partly due to electrochemical reactions of the substance itself or of impurities. Therefore it is to be expected that the voltage at which injection phenomena begin depends on the chemical nature of the substances and of the impurities. Usually the latter are not known. The electrochemical behavior of the pure substances, however, can be studied by polarographic investigations.

Figure 11 shows the oxidation and reduction potentials of some typical liquid crystalline materials. Obviously, in accordance with observations of Vildebrand and Matschiner[89], the characteristic potentials to first order depend on the middle group and not on the p-substituents. With respect to

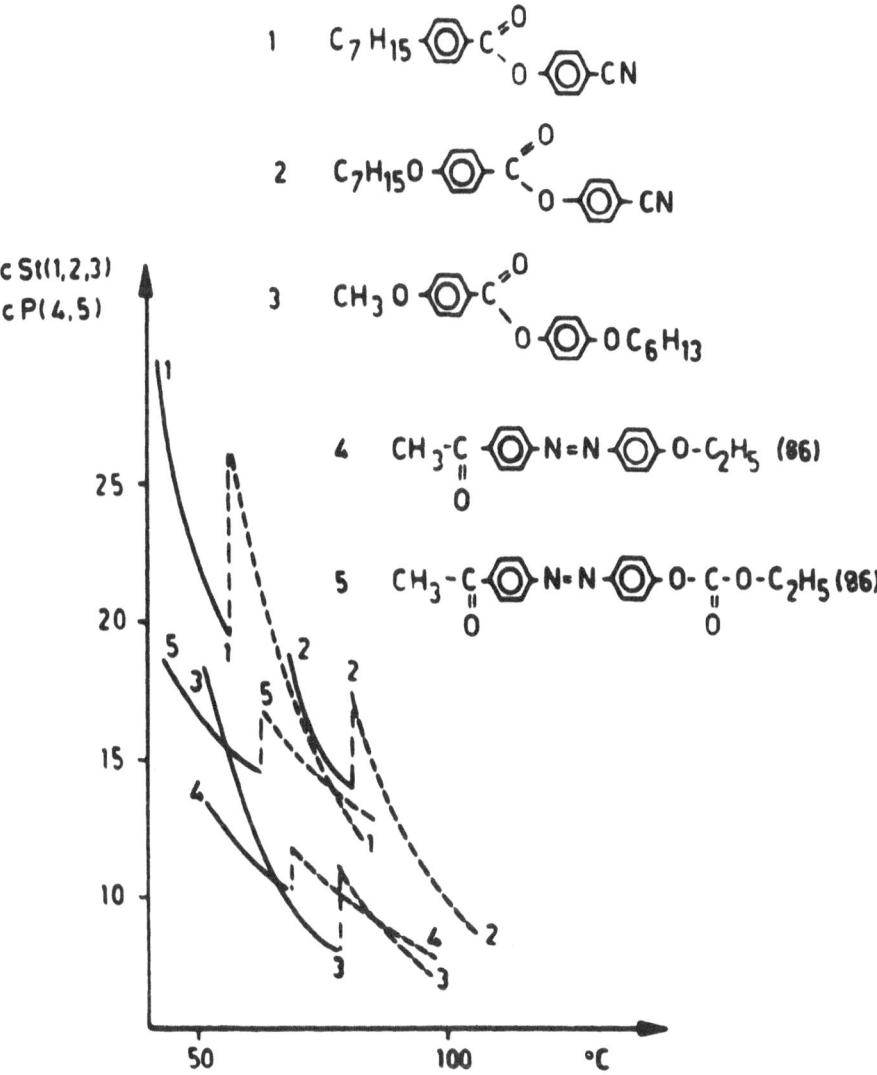

Fig. 8. Temperature dependence of the viscosities of substances with different substituents.

oxidation stability, we see a decreasing trend in the sequence: esters > azo, azoxy > tolane, azomethine. With respect to reduction stability the sequence is: tolane > azomethine, ester > azoxy > azo.

The polarographic investigations are done in a highly polar solvent with the addition of a strongly dissociating electrolyte. Therefore the results can be adopted for the pure liquid crystal, which is in general not very polar, only with some caution. Because of the high resistivity of the liquid crystals

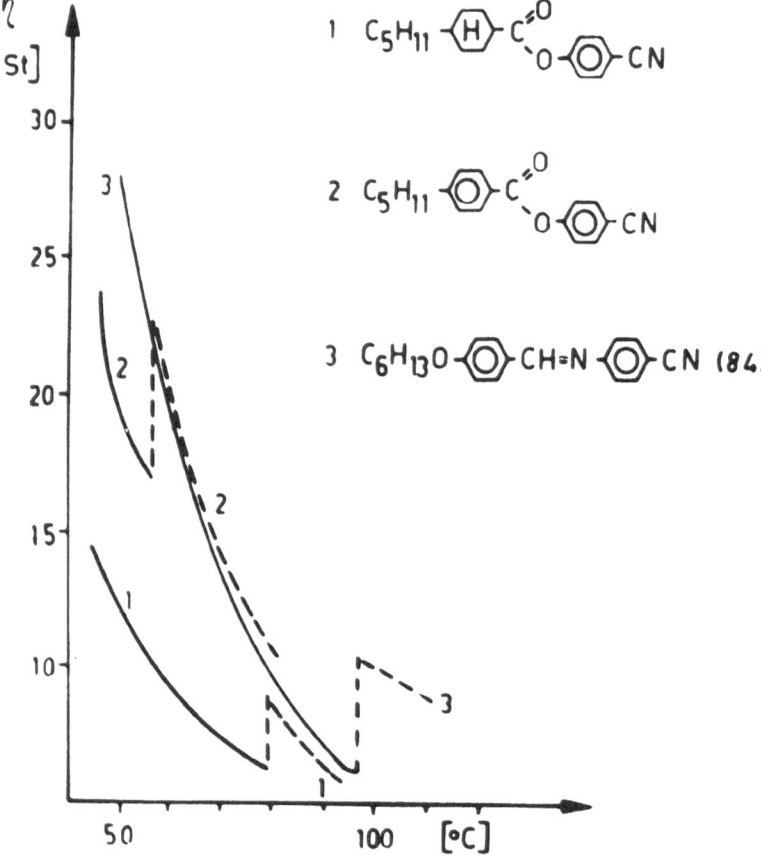

Fig. 9.   Temperature dependence of the viscosities of substances with different middle groups.

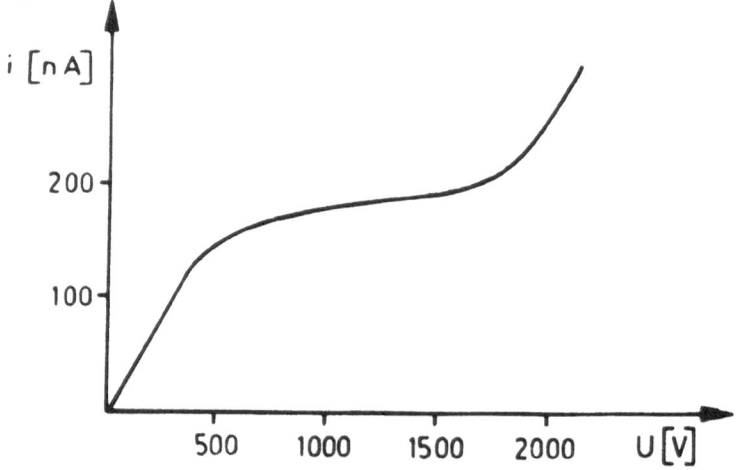

Fig. 10.   Steady-state current vs. applied voltage after Briere et al.[90] for electrodialysed MBBA: $50^\circ$ C (isotropic state); in the low voltage region $\sigma = 5 \times 10^{-11}$ $(\Omega cm)^{-1}$; electrode spacing 1 cm.

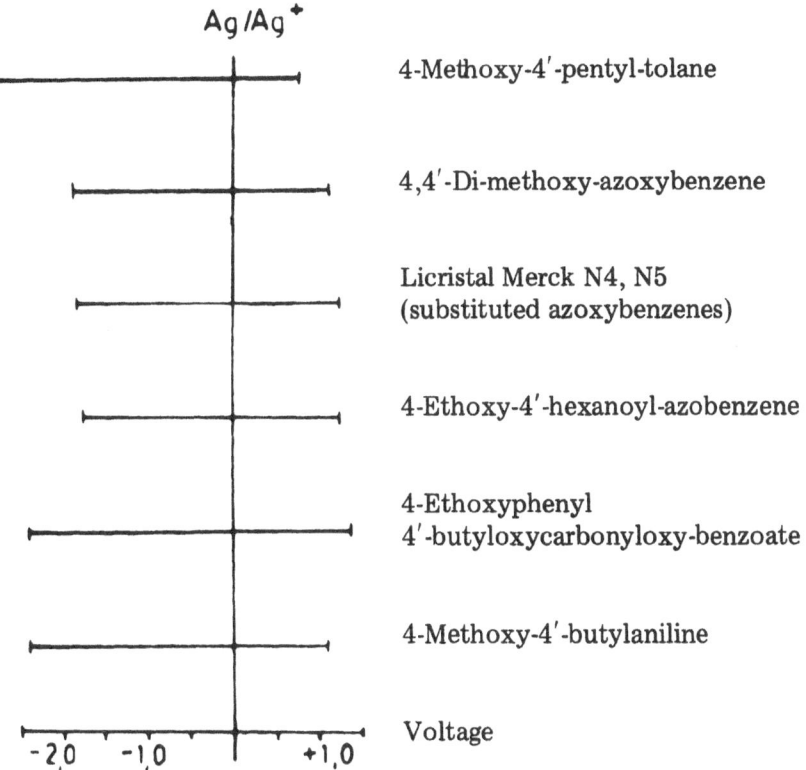

Fig. 11. Polarographic behavior of some liquid crystalline substances in solution ($CH_3CN$, added electrolyte: $Et_4N$ $ClO_4$). Halfwave potentials, left reduction, right oxidation. After Denat et al.[87].

even at high external voltages, the potential jump at the electrode can be too small for electrochemical processes. This is already indicated in Fig. 10. We have verified the distribution of potential directly (Fig. 12). In contrast to the results of Lu and Jones[92], we could not detect anomalously high field strengths near the electrodes. The potential at the lowest measurable distance from the electrodes (1 $\mu$m) was of the order 1 V, at an external field of 2000 V/cm.

The electrochemistry of liquid crystals is somewhat complicated and depends on the special nature of the chemical compound[88-90,93-96]. As Gaspard[97] points out, the temperature dependence of the electrical conductivity consists of a contribution due to ionic mobility

$$\mu = \mu_o \exp\left(\frac{-E_1}{RT}\right) \tag{6}$$

and a contribution from the increasing number of charge carriers due to dissociation

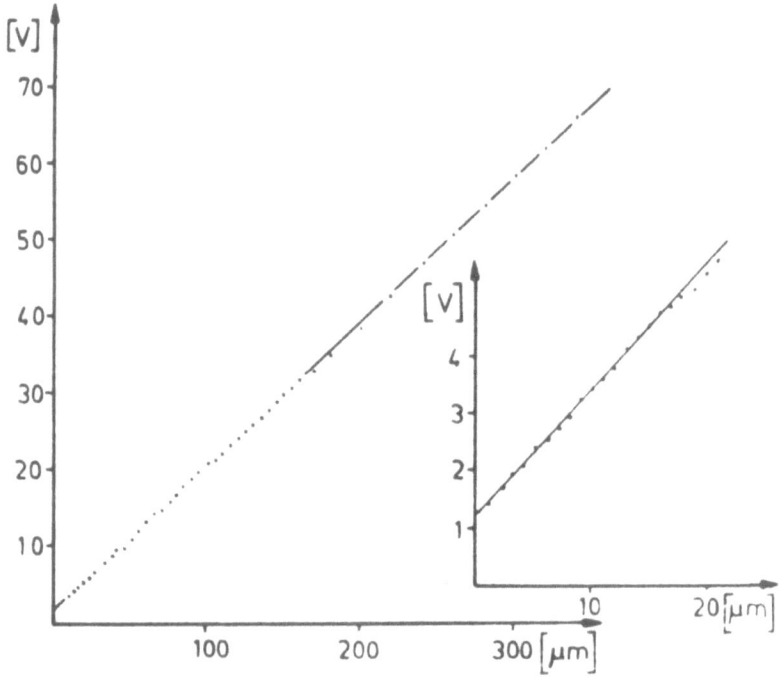

Fig. 12. Distribution of the potential in a nematic mixture of substituted phenyl benzoates[91]. dc electric field 2 kV/cm.

$$n = n_o \exp \left(\frac{-E_2}{RT}\right).\qquad(7)$$

The activation energy of the ionic mobility $E_1$ according to the rule of Walden should coincide with the activation energy of the viscosity, whereas the activation energy of the conductivity is

$$W = E_1 + E_2.\qquad(8)$$

For the nematic 4-n-octyloxyphenyl 4'-n-pentyloxybenzoate we have found $W = 16.3$ kcal/mole and from the viscosity measurement $E_1 = 9.2$ kcal/mole. This difference could be explained by the concept of temperature dependent dissociation equilibrium. However, the activation energy of viscosity in various substances in the isotropic state is higher than the activation energy of conductivity. For the substance mentioned above, $E_1 = 9.2$, $W = 6.2$. Thus, an additional influence is indicated. Heppke and Schneider[101] try to explain the remarkable decrease of conductivity in the homologous series of the p-alkyloxyazoxybenzenes by a decreasing dissociation of the electrolytes due to the lowering of the dielectric constants within the series. An effect of this kind can also be expected with increasing temperature, but it should not overwhelm the ordinary temperature dependence of dissociation. Therefore this problem seems to be unresolved.

The electrical conductivity of liquid crystals exhibits anisotropic behavior. Long experience has shown that the conductivity parallel to the director alignment $\sigma_\parallel$ is larger than the conductivity perpendicular to the director $\sigma_\perp$ in ordinary nematic liquids. Recently several substances have been investigated which possess a smectic modification at lower temperatures. In the temperature region just above smectic-nematic transition and in some cases up to higher temperatures, $\sigma_\parallel$ is smaller than $\sigma_\perp$ in these substances[98-101].

## 7. DOUBLE REFRACTION

All optical properties of liquid crystals are associated with double refraction. Unfortunately there is little data available which permits a comparison with the molecular structure. The double refraction of a given compound depends on its anisotropy of polarizability, the degree of order and the density. The degree of order, the density and, considering the flexibility of alkyl chains, also the anisotropy of polarizability decrease with growing temperature. Therefore a strong temperature dependence of double refraction results, and a comparison should be made only at reduced temperatures. Within homologous series the double refraction decreases with increasing length of the alkyl chain[103,108]. The influence of the anisotropy of polarizability may be inferred from Table 10, where examples 1-4 show a decreasing trend of the clearing points and also of the double refractions. According to the theory of Maier and Saupe[109], the clearing points are determined by the anisotropies of polarizability. From these examples we may conclude that the values of the double refraction correspond to the clearing points and therefore to the anisotropies of the polarizability of the compounds. As examples 5-9 (Table 10) indicate, this statement is not universally valid. The discrepancy can be explained, for instance on example 7, by the effect of the high dipole moment on the clearing point. It should be emphasized, however, that for the behaviour of the alkyloxybenzoic acid No. 8 and the cholesterylester No. 9 we do not have an explanation.

## 8. CONCLUDING REMARKS

For the dynamic scattering effect (DSM), the substance should have $\Delta\epsilon$ = 0.2 to 0.5 and good electrical conductivity, $\sigma = 10^{-8}$ to $10^{-11}$ $(\Omega\,\text{cm})^{-1}$. Especially for ac conditions the conductivity should not have too low values; otherwise there is danger of reaching the conductivity relaxation region at low frequencies, especially at lower temperatures, due to the natural decrease of conductivity. The threshold voltage for the onset of DSM is nearly independent of the substance. Low viscosities are advantageous for obtaining short rise and decay times.

The threshold voltage for the different field effects is given by

## TABLE 10

### Double Refraction 5° C Below the Clearing Point at $\lambda$ = 589 nm

| No. | Structure | $\Delta n$ | clear (°C) | Ref. |
|---|---|---|---|---|
| 1 | $CH_3O$—⬡—$CH=N-N-CH$—⬡—$OCH_3$ | 0.24 | 182 | 104 |
| 2 | $C_2H_5O$—⬡—$N = N$—⬡—$OC_2H_5$ (with $\uparrow O$) | 0.22 | 168 | 104 |
| 3 | $CH_3O$—⬡—$N = N$—⬡—$OCH_3$ (with $\uparrow O$) | 0.21 | 135 | 104 |
| 4 | $CH_3O$—⬡—$C \equiv C$—⬡—$C_5H_{11}$ | 0.20 | 59 | 112 |
| 5 | $CH_3O$—⬡—$CH = N$—⬡—$C_7H_{15}$ | 0.16 | 63 | 103 |
| 6 | $CH_3O$—⬡—$CH = N$—⬡—$C_4H_9$ | 0.16 | 44 | 105 |
| 7 | $C_8H_{17}O$—⬡—$\underset{\overset{\|}{O}}{C}$—$O$—⬡—$NO_2$ | 0.12 | 68 | 103 |
| 8 | $C_4H_9O$—⬡—$\underset{\overset{\|}{O}}{C}$—$OH$ | 0.12 | 160 | 104 |
| 9 | Cholesterylpelargonate | −0.015 | 92 | 106 |
|  |  | 0.03* |  | 106 |

*) calculated for the untwisted structure.

$$U_o = \left(\frac{4\pi K_i}{\Delta\epsilon}\right)^{1/2} . \tag{9}$$

$K_i$, depending on the particular effect, is one of the elastic constants or a combination of them. High values of $\Delta\epsilon$ and low values of $K_i$ are needed for technically interesting low threshold voltages. The rise and decay times may be expressed as[41]

$$\tau_r = \frac{4\pi\eta d^2}{\Delta\epsilon U^2 - 4\pi^3 K_i} \tag{10}$$

$$\tau_d = \frac{\eta d^2}{\pi^2 K_i} \tag{11}$$

where d is the thickness of the layer. Rise times as well as decay times are proportional to the viscosity, therefore low viscosities are needed. The rise time decreases with increasing $\Delta\epsilon$ and with decreasing elastic constants. However the decay time increases with decreasing $K_i$. There is little knowledge about the correlation between elastic properties and molecular structure. As Gruler[113] points out, the elastic constants in homologous series generally grow with increasing chain length. The same behavior has been found for viscosities[114]. It is an open question whether there is always a correlation. For a detailed consideration, of course, the different constants due to the anisotropy have to be taken into account.

Some important properties of the most utilized substance classes are listed in Table 11. Concerning the toxicity of liquid crystalline compounds, in a Merck leaflet some azoxy compounds and phenyl benzoates are claimed to have no toxic properties. Also cholesteryl esters may be assumed to be non-toxic because they are components of many pharmaceutical products.

There have been many efforts to achieve low threshold and driving voltages. Table 12 gives a short survey of mixtures with different chemical compositions. The threshold voltages depend on the particular surface conditions and therefore are not the best measure for comparing different substances. The better reproducible saturation voltages are to be preferred for this purpose.

## ACKNOWLEDGEMENTS

I am indebted to the VEB Werk für Fernsehelektronik Berlin for financial support of the work of the liquid crystal group of the University of Halle. Valuable discussions with Prof. Dr. H. Sackmann, Prof. Dr. H. Schubert, Dr. F. Kuschel, Dr. G. Pelzl, Dr. H. Kresse, Dr. H. Zaschke are acknowledged. Finally, I thank all members of our group who have participated in the experimental work.

TABLE 11

Properties of Some Substance Classes

| Substance class | chemicals | Stability to: | | | color | $\Delta\epsilon$ | $\eta$ | Useful for |
|---|---|---|---|---|---|---|---|---|
| | | dc | ac | light | | | | |
| R–⬡–CH=N–⬡–CN | − | − | + | ? | weak yellow | $>0$, high | high | FE |
| R–⬡–C(=O)–O–⬡–CN | + | + | + | + | no | $>0$, high | high | FE |
| R–⬡–C(=O)–O–⬡–⬡–CN | + | + | + | + | no | $>0$, high | ? | FE |
| R–(H)–C(=O)–O–⬡–CN | + | + | + | + | no | $>0$, high | low | FE |
| R–⬡–C(=O)–O–⬡–$C_2H_4CN$ | + | + | + | + | no | $>0$, middle | ? | FE |
| $R^1$–⬡–[pyrimidine N→O]–$R^2$ | + | + | + | + | no | $>0$, low | ? | FE, DSM |
| $R^1$–⬡–N=N(→O)–$R^2$ | + | limited | + | + | yellow | $<0$, low | low | FE, DSM |
| $R^1$–⬡–N=N–⬡–$R^2$ | + | − | + | − | brown | $\gtrless 0$, low | low | DSM |
| $R^1$–⬡–CH=N–⬡–$R^2$ | − | − | + | ? | no or weak yellow | $<0$, low | low | DSM |
| $R^1$–⬡–C(=CH–R)–⬡–$R^2$ | ? | ? | + | − | no | $<0$, middle | ? | DSM |
| $R^1$–⬡–C≡C–⬡–$R^2$ | + | ? | + | − | no | $\gtrless 0$, low | ? | FE, DSM |
| $R^1$–⬡–C(=O)–O–⬡–$R^2$ | + | + | + | + | no | $\gtrless 0$, low | low to middle | FE, DSM |
| $R^1$–⬡–CH=CH–⬡–$R^2$ | + | + | + | + | yellow | $<0$, middle | ? | DSM |

Symbols:   + stable,   − unstable,   FE field effects,   DSM dynamic scattering mode

## TABLE 12
### Characteristic Voltages of Some Mixtures in Twist Cells

| Mixture of | $U_O$ (threshold) | $U_s$ (90% saturation) |
|---|---|---|
| R—⬡—⬡—CN | 0.6 V | 1.3 V |
| R¹—⬡—C(=O)—O—⬡—R² <br> R¹—⬡(H)—C(=O)—O—⬡—CN | 0.5 V | 1.3 V |
| R¹—⬡(H)—C(=O)—O—⬡—OR² <br> R¹—⬡(H)—C(=O)—O—⬡—CN | 0.7 V | 1.5 V |

25° C; white light; 12 μm thickness, glass with obliquely evaporated tin dioxide.

## REFERENCES

1 W. R. Young, A. Aviram and R. J. Cox, J. Amer. Chem. Soc. 94 (1972) 3976
2 J. Malthete, M. Leclercq, J. Gabard, J. Billard and J. Jacques, Compt. Rend. 273 (1971) 265
3 H. Schubert and W. Weißflog (unpublished)
4 H. Kelker, B. Scheurle, R. Hatz and W. Bartsch, Angew. Chem. 82 (1970) 984
5 R. Steinsträsser and L. Pohl, Z. Naturforsch. 26b (1971) 577
6 J. van der Veen, W. H. de Jeu, A. H. Grobben and J. Boven, Mol. Cryst. Liq. Cryst. 17 (1972) 291
7 W. H. de Jeu and J. van der Veen, Phys. Lett. A44 (1973) 277
8 H. Schubert and H.-J. Deutscher (unpublished)
9 W. R. Young, A. Aviram and R. J. Cox, Angew. Chem. 83 (1971) 399
10 J. van der Veen and A. H. Grobben, Mol. Cryst. Liq. Cryst. 15 (1971) 239
11 I. Teucher, C. M. Paleos and M. M. Labes, Mol. Cryst. Liq. Cryst. 11 (1970) 187
12 J. van der Veen and Th. C. J. M. Hegge, Angew. Chem. Internat. Ed. 13 (1974) 344
13 H. Schubert and J. Pätzold (unpublished)
14 W. Weißflog, H. Schubert, D. Demus, B. Priesemuth and L. Vogel, DDR WP 106524 (1974)
15 B. P. Smirnov and I. G. Chistyakov, Isv. vyssh. uchebn. Zaved., Khim. Khim. Tekhnol. 13 (1970) 217
16 V. V. Titov, E. I. Kovshev, A. I. Pavluchenko, V. T. Lazareva and M. F. Grebenkin, J. Phys. Colloque 36 (1975) C1-387
17 W. Weißflog, H. Schubert, D. Demus and A. Wiegeleben, DDR WP 108023 (1974)

18  S. L. Arora, J. L. Fergason and T. R. Taylor, J. Org. Chem. 35 (1970) 4055

19  W. Weißflog, H. Schubert, H. Kresse and D. Demus, DDR WP 106933

20  W. Weißflog, H. Heberer, Ch. Mohr, H. Zaschke, H. Kresse, S. König and D. Demus DDR WP 116732 (1975)

21  D. Demus, L. Richter, C.-E. Rürup, H. Sackmann and H. Schubert, J. phys. Colloque 36 (1975) C1-349

22  R. Steinsträsser, Angew. Chem. 84 (1972) 636

23  J. P. van Meter and B. H. Klanderman, Mol. Cryst. Liq. Cryst. 22 (1973) 285

24  G. W. Gray, Mol. Cryst. Liq. Cryst. 7 (1969) 127

25  G. W. Gray, K. J. Harrison, J. A. Nash, J. Constant, D. S. Hulme, J. Kirton and E. P. Raynes in J. F. Johnson and R. S. Porter (ed.), 'Liquid Crystals and Ordered Fluids', Vol. 2, Plenum, New York (1974)

26  G. W. Gray, J. phys. Colloque 36 (1975) C1-337

27  H. Zaschke, H. Schubert, F. Kuschel, F. Dinger and D. Demus, DDR WP 95892 (1973)

28  K. Hanemann, H. Schubert, D. Demus and G. Pelzl, DDR WP 115283 (1975)

29  R. Champa, Mol. Cryst. Liq. Cryst. 19 (1973) 233

30  H. J. Deutscher, F. Kuschel, H. Schubert and D. Demus, DDR WP 105701 (1974)

31  J. A. Castellano, Ferroelectrics 3 (1971) 29

32  L. T. Creagh, Proc. IEEE 61 (1973) 814

33  G. W. Gray, 2. Freiburger Arbeitstagung Flüssigkristalle, Freiburg, April 1972

34  R. Steinsträsser and L. Pohl, Angew. Chem. 85 (1973) 706

35  D. Demus and F. Kuschel, Liquid Crystal Conference, Uniejew (Poland) October 1973 (in press)

36  D. Demus, Z. Chem. 15 (1975) 1

37  D. Demus, H. Demus and H. Zaschke, 'Flüssige Kristalle in Tabellen', VEB Deutscher Verlag für Grundstoffindustrie, Leipzig (1974)

38  K. Hanemann, H. Schubert, U. Bargenda and D. Demus, DDR WP 115829 (1975)

39  J. A. Castellano, RCA Rev. 33 (1972) 296
    J. A. Castellano and M. T. McCaffrey, DOS 2.121.085

40  R. A. Soref in W. A. Albers jr. (ed.), 'The Physics of Optoelectronic Materials', Plenum, New York (1971)

41  L. M. Blinov, Uspekh. fiz. nauk 114 (1974) 67

42  A. P. Kapustin, 'Elektrooptische und akustische Eigenschaften flüssiger Kristalle', Nauka, Moskau (1973)

43  A. Saupe, Ann. Rev. Phys. Chem. 24 (1973) 441

44  W. Helfrich, Mol. Cryst. Liq. Cryst. 21 (1973) 187

45  L. Kofler and A. Kofler, 'Thermo-Mikro-Methoden zur Kennzeichnung organischer Stoffe und Stoffgemische', Verlag Chemie, Weinheim (1954)

46  A. R. Ubbelohde, 'Melting and Crystal Structure', Clarendon, Oxford (1965)

47  E. C.-H. Hsu and J. F. Johnson, Mol. Cryst. Liq. Cryst. 20 (1973) 177

48  E. C.-H. Hsu and J. F. Johnson, Mol. Cryst. Liq. Cryst. 27 (1974) 95

49  D. Demus, Ch. Fietkau, R. Schubert and H. Kehlen, Mol. Cryst. Liq. Cryst. 25 (1974) 215

50  M. Domon and J. Billard, in S. Chandrasekhar (ed.), 'Liquid crystals', Pramana, India (1975)

51  D. S. Hulme, E. P. Raynes and K. J. Harrison, J. Chem. Soc. Chem. Comm. (1974) 98

52  G. W. Gray, K. J. Harrison and J. A. Nash, J. Chem. Soc. Chem. Comm. (1974) 431

53  G. W. Smith, Mol. Cryst. Liq. Cryst. (in press)

54  D. Demus, Z. Naturforsch. 22a (1967) 285

55  L. Pohl and R. Steinsträsser, Z. Naturforsch. 26b (1971) 87

56  D. Demus, R. Schubert and Ch. Fietkau (unpublished)

57  A. Boller, H. Scherrer, M. Schadt and P. Wild, Proc. IEEE 60 (1972) 1002

58  D. Demus, M. Süße, L. Vogel and G. Pelzl (unpublished)

59  J. Billard, J. C. Dubois and A. Zann, J. Phys. Colloque 36 (1975) C1-355

60  J. van der Veen and W. H. de Jeu, Mol. Cryst. Liq. Cryst. 27 (1974) 251

61  J. P. van Meter, R. T. Klingbiel and D. J. Genova, Solid State Comm. 16 (1975) 315

62  R. T. Klingbiel, D. J. Genova, T. R. Criswell and J. P. van Meter, J. Am. Chem. Soc. 96 (1974) 7651

63  W. H. de Jeu and Th. W. Lathouwers, Z. Naturforsch. 29a (1974) 905

64  W. H. de Jeu and Th. W. Lathouwers, Z. Naturforsch. 30a (1975) 79

65  H. Kresse, D. Demus and Chr. Krinzner, Z. phys. Chem. 256 (1975) 7

66  A. Ashford, J. Constant, J. Kirton and E. P. Raynes, Electron. Lett. 9 (1973) 118

67  W. H. de Jeu and Th. W. Lathouwers, Mol. Cryst. Liq. Cryst. (in press)

68  H. Kresse and P. Schmidt, Z. phys. Chem. (in press)

69  W. Maier and G. Meier, Z. Naturforsch. 16a (1961) 262

70  H. van der Veen, W. H. de Jeu, A. H. Grobben and J. Boven, Mol. Cryst. Liq. Cryst. 17 (1972) 291

71  R. Steinsträsser and L. Pohl, 5th International Liquid Crystal Conference, Stockholm, June 1974

72  H. Kresse and P. Schmidt (unpublished)

73  W. Maier and A. Saupe, Z. Naturforsch. 14a (1959) 882 and 15a (1960) 287

74  W. Maier and A. Saupe, Z. Naturforsch. 16a (1961) 1200

75  H. Weise and A. Axmann, Z. Naturforsch. 21a (1966) 1316

76  A. Axmann, Z. Naturforsch. 21a (1966) 290, 615

77  H. Kresse, P. Schmidt and D. Demus, Phys. Status Solidi a 32 (1975) 315

78  W. H. de Jeu, C. J. Gerritsma, P. van Zanten and W. J. A. Goossens, Phys. Lett. A 39 (1972) 355

79  G. Baur, A. Stieb and G. Meier, 16th National Meeting of the American Chemical Society, Chicago, 1973

80  H. K. Bücher, R. T. Klingbiel and J. P. van Meter, Appl. Phys. Lett. 25 (1974) 186

81  H. Kresse, F. Kuschel and D. Demus, DDR WP 107563 (1974)

82  H. Kresse, F. Kuschel and D. Demus, DDR WP 107561 (1974)

83  E. P. Raynes, I. A. Shanks, Electron. Lett. 10 (1974) 114

84  Ch. Gähwiller, Mol. Cryst. Liq. Cryst. 20 (1973) 301

85  P. G. de Gennes, 'The Physics of Liquid Crystals', Clarendon, Oxford (1974)

86  R. S. Porter and J. F. Johnson in F. Eirich (ed.), 'Rheology IV', J. Wiley, New York (1968) 301

87  A. Denat, B. Gosse and J.-P. Gosse, J. Chim. Phys. 2 (1973) 327

88  G. H. Heilmeier, L. A. Zanoni and L. A. Barton, Proc. IEEE 56 (1968) 1162

89  J. Vildebrand and H. Matschiner (unpublished)

90  G. Briere, R. Herino and F. Mondon, Mol. Cryst. Liq. Cryst. 19 (1972) 157

91   G. Spott, W. Thiel and D. Demus, Wiss. Z. Univ. Halle (in press)

92   S. Lu and D. Jones, Appl. Phys. Lett. 16 (1970) 484

93   A. Denat, B. Gosse and J. P. Gosse, Chem. Phys. Lett. 18 (1973) 235

94   A. Denat and B. Gosse, Chem. Phys. Lett. 22 (1973) 91

95   A. Denat, B. Gosse and J. P. Gosse, J. Chim. Phys. 2 (1973) 319

96   A. Lomax, R. Hirasawa and A. J. Bard, J. Electrochem. Soc. 119 (1972) 1679

97   F. Gaspard, R. Herino and F. Mondon, Mol. Cryst. Liq. Cryst. 24 (1973) 145

98   F. Rondelez, Solid State Commun. 11 (1972) 1675

99   Sh. Kh. Kuvatov, A. P. Kapustin and A. N. Trofimov, Shurn. eksp. teor. fiz. 19 (1974) 89

100  G. Heppke and F. Schneider, Z. Naturforsch. 29a (1974) 1523

101  G. Heppke and F. Schneider, Z. Naturforsch. 30a (1975) 316

102  'Nematic Phases Licristal', leaflet of E. Merck, Darmstadt (1974)

103  G. Pelzl, R. Rettig and D. Demus, Z. phys. Chem. 256 (1975) 305

104  W. Maier in Landolt-Börnstein, 'Zahlenwerte und Funktionen aus Physik-Chemie - Astronomie - Geophysik - Technik', Vol. 2, Springer, Berlin (1960) 533

105  M. Brunet-Germain, Compt. Rend. 271 (1970) 1075

106  G. Pelzl and H. Sackmann, Z. phys. Chem. 254 (1973) 354

107  G. Pelzl and H. Sackmann, Mol. Cryst. Liq. Cryst. 15 (1971) 75

108  G. Pelzl and H. Sackmann, Symposia of the Faraday Society, 5 (1971) 68

109  W. Maier and A. Saupe, Z. Naturforsch. 13a (1958) 564, 14a (1959) 882 and 15a (1960) 287

110  C. S. Bak, K. Ko and M. M. Labes, J. Appl. Phys. 46 (1975) 1

111  H. Sorkin and A. Denny, RCA Review 34 (1973) 308

112  G. Labrunie and M. Bresse, Compt. Rend. 276B (1973) 647

113  H. Gruler and G. Meier, Mol. Cryst. Liq. Cryst. 23 (1973) 261

114  J. Wulf and D. Demus (unpublished)

## DISCUSSION

### E. Guyon (Université de Paris-Sud)

I am not sure that a systematic study of the viscosity properties of liquid crystals as a function of material can be done by measuring a single viscosity term. It is not clear that the behavior of the different Leslie terms will show the same type of variation, for instance with temperature. This is especially questionable in the case of the two viscous torques. In large shears, one should make clear that no tumbling effect takes place and that one does not measure a viscosity largely conditioned by the motion of disclinations. It would be interesting to have systematic measurements of individual viscosity terms for comparison with specific predictions of various molecular models.

### D. Demus

I agree in principle. We used a simple measurement because no other technique was available, hoping that a useful correlation to the individual coefficients exists, at least far from phase transitions.

F. J. Kahn (Hewlett-Packard)

We too have found it useful to make a single viscosity measurement in an attempt at a practical comparison of different mixtures. A theoretical analysis of the relationship between the measured quantity and the individual viscosities would be helpful.

H. Kelker (Hoechst)

Would you agree that as a rule the melting point minimum in a homologous series occurs for the C4 compound?

D. Demus

The minimum seems to fall in the range C4 to C8.

M. L. Hitchman (RCA Zurich)

In the field of polymer chemistry it is possible to predict, often with remarkable accuracy, the physical properties of polymeric materials on the basis that each functional group in the molecular structure actually performs a function that is reflected in the macroscopic properties (cf. D. W. van Krevelen, "Properties of Polymers", Elsevier, 1972). Do you know if any similar correlation has been attempted in relation to liquid crystal materials and their properties?

D. Demus

There are relations between the molecular structure and the clearing points, which for special structures allow the rough calculation of clearing points with the aid of substituent parameters. As I have tried to show in my paper, there are connections of the molecular structure with the dielectric and rheological properties and the double refraction. In the case of transition enthalpies there is a complicated situation. In most homologous series the transition enthalpies show an increasing trend, but there are also some series with a decreasing trend. Therefore the connections between molecular structure and transition enthalpies may be valid for one series or one substance class, but are not general.

# ANCHORING PROPERTIES AND ALIGNMENT OF LIQUID CRYSTALS

E. GUYON and W. URBACH*
Université Paris-Sud, Orsay, France

## SUMMARY

The anchoring properties of liquid crystals in contact with a solid substrate have received considerable attention, from the early work on surface defects to the most recent theoretical and experimental studies of surface disclinations. From these analyses, as well as from the elastic behavior of very thin films, the notions of surface energy and of an extrapolation distance giving the spatial range of surface effects are introduced. A schematic description of the various configurations of alignment, as related to various electrooptic devices, is given. The possibility of orientations intermediate between the planar and homeotropic is discussed. The more important techniques of alignment are briefly reviewed. Finally, some limiting mechanisms of orientation are discussed: the mechanical one involves the bulk elastic energy on deformed surfaces; chemical mechanisms consider the polar and dispersive contributions to the alignment. A conclusion is that much more remains to be done in a systematic study of the mechanisms and on the evaluation of surface energy parameters.

## 1. INTRODUCTION

The study of the interaction of a liquid crystal (LC) with a solid substrate has a long history, initiated by the careful observations by Grandjean[1] in 1917. Surprisingly, the preparation of single domain samples — single crystals — by inducing a bulk alignment through surface treatment was not a major concern for experimentalists for a long time, probably because the

*) Also Lab. de Biophysique, C.H.U. Cochin, 75 Paris, France

erroneous "swarm" model, which also slowed down the development of the continuum theory, dealt well with polydomain samples. Zocher[2] and Chatelain[3] were able to align nematics in the planar state[4] wherein the average molecular axis (the director field $\vec{n}$, $|\vec{n}| = 1$) lies along x in the plane of a glass substrate which had been previously rubbed on paper along the x direction. Homeotropic alignment with $\vec{n}$ perpendicular to the substrate (along z) was also obtained by deposition of the LC on very clean glass plates. However the exact mechanisms of these two techniques were not understood and the conditions for obtaining alignment depended crucially on a "tour de main".

Only recently has careful consideration been given to the two fundamental problems:

  — inspection of the physicochemical mechanisms
  — quantitative evaluation of the energies involved in the "anchoring".

We will focus our attention in this review on these two aspects. The technical aspects will be treated briefly with a comparative study of the ways of preparation of surfaces for electrooptic devices. For a more exhaustive treatment it would be necessary to consider the choice of LC material, dopants, impurities, sealants, external agents (temperature, effects of light), together with the substrate; excellent recent review articles already exist.[5-7] We hope that the present approach can be of some help in:

  — the evaluation of new, more reliable and durable orientation techniques (in particular for the case of smectic phases where a lot remains to be explored)
  — thinking of devices where new alignment conditions (weak anchoring, conical or degenerate alignment) could be used.

We will begin by introducing briefly the conditions of orientation for the most classical electrooptic devices. Largely inspired by the recent work of Kleman's group in Orsay, we will then describe the parameters which characterize the alignment of a liquid crystal (mostly nematics) on a substrate. The next section will briefly review the alignment techniques. The last section will discuss some limiting mechanical and chemical mechanisms of orientation.

## 2. SOME FUNDAMENTAL PROPERTIES OF DEVICES

### 2.1. Geometry

Liquid crystals are strongly anisotropic. Most phases are uniaxial. The ellipsoid of indices has a rotational symmetry about its long (extraordinary) axis which is parallel to $\vec{n}$. In MBBA, $\nu_e \cong 1.7$; $\nu_o \cong 1.5$. The geometry is defined in Fig. 1 where $\vec{n}$ is characterized by the polar angles $\phi$ and $\theta$. The following orientations can be obtained:

  — homeotropic: $\vec{n}$ (and extraordinary axis) perpendicular to the substrate ($\phi = 0$), extinction between crossed polarizers.

— planar: $\vec{n}$ in the plane of the substrate, along x ($\phi = \pi/2$, $\theta = 0$). Four positions of extinction between crossed polarizers are found.

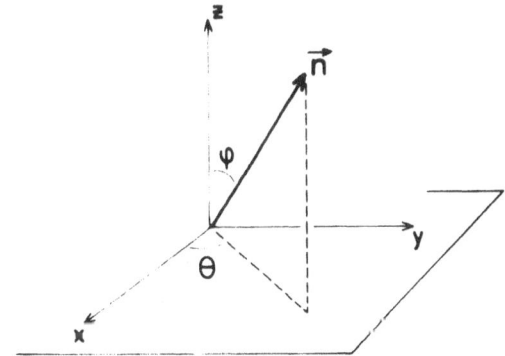

Fig. 1. Coordinate geometry: xy is the plane of the substrate.

— oblique (or tilted): $0 < \phi < \pi/2$, $\theta$ = constant. The observations between crossed polarizers are the same as in the planar case.
— degenerate: planar, $\phi = \pi/2$, degenerate in $\theta$; or conical, $0 < \phi < \pi/2$, degenerate in $\theta$. This is obtained in particular at the free surface[8] or at the nematic-isotropic interface[9] of MBBA.

In cholesterics, the planar alignment causes the axis of the helix to be perpendicular to the substrate whereas, with a homeotropic alignment of the molecules, the helix axes are randomly distributed in the plane xy and a focal conic structure is seen. In smectics A, where the director is perpendicular to the layers, the homeotropic alignment has layers parallel to the planes and is usually quite transparent as the fluctuations of orientation are limited by the rigidity of the planes. A planar molecular alignment with planes perpendicular to the plates can also be obtained. It is also possible to obtain single-domain samples of the more complex smectic phases. In all cases we have found that the ordering depends on the possibility of cooling from a well ordered higher temperature mesophase.

## 2.2. Optical Measurements

A polarizing microscope can be used to observe the uniformity of alignment and the absence of defects. However, it is an insufficient test, in particular for the oblique and degenerate cases (see above). Observations of the conoscopic images formed in a monochromatic beam converging in the sample is needed[10] (Fig. 2). In planar samples, the ambiguity in the direction of $\vec{n}$ between the two axes of the sets of hyperbolas can be resolved by observing the deviation of the image of a thick sample (d $> 100~\mu$m) by a small magnet or from the direction where the "swarming" observed with a single polarizer is most visible.

It is well known that nematics scatter light strongly, especially at small angles, because of the long wavelength ($\lambda = 2\pi/q$) spontaneous fluctuations of orientation which are large compared with the density fluctuations. (The

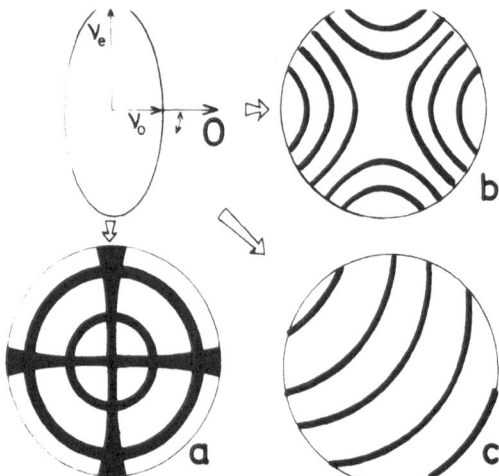

Fig. 2. Ellipsoid of indices for a positive $(\nu_e > \nu_o)$ uniaxial liquid crystal. The effect of fluctuation of orientation around the planar and homeotropic alignment is shown. The typical conosocopic images with average beam direction perpendicular to the plates are shown for homeotropic (a), planar (b) and oblique (c) orientation.

critical opalescence observed at the critical point of an ordinary fluid implies a diverging compressibility.) A careful analysis of the scattering cross section $\sigma$ (which, by angular integration, gives the absorption coefficient) has recently been given by Bouchiat and Langevin.[11] The anisotropy of $\sigma$ stems from the optical anisotropy and from that of the elastic constants. For simplicity, we will neglect this last anisotropy factor in the following and consider a single elastic constant K ($\sim 10^{-6}$ erg/cm).

The thermal fluctuations are evaluated from the ratio $k_B T/Kq^2$. However experiments lead to much larger values which indicate that the static contribution due to surface roughness is dominant. The analysis can be carried out by replacing $k_B T$ by an expression f(q) characteristic of the surface irregularities. A control of the function f(q) is a major aim in order to attain LC cells of good optical quality. In practice, homeotropic samples can be obtained on substrates of better smoothness and are more transparent. In planar samples, the "swarming" is more visible when the polarization is along the director (along x), as some selection rules[12] prevent the divergent contribution of low q with light polarized along y (normal incidence). Another contribution arises from a consideration of the ellipsoid of indices (Fig. 2). In the presence of a variation of $\vec{n}$ in the oe plane, the variation of $\nu$ is larger around the planar orientation because the radius of curvature for the polarization along the extraordinary axis e is smaller. This also explains why the swarming effect is more important for an oblique orientation, as any fluctuation of $\vec{n}$ implies, in this case, a first order variation of $\nu$ rather than second order.

While the advantage of homeotropy in devices is clear, this discussion indicates the need for a definition of orientation beyond "it aligns or not".

In particular, with large anchoring energy treatments to be defined next, very good transparency is achieved in planar films.

## 2.3. Field Effect Devices

As a classical example of the field effect used in many devices, we have given in the appendix a simplified solution of the Freedericksz transition[13,14] of a planar nematic film of positive dielectric constant anisotropy ($\epsilon_a = \epsilon_{\parallel} - \epsilon_{\perp} > 0$) in a vertical electric field. The calculation can be extended immediately to our homeotropic geometry with $\epsilon_a < 0$. The existence of a sharp "second order" onset of distortion at a critical voltage $V_c$ comes from the orthogonality of the initial alignment with the field. Any misalignment (in particular for poorly aligned samples) will suppress the threshold and cause rounded transitions, which are to be avoided in devices.[14] From equation (A7) we find that

$$V_c = E_c d \cong \pi (4\pi K/\epsilon_a)^{1/2} (1 - 2K/W_1 d). \qquad (1)$$

This equation introduces the notion of a surface energy $W_1$ and of weak anchoring.[14b] We give a plausible angular dependence of a surface energy

$$w_1\big|_{erg/cm^2} = W_1 \sin^2 \Delta\phi, \qquad (2)$$

where $w_1$ represents the energy to rotate the director from its preferred orientation in the vertical plane by an angle $\Delta\phi$. In the strong anchoring limit ($W_1 \to \infty$), the planar boundary condition reads

$$\phi(\pm d/2) = 0. \qquad (3)$$

Furthermore the solution of the linear Freedericksz problem just above threshold satisfies the condition

$$\frac{d\phi}{dz}\bigg|_{\pm d/2} = \mp \frac{\phi}{b}. \qquad (4)$$

The extrapolation length

$$b = K/W_1 \qquad (5)$$

will be used often in the following to characterize the anchoring energy. In Fig. 3 we see that it measures the continuation of variation of the angle $\phi(z)$ outside the film: indeed, expression (1) can be transformed into

$$E_c(d+2b) \cong E_{c_\infty} d .$$

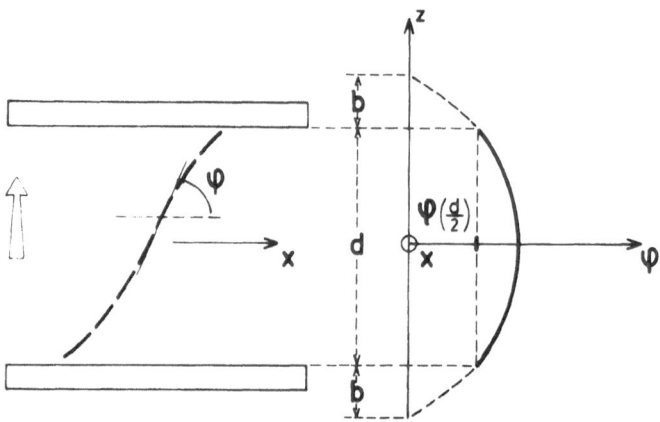

Fig. 3. Planar nematic with $\epsilon_a > 0$ in a perpendicular electric field. Above the critical threshold a distortion $\phi(z)$ develops. Note the finite angle at the boundaries and the extrapolation length b due to a weak anchoring condition (b = 0 for strong anchoring).

Compared with the strong anchoring limit for b = 0, the weak anchoring modification $E_c$ comes from a replacement of the thickness d by d+2b. The variation $V_c(d)$ gives access, in principle, to a determination of $W_1$. However, it would require the use of extremely thin films if strong anchoring were to be measured: for $W_1 = 1$ erg/cm$^2$, b ~ 50 to 100 Å.

Another energy

$$w_2 = W_2 \sin^2 \Delta\theta \tag{6}$$

characterizes the rotation from the preferred polar angle $\theta = 0$ for a planar or oblique alignment. It can be measured from the Freedericksz transition of a planar layer with diamagnetic susceptibility anisotropy $\chi_a > 0$ in a magnetic field along y. In a recent experiment,[15] a planar film contained between a treated plate and an untreated one (weak anchoring) was twisted by rotating one of the plates by $\pi/2$. $W_2$ was measured by studying the variation of the twist angle when a magnetic field was applied along the strong anchoring direction. A small value $W_2 \cong 5 \times 10^{-2}$ cgs was obtained.

In practice, the homeotropic alignment used in field-effect devices with $\epsilon_a < 0$ can be obtained without any surface treatment on plates freshly covered with a transparent conducting tin oxide (NESA) or indium oxide film. However, a good homeotropic alignment will also imply the formation of domains at $\phi$ and $-\phi$ which are degenerate in $\theta$ because there is no preferred direction in the xy plane of the film: one observes the formation of umbilics.[18] The width of the walls separating the domains increases as the voltage approaches $V_c$. At $V_c$ it is much larger than the film thickness. To prevent these domains it is desirable to have a uniform tilt of the homeotropic alignment at the plates, small in order to retain most of the sharp transition. By depositing a surfactant giving a homeotropic alignment on a rubbed plate, only two opposite directions of bend along the direction of rubbing will be obtained[19] when a voltage $V > V_c$ is applied. Some surfactants[20] lead to a small initial tilt of the alignment. We have also been able to obtain a tilt $\phi_0 \sim 5$ to $15°$ along a well defined direction by deposition of a homeotropic surfactant on a grazing incidence evaporated film (see 4.2).

A small initial distortion $\phi_0$ is also needed to get a fast rise time when the voltage is applied if the zero state corresponds to no applied voltage. The increase of $\phi$ follows a law: $\phi = \phi_0 \exp(t/\tau)$, where the time constant $\tau$ is a function of the applied voltage and becomes long near the second order transition at $V_c$[21]. The prefactor angle $\phi_0$ also influences the rise time. In some devices a residual distortion is kept in the unactivated state by maintaining an applied voltage; this involves little power consumption as high resistivity materials are used. However, in display devices where the change in transmitted intensity between crossed polarizers or in reflected intensity between a mirror and circular polarizer is used, this reduces the contrast ratio. A good compromise seems to be an initial angle $\phi_0 \sim 5°$.

The planar alignment is used in particular for the field effect on a twisted sample with $\epsilon_a > 0$[22]. The directions of planar alignment on the two plates make a $\pi/2$ angle. This leaves the possibility of two signs of the chirality of the nematic, unless a small amount of cholesteric material is added. The separation between domains of reverse twist is seen as lines whose width is measured by an extrapolation length b which can give access to the energy $w_2(\theta)$ (see 3.3). A photograph (Fig. 4) showing both field induced lines and surface lines and a discussion of the electrooptic quality in conjunction with surface preparation in these devices are given by Hareng[23,24].

Cholesterics can be used in a field effect display by inducing the distortion from the planar to the focal conic texture in materials with $\Delta\epsilon > 0$[25]. Smectics are generally unsuitable for field effects since the rigidity of the layers strongly limits the possible distortion in a field[26]. It is important to note that field effects require an electric field frequency $\omega \sim 10^3$ to $10^5$ Hz to prevent the convective mechanisms used in the display to be discussed next. A field effect can also be used to restore the order of a layer (to "erase") when another distortion mechanism is used. By using a superposition of two fields at different frequencies, it is possible to obtain at the same time the alignment and the distortion effect. This is also possible in materials where the sign of $\epsilon_a$ changes with frequency[16].

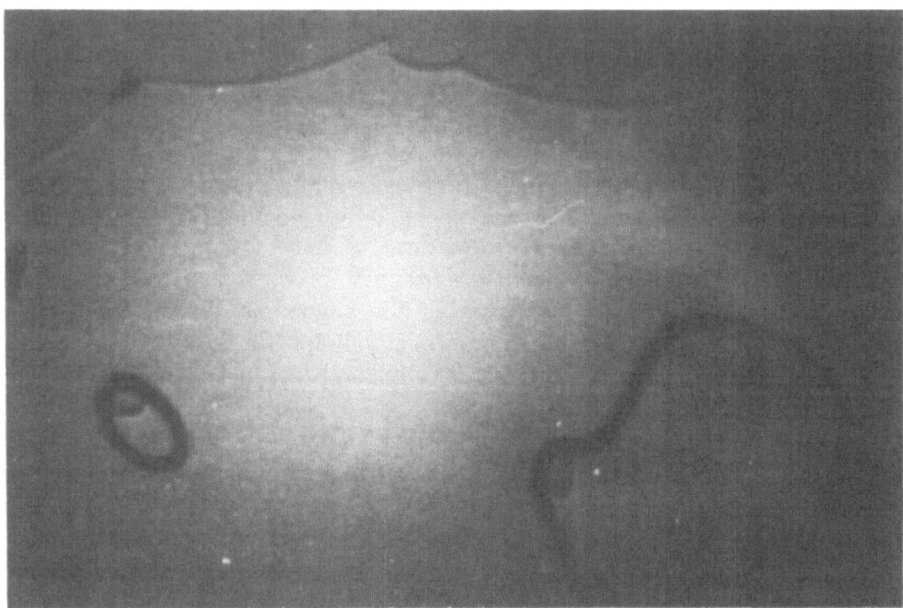

Fig. 4. Twisted planar sample ($\epsilon_a > 0$) in a vertical electric field[24]. Broad lines correspond to $\phi$ type defect lines and have a field dependent width. Narrow lines are $\theta$ type surface disclination lines.

## 2.4. Dynamic Scattering Mode (DSM) Devices

The application of a large enough electric field of low frequency to a planar nematic film of negative $\epsilon_a$ induces a strong distortion of the structure and, consequently, a large increase of turbidity. Memory effects can also be obtained if a cholesteric guest is added[28]. We will not discuss the mechanisms which have received considerable theoretical attention and which involve convective motions of ions[29]. Because of the greater transparency in the undistorted state, a homeotropic geometry is used. In this case, the DSM involves first a field effect which tends to orient the molecules perpendicular to the field and then the turbulent motion. In multiplexed systems which involve the continuous application of a voltage, the planar alignment is preferred because it has a single sharp threshold.

We will not consider other classes of displays where LC's are used as detectors of sound waves, mechanical stresses, temperature changes or chemicals (with cholesterics). Of particular interest is thermal addressing wherein cholesterics or smectics are disoriented by the local heating of a laser beam. In all these cases, similar orientation problems arise.

## 3. MEASUREMENT OF SURFACE ENERGY[30-35]

In Fig. 5 we consider a planar sample with a 180° Bloch-like torsion wall.[30,34,35] We assume for the moment that the distortion is uniform through the thickness. This corresponds to the case of thin samples with d < b. The bulk elastic free energy per unit length of the wall is given by

$$F_B \sim \frac{\pi^2 d}{2e} K, \qquad (7)$$

and clearly decreases if the thickness e of the wall increases. On the other hand, assuming a linear twist rotation of $\vec{n}$ through the wall, the surface energy at the two surfaces is given by

$$F_S = 2 \int_o^e W_1 \sin^2\phi \, du = W_1 e. \qquad (8)$$

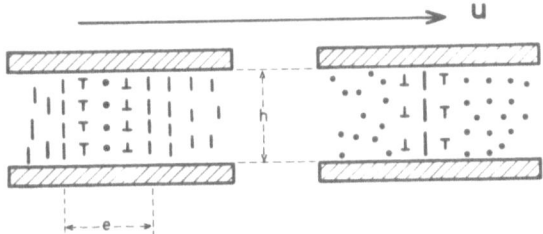

Fig. 5. Schematic representation of Bloch walls implying a twist uniform in thickness in planar and homeotropic liquid crystals. The thickness of the wall e increases as the anchoring strength decreases.

The total free energy $F_W = F_B + F_S$ is a minimum when

$$\delta F_W / \delta e = 0$$

which yields

$$e = \frac{\pi}{2} \sqrt{\frac{2dK}{W_1}} = \frac{\pi}{2} \sqrt{2bd} \qquad (9)$$

and

$$F_W = \sqrt{2dKW_1}. \qquad (10)$$

Another possibility is the occurence of a pair of surface lines.[31,32,33] In this case the variation of orientation with z must be considered. One can carry out a calculation of the elastic energy[32] similar to the case of a bulk disinclination line. (Note that a surface line with s = 1/2 where the director rotates by $\pi$ corresponds to a coreless s = 1 bulk line and its energy can also be described by a continuum model without a core singularity.) An order of magnitude for the line energy per unit length is

$$E_L = \pi K . \tag{11}$$

The existence of one or the other type of defect is determined by the ratio

$$\frac{F_W}{2E_L} = \sqrt{\frac{2dW_1}{K}} = \sqrt{\frac{2d}{b}} . \tag{12}$$

As was intuitively obvious, walls which do not involve a distortion along z are obtained for thin samples (d < b), whereas surface defects with a distribution of orientation through the wall thickness are seen when d > b. Detailed observations and measurements in these two limits have been carried out[34].

In thick planar films (d $\sim$ 100 $\mu$m) and with strong anchoring conditions, a detailed optical analysis of the distribution of director orientation $\theta(u)$ as one crosses a surface disclination perpendicular to u has led Ryschenkow[34] to a measure of the angular distribution of energy $w_2(\theta)$. His work follows the analysis by Kleman and Vitek[32]. They use the form of a surface boundary condition similar to that obtained in the Rapini calculation[14] (eq. A3) to express the balance between the torque $dw_2/d\theta$ exerted on the director at the surface and the elastic torque arising from the measured distribution of orientations along u. By numerical integration one gets $w_2(\theta)$. Figure 6 gives the form of the results obtained for a substrate covered with an obliquely evaporated SiO film. The variation is not too far from the $\sin^2\theta$ distribution (equation 6). The radii of curvature near the extrema could, in principle, also be evaluated from the Freedericksz measurement of the extrapolation length on a twisted nematic with $\theta = \pi/2$. However, the values of $W_2 \sim 1$ erg/cm$^2$ correspond to a very small b $\sim$ 30 Å and the measurement would be impossible in practice.

The other limit was obtained for thin LC layers ($\sim$ 10 $\mu$m) on a substrate with weak anchoring[35]: a carbon film was formed on glass by burning paper under it. At room temperature, MBBA gives a degenerate conical orientation. Figure 7 shows a wall with variation of twist across it. From the width of the wall e $\sim$ 20 $\mu$m and using (9), one gets an anchoring energy $W_1 \sim 10^{-4}$ erg/cm$^2$. The difference in order of magnitude between these two extreme cases points out the need for consideration of the surface energy parameter in technical discussions on alignment in liquid crystal displays.

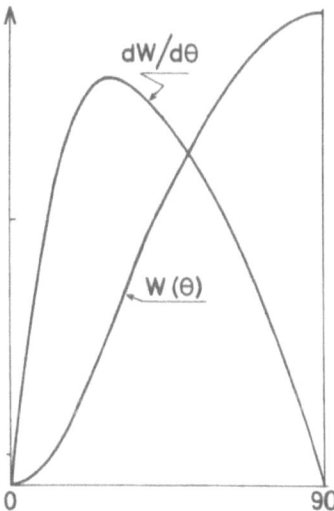

Fig. 6. Angular variation of the surface torque $dw/d\theta$ from the profile of orientation about a surface disinclination line, and corresponding surface energy dependence $w(\theta)$. $\theta = 0$ is the easy orientation axis.

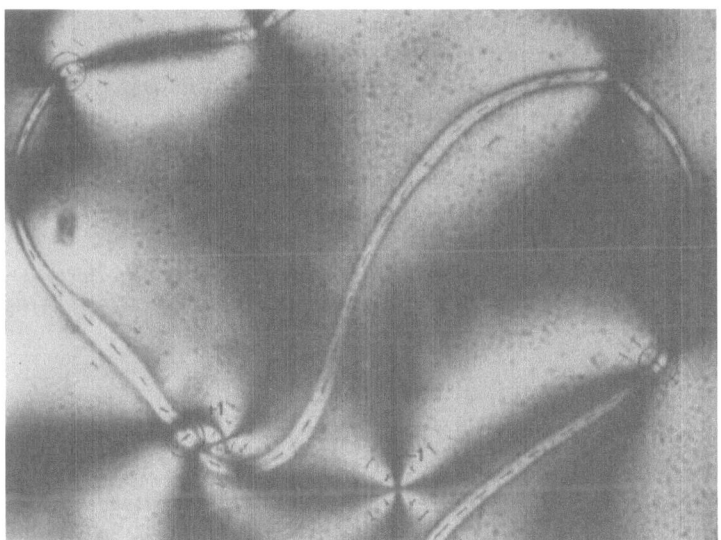

Fig.7. Photograph[34,35] shows Bloch walls obtained with low anchoring energy conical alignment on carbon. It involves a twist of the director which is planar in the middle of the wall. Inversion points (circles) and classical nematic nuclei can also be seen.

## 4. METHODS OF ALIGNMENT[5]

We shall not try to give a detailed list of techniques used to obtain the various possible orientations. In any case, such a discussion should be carried out in conjunction with the LC materials used. Nor will we discuss crucial parameters such as simplicity of application, cost and lifetime. However, the techniques reported here all led to a stable alignment during the course of our experiments lasting several months.

### 4.1. Chemical Effects

Homeotropic alignment. We have mentioned the homeotropic alignment on very clean glass. Creagh and Kmetz[19] have done a careful study of chemically cleaned surfaces controlled by Auger spectrometry and pointed out the role of contaminents (see also Ref. 36). In practice, a glass substrate covers rapidly with a film of adsorbed water. It is also likely that when an LC is deposited, a monomolecular LC layer strongly binds by polar forces to the substrate and the subesequent alignment of the LC must be considered on this first layer substrate. This point is suggested by dynamic observations during the filling of a cell where wetting of the substrate — which, as we will see, is often associated with a planar alignment — is followed by dewetting[5,19].

More reliable and controlled techniques involve the use of surfactants in the bulk of the LC or on the substrate. Haas et al.[37] obtained homeotropy by adding 0.5% of Versamid 100 in the bulk of Schiff base LC's (a derivative '360' is indicated[24] as giving very long lifetime alignment). However, it seems that bulk additives should be used with care as the LC properties (conducitivity, elasticity, etc.) can vary rapidly with small concentrations of additives and with the time evolution of their concentration. Surfactant molecules deposited directly on the substrate with a hydrophylic head sticking to the polar substrate and a hydrophobic tail pointing away are good candidates to reduce the surface energy. A classical example is barium stearate deposited as a 25 Å thick Langmuir film. Hexadecyltrimethyl ammonium bromide (HMAB) which also can be added in the bulk, was first used by Haller and Huggins[38]. The substrate is treated by dipping in a saturated water or chloroform solution. The aliphatic chains point out of the substrate and induce the homeotropic alignment. One can improve the binding of surfactants by using molecules having severable groups. Kahn[39] has developed the use of the "surface coupling" organosilanes ($R \; Si(OCH_3)_3$) to produce various types of alignment. When R is a long ($C_{18}$) hydrocarbon chain, homeotropy is obtained. The stability of the surface layer can be increased by hydrolysis and subsequent drying. One gets a two dimensional polymeric network which appears to be very stable with time and temperature. Polymerization in a gaseous plasma has also been used successfully[40].

Planar alignment. A connection between these two limiting types of alignment was obtained in experiments using surface deposited HMAB[41]. Figure 8 shows the surface density $\delta/cm^2$ of a $C^{14}$ marked HMAB film obtained by slowly pulling a glass plate from an aqueous solution of various concentrations c. For large c, homeotropy is attained in agreement with Haller's findings. The upper plateau corresponds to a close packing of HMAB molecules. The lower plateau corresponds to a less dense packing and is consistent with HMAB molecules lying flat on the substrate. In this range of concentrations, MBBA aligns in the planar state. The director is along the axis of pulling of the plates. The weak effect due to the flow action on the HMAB molecules can provide a preferential alignment of these molecules. This is possibly a cooperative effect relating to quasi long range order in 2D systems. No anchoring energy data has been reported in this case. It could probably be correlated with the amount of alignment of the HMAB molecules along the axis of pulling.

In practice, planar alignment using chemical techniques, can be obtained with surfactants having two polarizable heads. Organosilanes have again been used successfully, where R contains a short $(C_3)$ carbon chain with an amino end group which can attach to the substrate. Since a non degenerate planar alignment is generally required, rubbing of the substrate after deposition of the surfactant (but before polymerization) or physical treatment of the surface (see 4.2) before deposition are used.

Other types of alignment. Intermediate oblique alignment can be obtained by using a first molecular layer which lies at an angle to the polar substrate[20]. Finally, we recall the Ryschenkow technique where, by depositing

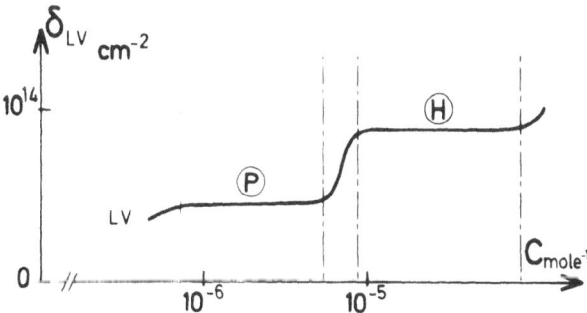

Fig. 8. Surface density (in molecules/cm$^2$) of HMAB molecules adsorbed on glass, plotted as a function of the molar concentration of the solution. Orientation of liquid crystal on the two plateaus is indicated and corresponds to that of the first HMAB layer: P = planar, H = homeotropic.

a thin amorphous carbon film formed by burning paper, a degenerate conical alignment (with a $\phi$ angle varying rapidly with T) was obtained with MBBA.

## 4.2. Mechanical Effects

The classical preparation of planar samples developed by Chatelain[3] and Zocher[2] involves the directional rubbing of a glass substrate on paper. It is likely that the dominant mechanism in this case is the alignment of organic contaminants present on the glass.[19] It is paradoxical that not more is known about the mechanism of this famous technique! On the other hand, controlled and systematic experiments have been reported where the substrate was directionally scribed, deformed and etched to obtain planar alignment[36,42]. An experiment[43] showing planar alignment on the replica of a rubbed surface strongly suggests the existence of a physical mechanism. However, the Creagh and Kmetz[19] experiment shows that, on very clean scribed glass plates, homeotropy is obtained as a result of the competition between chemical and physical mechanisms.

A very reliable technique used initially by Janning[44] involves the vacuum deposition of a thin film under oblique incidence. The geometry is indicated in Fig. 9. We have studied this technique in detail[45,46,37] as a func-

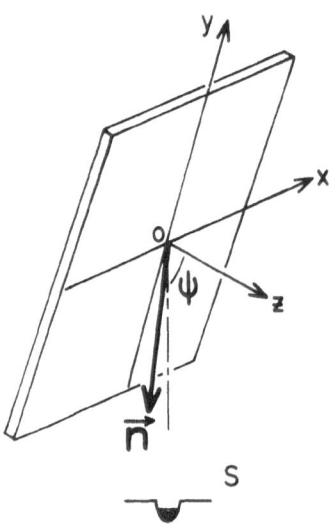

Fig. 9.  Geometry of the oblique incidence evaporation technique. Orientation of liquid crystal along x or in the yz plane in a direction close to that indicated for the evaporation beam is possible.

tion of the angle of incidence of the beam $\psi$. Films of gold or SiO evaporated at normal incidence ($\psi = 0$) give a degenerate planar alignment. (This does not imply a weak alignment: such samples show uniform alignment by "plages" and twisting the LC director field or rotation of the plates does not give any indication of weak anchoring.)

If the incidence angle is larger than 45°, a planar orientation is obtained where the director is along the direction x perpendicular to the plane of incidence of the beam. The mechanism, developed in the case of oblique incidence magnetic film, involves the self-shadowing of evaporated grains whereby long chains of the evaporated film form along $x$[48]. Figure 10 shows an electron micrograph of an SiO film with $\psi = 60°$. Grazing X-ray measurements on a similar film indicate a modulation $\Delta D \sim 100$ Å for an average film thickness $D \sim 200$ Å[49]. As indicated in the discussion of Fig. 6, this alignment technique leads to a very strong anchoring and we have been able to align in the planar state practically all mesophases we have studied. Lifetime exceeds two years, and temperature cycling tests up to 250° C have been performed with success.

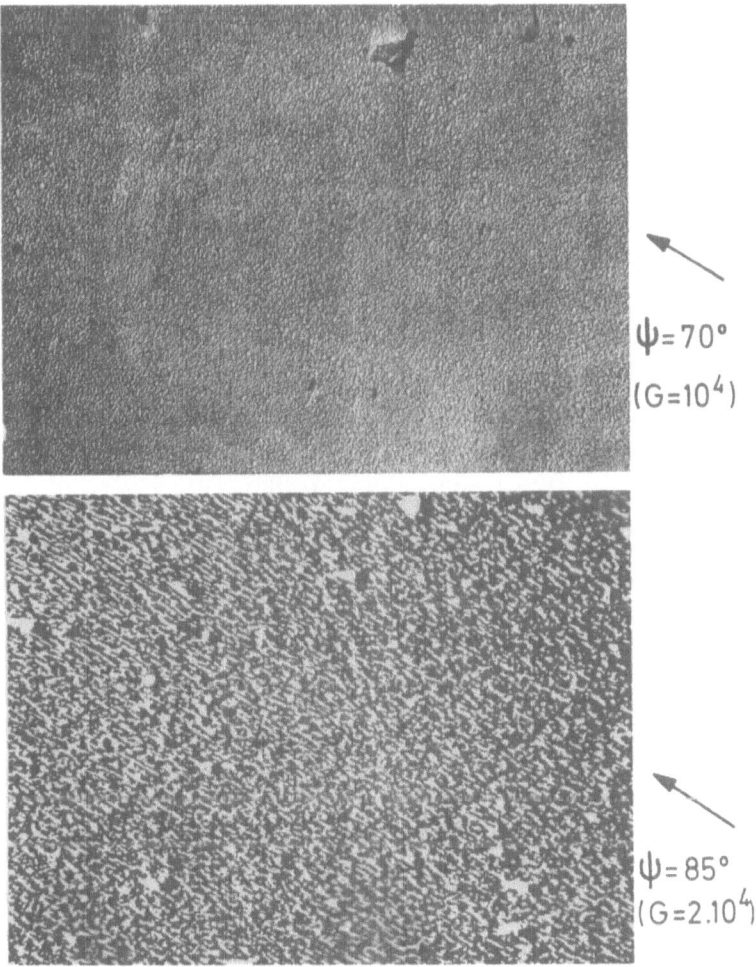

Fig. 10. Electron micrographs (magnification G) of obliquely evaporated Au films. For 70° incidence angle, channels roughly perpendicular to direction of evaporation beam (marked by arrow) can be seen. For 85° grazing incidence angle, the elongated shape of the grains dominates, the alignment being roughly along the grains (courtesy of Mmes Leger and Lottin).

For grazing incidence angle ($\psi > 80°$) the prolate shape of the grains in the direction of the beam dominates (Fig. 10). This induces nondegenerate oblique alignment of nematics with the director in the yz incidence plane at an angle $\phi \sim 80°$. Apart from the existence of these limiting $\psi$ angle values, the technique does not appear to depend crucially on the conditions of evaporation: vacuum pressure, values of D or $\psi$. In particular, we have found no systematic decrease of the anchoring energy measured from the width of surface lines as discussed in Section 3, when $\psi$ decreases towards 40°. A technique which would lead to a continuously variable anchoring energy is yet to be found.

Together with Proust and TerMinassian, we have also studied the additional effect of an HMAB film deposited on evaporated substrates. Measurements of surface coverage indicate a uniform deposition on the oblique film[50] except in the case of extreme grazing incidence. In the intermediate angle region ($45° < \psi < 85°$), homeotropy is obtained if a dense HMAB film is deposited. However, for the largest angles ($\psi > 85°$), a small tilt of the director from homeotropy ($\phi \sim 5°$ to 15°) is obtained along the direction of formation of the grains. Although more remains to be done for a systematic study of this last effect, it might be usable if a small deviation from homeotropy is needed in devices as discussed in 2.3.

Although the foregoing discussion was carried out for nematics, it can be extended to the cholesteric and smectic phases. However, in the smectic case, a strong anchoring (as defined with nematics) is needed to get a planar alignment. The oblique evaporation technique works well if the alignment of the smectic is obtained by cooling from a nematic phase. With grazing evaporation followed by HMAB deposition, we have also been able to obtain single crystal smectic C materials with the plane of the layer parallel to the substrate and with the axis of $\vec{n}$ towards the tilt direction $\phi$. Careful cooling from a homeotropic smectic A phase can also give good smectic C crystals if shear is applied by sliding one glass plate with respect to the other to align the axis of $\vec{n}$ along the direction of motion.

## 5. MECHANISMS

### 5.1. Physical Mechanism

It is qualitatively obvious that, due to the second order (orientational) elasticity of nematics, the energy of an LC deposited on a grooved surface (we assume z = A sin qy) can be minimized by aligning $\vec{n}$ along x. Independently several authors[36,42,51] have obtained expressions for the elastic surface energy. One assumes that, at the interface, $\vec{n}$ is locally parallel to the substrate. If one neglects the anisotropy of the elastic constants, the angular distribution of the director in the yz plane would be given from the solution of the Laplace equation

$$K\nabla^2\phi = 0, \tag{13}$$

where $\phi$ is the angle with the average y direction. The free energy difference between this orientation and the preferrred one, with $\vec{n}$ along the grooves (along x), can be obtained simply from a dimensional consideration and is given by

$$W_2 = \frac{K}{4} A^2 q^3. \tag{14}$$

In the oblique incidence experiment with $A = \Delta D \sim 100 \text{ Å}[44]$ and $q^{-1} = 200$ to $400 \text{ Å}$, one gets $W_2 \sim 1 \text{ erg/cm}^2$ in good agreement with the measurement of Fig. 6. This value is typically 10 times larger than that calculated by Berreman on glass substrates polished with a diamond paste[42] ($A \sim 10 \text{ Å}$; $q^{-1} = 200 \text{ Å}$). We have solved (13) in a similar manner to explain the oblique orientation on films with grazing incidence, using a model of asymmetric sawtooth substrates[45].

Another mechanism[34] involves the consideration of a local surface energy, $w_{\ell/cm^2} = W_{\ell} \sin^2\phi$ rather than the elasticity. $\phi$ refers to the direction with the local normal to the grooved surface. The molecules are assumed to remain everywhere parallel to the average direction. One obtains

$$w_2(\theta) = \frac{W_{\ell}}{2} A^2 q^2 \sin^2\theta. \tag{15}$$

This model, which gives better consideration to the LC-substrate interaction mechanism, can possibly apply for weak anchoring substrates.

The elasticity equation (13) implies that the distortion of $\vec{n}$ will be damped exponentially over a thickness of the order of $q^{-1}$. Using this result and the data in the evaluation of (14), we reach an energy density $\sim Wq$ or $\sim 10^{-3} \text{ eV/particle}$. This figure can be used to put boundaries on the possible available anchoring energies. The smectic case is markedly different from the nematic and the anisotropic elasticity plays a dominant role, the elasticity being first order for a displacement perpendicular to the layers[52]. In particular, if a smectic is deposited on a grooved surface with the planes of the layers following the ripples, the damping of the oscillation of the planes will take place over a much larger distance[53]

$$\ell \sim q^{-1} (aq)^{-1}$$

where a is of the order of an atomic dimension: $\ell \sim 1 \text{ mm}$ for an undulation of wavelength $10 \ \mu m$. The long range memory of surface defects explains many difficulties in the alignment of smectics and in particular why the condition of planarity of the substrates is much more stringent for smectics than for nematics. Float glass or fire polished glass can be used which are cheaper than optically polished glass.

## 5.2. Chemical Mechanisms

The observed correlation between wetting and planar alignment, unwetting and homeotropy led Creagh and Kmetz[19] to a simple general criterion for the alignment of various nematics. The substrates are characterized by a critical surface tension $\gamma_c$ obtained by measuring the contact angles $\theta$ of different liquids of surface tension $\gamma_L > \gamma_c$ and extrapolating to $\theta = 0$ (limit of wetting). The term $\gamma_c$, proportional to the surface tension of the solid, characterizes the interaction with the LC whereas $\gamma_L$ measures the cohesion of the LC. The difference $\gamma_c-\gamma_L$ in erg/cm$^2$ can be used as a measure of the surface energy difference between homeotropic and planar alignment. If $\gamma_L-\gamma_c > 0$ (no wetting), homeotropic alignment is predicted. A planar alignment should be obtained when $\gamma_L-\gamma_c < 0$, for example with metal substrates of large cohesive energies and large $\gamma_c$. The case of clean glass with a high value of $\gamma_c$ which gives homeotropy does not fit the description. However, we have mentioned the difficulties with adsorption of water or other impurities, or even that of a first LC layer, which all would lower $\gamma_c$. In contradiction with the Creagh & Kmetz classification, Haller[54] has obtained different alignments on the same substrate for LC's which had practically the same surface tension.

It seems necessary to separate, in a detailed analysis, the various contributions: steric, dipolar (p) and dispersive (d). A systematic study of the p and d terms has been carried by Proust and TerMinassian[55] in the case of MBBA deposited on an HMAB film. The electric dipole of MBBA being nearly perpendicular to the molecular axis ($\epsilon_a < 0$), the molecules like to lie in a planar configuration on the substrate. When the HMAB surface coverage $\delta$ increases (Fig. 8) this dipolar contribution, and consequently the surface energy measured by the contact angle, is reduced. The authors assume that only the dispersive contribution is left when there is saturation of HMAB. This reduction would explain the homeotropy obtained for large $\delta$ values. We see that this model differs from the empirical Creagh & Kmetz classification. In particular, the important role of the sign of $\epsilon_a$ is not apparent in their model. The difference in the energies found for the planar and homeotropic cases, desired for comparison of the chemical effects with the elastic mechanism, is of the order of 1 erg/cm$^2$, but since it is obtained by subtraction of much larger terms, this result should be used with care.

The forces involved in the Proust & TerMinassian analysis are all short range. If one now considers the case of an anisotropic substrate (large dielectric anisotropies can be obtained in paraelectric crystals), the range of Van der Waals interaction increases as $z^{-3}$ for a non-retarded interaction and as $z^{-4}$ for a retarded interaction. The problem of a nematic layer deposited at a distance $\Delta$ from such a substrate was considered by Smith and Ninham.[56] The separation $\Delta$ could be due to an isotropic slab giving no preferred orientation. If the alignment at the other LC surface ($z = \Delta+d$) is rigidly fixed along a direction different from the one preferred by the anisotropic Van der Waals mechanism, the distribution of orientations across the thickness of the LC should lead to an evaluation of the Van der Waals force constant. The continuum elasticity in some sense replaces the torsional pendulum involved in classical Van der Waals force measurements. This problem was reconsidered in more detail by Dubois-Violette and de Gennes[57] for variable $\Delta$ (retarded to non-retarded interaction) and for different boundary conditions at the surface $z = \Delta$. An explicit expression for the extrapolation length b is obtained.

Although this example may look rather academic, it is a good example of a possible fundamental approach to surface anchoring in nematics leading to comparisons between experiment and theory without the need for a detailed molecular description or knowledge of the substrate topology (a perfectly planar surface is assumed). An amusing "local Freedericksz" transition has been described in the case where the long range Van der Waals forces favor homeotropy whereas the short range anchoring favors planar alignment.[58] Planar-to-conical and conical-to-homeotropic transitions are predicted. This result may possibly be relevant in the description of the Ryschenkow experiments on carbon film: as T increases, the MBBA alignment varies from nearly planar to homeotropic. The temperature variation of the Van der Waals forces should be like that of the order parameter S whereas short range forces vary as $S^2$ because they are governed by the elasticity.

# 6. CONCLUSION

In this review, we have attempted to give indications on orientation mechanisms as well as some quantitative estimates of the anchoring properties. More systematic data as a function of the LC itself are clearly desirable. The effect of degradation of the LC (hydrolysis or UV interaction) and of the substrate (chemical or electrochemical[59]) should be considered in more detail. The anchoring energy also controls in practice the electrooptic qualities of the devices (sharpness of transitions, transparency). Finally, the description of new types of alignment (conical, degenerate) as well as the possibility of weak anchoring can possibly lead to new applications.

We quote here two examples of possible effects. Ryschenkow, Pikin and Urbach[60] have studied electrohydrodynamic instabilities of nematics in an oblique geometry obtained from grazing incidence Au evaporation. Instead of the rolls (vortices) perpendicular to the average molecular orientation seen in the planar Williams problem, the rolls are along the $\vec{n}$ direction at threshold. The difference comes from the reduced symmetry between $\vec{n}$ and the rolls circulating in a vertical plane. Kleman and Pikin[61] have also studied theoretically some possible hydrodynamic instabilities obtained in a shear flow with conical orientation. Depending on the relative values of the surface energies $W_1$ and $W_2$, different instabilities are expected with variation of $\theta$ or $\phi$. This study could give access to new measurements of surface energy.

The possibility of weak anchoring conditions has been partly exploited; Helfrich has proposed some measurements on flexoelectricity in nematics. In a first experiment,[62] a homeotropically aligned sample was placed in a field parallel to the plate. Although the dielectric effect was stabilizing ($\epsilon_a < 0$), a distortion $\phi \propto z$ due to flexoelectricity was observed. The $\phi(z)$ dependence implies non-zero boundary conditions and weak anchoring for the lecithin agent used. Helfrich[63] has also proposed a polarity dependent flexoelectric device based on the existence of two boundaries of different anchoring energies (otherwise the asymmetric flexoelectric contributions cancel due to the integration over thickness).

One can also think of memory effects where the transient application of an external agent (for example the temperature, which by itself can also change the alignment condition) can change the surface orientation and lead to memory effects if surface defects are present. This would also take place with strong anchoring conditions if several energy minima were obtained for different directions of alignment. Indications that such effects are present have been obtained.[23]

## ACKNOWLEDGMENTS

We have had important discussions with M. Kleman, C. Williams, G. Ryschenkow on the properties of surface defects and with L. TerMinassian and J. Proust on chemical surface effects. The support of DGRST contract 73-7-1483 is acknowledged.

## REFERENCES

1   F. Grandjean, Bull. Soc. Fr. Min. 39 (1916) 164

2   H. Zocher and G. K. Cooper, Z. Phys. Chem. 132 (1928) 295

3   P. Chatelain, Bull. Soc. Fr. Min. 66 (1943) 105

4   One uses also the term "homogeneous" to characterize this alignment. We feel that it is misleading as it also characterizes a single domain sample and will not use it.

5   L. A. Goodman, RCA Rev. 35 (1974) 347 and F. J. Kahn, G. N. Taylor and H. Schonhorn, Proc. IEEE 61 (1973) 823

6   L. T. Creagh, Proc. IEEE 61 (1973) 814

7   J. Borel, G. Labrunie and J. Robert, J. Phys. 36 (1975) C1-215

8   M. A. Bouchiat and D. Langevin, Phys. Lett. A 34 (1971) 331

9   R. Vilanove, E. Guyon, C. Mitescu and P. Pieranski, J. Phys. 35 (1974) 153

10  N. H. Hartshorne and A. Stuart, "Crystals and the Polarizing Microscope", Arnold, London (1970)

11  D. Langevin and M. A. Bouchiat, J. Phys. 36 (1975) CI-197

12  Orsay group on liquid crystals, Mol. Cryst. Liq. Cryst. 13 (1971) 187

13  V. Freedericksz and V. Zolina, Trans. Far. Soc. 29 (1933) 919 and H. Gruler and G. Meier, Mol. Cryst. Liq. Cryst. 16 (1972) 299

14a A. Rapini, These de 3eme cycle, Orsay (1970)

14b A. Rapini and M. Papoular, J. Phys. 30 (1969) C4-54

15  J. Sicart, private communication

16  H. K. Büchner, R. T. Klingbiel and J. P. VanMeter, Appl. Phys. Lett. 25 (1974) 186

17  e.g. F. J. Kahn, Appl. Phys. Lett. 20 (1972) 199 and W. Helfrich, Mol. Cryst. Liq. Cryst. 21 (1973) 187

18  A. Rapini, L. Leger and A. Martinet, J. Phys. 36 (1974) C1-189 and L. Leger, Mol. Cryst. Liq. Cryst. 24 (1973) 33
19  L. T. Creagh and A. R. Kmetz, Mol. Cryst. Liq. Cryst. 24 (1973) 59
20  J. F. Dreyer, private communication
21  P. Pieranski, F. Brochard and E. Guyon, J. Phys. 33 (1972) 681 and 34 (1973) 35
22  D. W. Berreman, this book
23  M. Hareng, private communication
24  J. C. Dubois, M. Gazard, A. Zann, M. Hareng and J. J. Metzger, Compte rendu de fin d'etudes DGRST (73-7-1979)
25  E. P. Raynes, this book
26  A. Rapini, J. Phys. 33 (1972) 237
27  G. H. Heilmeier, L. A. Zanoni and L. A. Barton, Proc. IEEE 56 (1968) 1162
28  G. H. Heilmeier and J. E. Goldmacher, Proc. IEEE 57 (1969) 34
29  W. Helfrich, J. Chem. Phys. 51 (1969) 4092
30  P. G. de Gennes, "The Physics of Liquid Crystals", Oxford University Press (1974) Chapt. 4
31  C. Williams and M. Kleman, Phil. Mag. 28 (1973) 725
32  V. Vitek and M. Kleman, J. Phys. 36 (1975) 59
33  C. Williams, V. Vitek and M. Kleman, Solid State Commun. 12 (1973) 581
34  G. Ryschenkow, These de 3eme cycle, Orsay (1975)
35  G. Ryschenkow and M. Kleman, to be published in J. Chem. Phys.
36  U. W. Wolf, W. F. Greubel and H. H. Krüger, Mol. Cryst. Liq. Cryst. 23 (1973) 187
37  W. Haas, J. Adams and J. B. Flannery, Phys. Rev. Lett. 25 (1970) 1326
38  I. Haller, J. Chem. Phys. 57 (1973) 1400
39  F. J. Kahn, Appl. Phys. Lett. 22 (1973) 111 and 386
40  J. C. Dubois, M. Gazard and A. Zann, Appl. Phys. Lett. 24 (1974) 297
41  J. E. Proust, L. TerMinassian-Saraga and E. Guyon, Solid State Commun. 11 (1972) 1227
42  D. W. Berreman, Phys. Rev. Lett. 28 (1972) 1683
43  J. F. Dreyer, Proceedings 3rd Internat. Liquid Crystal Conf. Berlin (1970)
44  J. L. Janning, Appl. Phys. Lett. 21 (1972) 173
45  E. Guyon, P. Pieranski and M. Boix, Lett. in Appl. & Eng. Science 1 (1973) 19
46  W. Urbach, M. Boix and E. Guyon, Appl. Phys. Lett. 25 (1974) 479
47  See also G. Dixon, T. Brody and W. Hester, Appl. Phys. Lett. 24 (1974) 47
48  D. O. Smith, M. S. Cohen and G. P. Weiss, J. Appl. Phys. 31 (1960) 1755
49  P. Croce and L. Nevot, private communication and J. Appl. Cryst. 7 (1974) 125
50  M. Clavillier, private communication. The surface area was evaluated from the charge involved in an anodic oxidation of the metallic film
51  P. G. de Gennes, Orsay lectures (1972)
52  Ref. 30, chapt. 7
53  G. Durand, C. R. Acad. Sc. 275B (1972) 629
54  I. Haller, Appl. Phys. Lett. 24 (1974) 349
55  J. E. Proust and L. TerMinassian-Saraga, J. Phys. 36 (1975) C1-77
56  E. R. Smith and B. W. Ninham, Physica 66 (1973) 111
57  E. Dubois-Violette and P. G. de Gennes, to be published in J. Colloid & Interface Sci.
58  E. Dubois-Violette and P. G. de Gennes, J. Phys. Lett. 36 (1975) L-255
59  A. Sussman, Appl. Phys. Lett. 21 (1972) 126

60  G. Ryschenkow, P. Pikin and W. Urbach, to be published in J. Phys.
61  M. Kleman and P. Pikin, to be published
62  D. Schmidt, M. Schadt and W. Helfrich, Z. Naturforsch. A 27 (1972) 277
63  W. Helfrich, Appl. Phys. Lett. 24 (1974) 451
64  J. Nehring, A. R. Kmetz and T. J. Scheffer, to be published in J. Appl. Phys. 47
    (1976)

## APPENDIX

We consider a planar ($\vec{n} \, // \, x$) nematic sample in an electric field $\vec{E}$ along z. The elastic energy can be expressed as

$$F_1 = \frac{K}{2}(d\phi/dz)^2$$

if we retain only a possible bend of the molecules along z. The electrostatic contribution is:

$$F_2 = -\frac{1}{4\pi} \int \vec{D}\, d\vec{E} = \frac{-\epsilon_\perp E^2 - \epsilon_a (\vec{n}\vec{E})^2}{8\pi} \, .$$

The second term tends to align the molecules parallel to E if $\epsilon_a > 0$. We also add a finite surface energy

$$w_1 = W_1 \sin^2\phi \qquad \text{at both boundaries } z = \pm d/2.$$

The total free energy, in the limit of small distortions, is written

$$F = \int_{-d/2}^{d/2}[K(d\phi/dz)^2 - \epsilon_a E^2 \phi^2/4\pi]dz + W_1[\sin^2\phi(d/2) + \sin^2\phi(-d/2)]. \quad (A1)$$

We minimize F with respect to variations of $\phi(z)$. The equilibrium configuration which expresses the balance of the elastic and electric torques is given by

$$K\, d^2\phi/dz^2 + \frac{\epsilon_a E^2 \phi}{4\pi} = 0. \qquad (A2)$$

An additional condition (coming from integration by parts) is

$$2K(d\phi/dz)_{d/2} + W_1 \sin 2\phi(d/2) = 0 \qquad (A3)$$

with a similar condition (with a minus sign) at $-d/2$. The form of this boundary condition, obtained first by Rapini,[14b] expresses the equilibrium between the surface anchoring torque $dw/d\theta$ and the elastic torque at the surface. We look for solutions of (A2) of the form

$$\phi = \cos qz \qquad (A4)$$

with

$$Kq^2 - \frac{\epsilon_a E^2}{4\pi} = 0. \tag{A5}$$

The boundary condition (A3) leads to

$$q = \pi/(d+2b) \tag{A6}$$

with $b = K/W_1$. The definition of b as an extrapolation length is fairly obvious from Fig. 2. Finally, using (A5) and (A6), one gets the critical field condition

$$E_c = \frac{\pi}{d+2b} \sqrt{\frac{4\pi K}{\epsilon_a}}. \tag{A7}$$

The distortion above threshold is found to depend on the ratios of elastic constants as well as on the value of $W_1$.[14a] Experimentally sharper transitions are found for low anchoring energies. This can be useful for the evaluation of $W_1$.

After completing this review, we received a detailed analysis[64] of the possible technical use of small anchoring for obtaining saturation of distortion as close as possible to threshold. This result is certainly in line with our consideration of the anchoring strength factor.


## DISCUSSION

W. Helfrich (Freie Universität Berlin)
How do you obtain conical alignment and do you have a physical explanation for it?

E. Guyon
Ryschenkow obtains conical alignment very consistently by burning paper under his glass plates. Practically complete azimuthal degeneracy is shown for instance by shear flow experiments. De Gennes and Dubois-Violette explain the tilt angle in terms of a competition between long range alignment forces from the substrate and short range forces from the surface layer. The balance between these forces shifts with temperatures as the elastic and dielectric constants change.

W. G. Freer (Rank)
You state that the anchoring energy for oblique evaporation is two orders of magnitude larger than that for Chatelain polishing. Is this due to a lower density of anchoring points in the latter case?

E. Guyon

The anchoring energy of glass (contaminated by finger oil?) rubbed on paper according to the Chatelain technique has been evaluated in Ref. 34. However we have not studied the structure of these polished surfaces.

A. R. Kmetz (Brown Boveri)

Using electron microscopy on replicas of Chatelain-polished substrates, Creagh and I confirmed that grooves in the organic contaminant layer were essential to uniform alignment, but we did not make quantitative measurements of their depth and spacing.

E. Guyon

Nevertheless the oblique evaporation technique is far more reproducible and durable for planar alignment. The Chatelain technique is seldom effective with smectics, but evaporation works with nearly all smectics.

F. J. Kahn (Hewlett-Packard)

Since your measurements of anchoring energy for Chatelain polishing involved a rubbed surface whose chemical nature was very uncertain, it would be interesting to repeat the experiments with a chemically well defined surface.

E. Guyon

In the experiments using HMAB surfactant the number of molecules adsorbed on glass was determined from $C^{14}$ counting, and the surface area and structure could be obtained independently. We have seen different tilt angles as a result of competition between chemical and elastic mechanisms, but anchoring energy was not measured in that experiment. It will probably take a few years before the chemical aspects of alignment are well understood.

J. P. Hurault (LEP)

Can you comment on analytical methods which will be able to give information on the anchoring conditions on a microscopic scale?

E. Guyon

The experiment described by Smith and Ninham and redeveloped by De Gennes and Dubois-Violette is at least one case where a microscopic physical model can be experimentally tested.

SHORT COMMUNICATION

# MEASUREMENT OF ALIGNMENT TILT IN

# TWISTED NEMATIC DISPLAYS

K. TORIYAMA and T. ISHIBASHI

Electron Tube Division, Hitachi Ltd., Mobara City, Japan

The development of a twisted nematic display for an electronic watch is a problem in total system design involving materials, fabrication and electronics. In this short communication we concentrate on the technology of molecular alignment which has much influence on display performance. A technique suitable for mass production is required which reliably provides uniform quiescent alignment and freedom from induced domains of reverse tilt. To develop such a process, we have investigated the oblique evaporation method of alignment[1,2] by measuring the direction of molecular orientation with a "magneto-capacitive" method.

If a magnetic field is applied to the twisted nematic device, the molecules reorient to a new state under the combined influence of the magnetic field and the intermolecular forces. Typical models for molecular orientation in a field are shown in Fig. 1. From the symmetry of the twisted structure, we consider variation of the magnetic field direction in two planes. Molecules in the center of the device are parallel to the A-A' plane and are perpendicular to the B-B' plane in the absence of field.

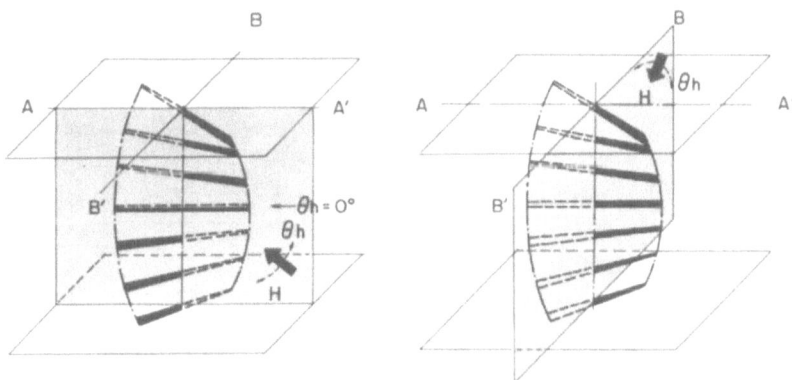

Fig. 1.  Definition of plane of magnetic field ($\vec{H}$): A-A' plane parallel to center molecules; B-B' plane perpendicular to center molecules.

We infer the state of liquid crystal orientation from observation of the capacitance C and its derivative $dC/d\theta_h$ with the direction angle $\theta_h$ of the applied field. Measurements of the ratio of capacitance $C_h$ in a 9 kGauss field to the capacitance $C_o$ in zero field are shown in Fig. 2 as a function of the field direction in the two planes. Results are given for samples oriented by SiO evaporation at two different angles of beam incidence $\theta$.

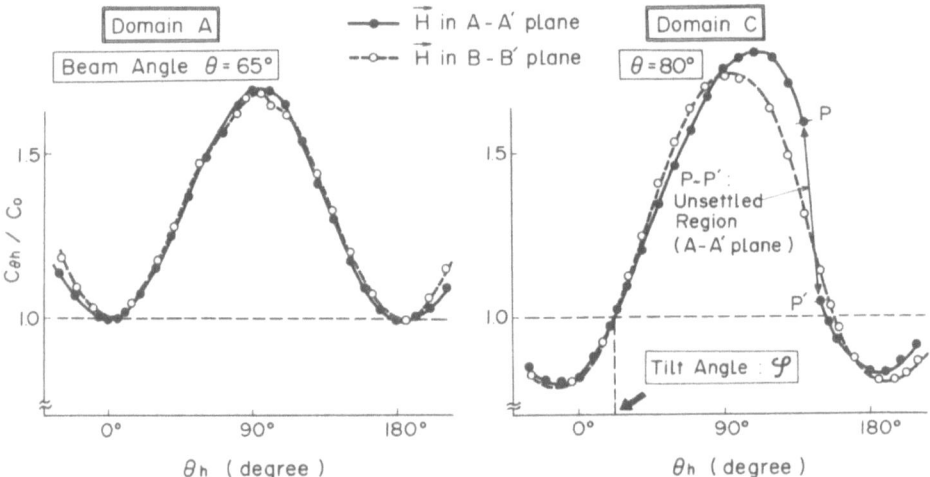

Fig. 2. Magneto-capacitive curve ($C_h/C_o$ vs $\theta_h$) in TN display devices with substrate interfaces of Domain A and Domain C. Strength of applied magnetic field ist 9 kGauss.

If we can assume complete and pure dissolution of the liquid crystal twist under the influence of the magnetic field, **Fig. 3** illustrates that the

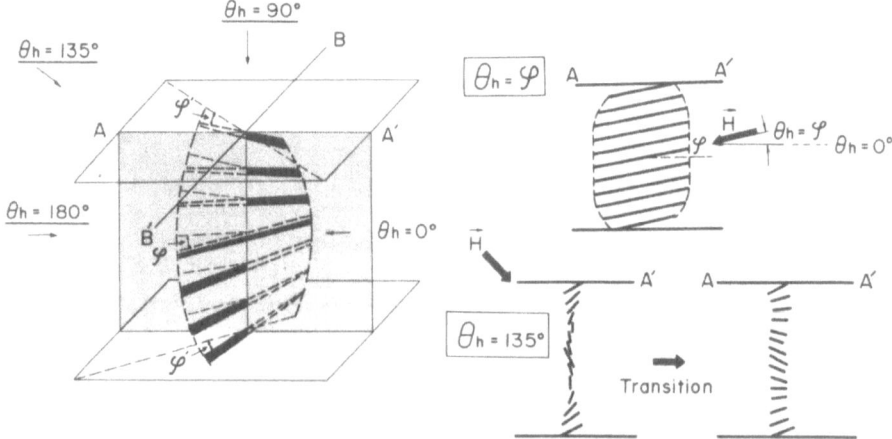

Fig. 3. Model for tilt orientation of liquid crystal molecules in twisted nematic devices with substrate interface of Domain C.

angle of the field in the A-A′ plane for which $C_h = C_o$ corresponds to the average molecular tilt angle $\varphi$ with respect to the surface planes. Using this measurement technique, we find that molecular alignments produced by various angles of beam incidence fall into three domains:

A) $50° < \theta < 73°$ — alignment in the substrate plane ($\varphi = 0$) perpendicular to the projection of the beam.
B) $73° < \theta < 77°$ — no preferred orientation.
C) $\theta \geqslant 78°$ — alignment in the plane of beam incidence with $15° \leqslant \varphi \leqslant 30°$.

Table 1 shows that different tilt angles $\varphi$ were measured on comparable Domain C substrates for different liquid crystal materials.

TABLE 1

Liquid Crystal Materials and their Tilt Angle ($\varphi$) for the Device with Substrate Interface of Domain C

| Liquid Crystal Mixtures | $\varphi$ |
|---|---|
| **LC − I:** | 22° |

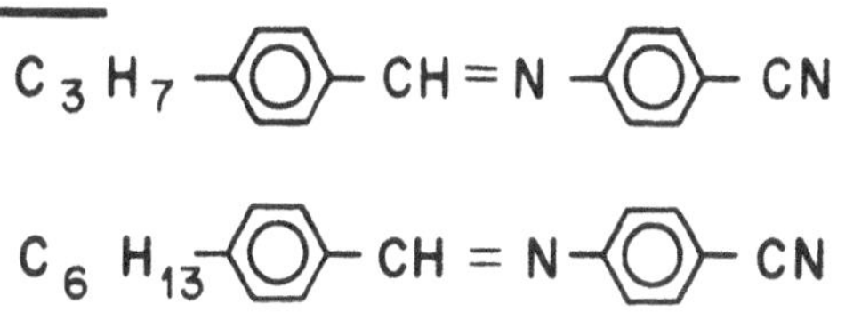

$$C_3H_7 - \bigcirc - CH = N - \bigcirc - CN$$

$$C_6H_{13} - \bigcirc - CH = N - \bigcirc - CN$$

1:2 mole ratio

| **LC − II:** 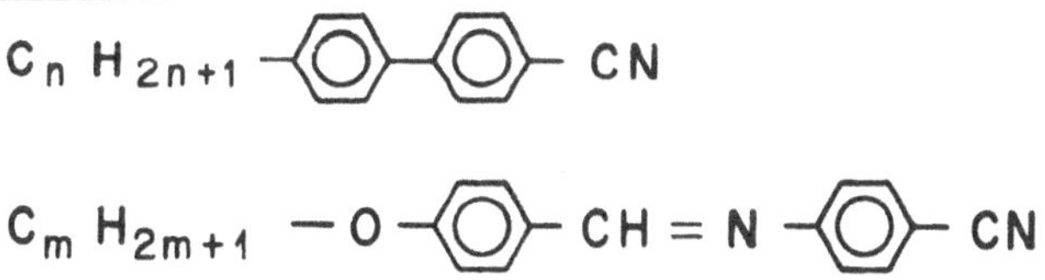 | 23° |

$$C_nH_{2n+1} - \bigcirc - \bigcirc - CN$$

$$C_mH_{2m+1} - O - \bigcirc - CH = N - \bigcirc - CN$$

| **LC − III:** | 27° |

$$C_7H_{15} - \bigcirc - \bigcirc - CN$$

$$C_mH_{2m+1} - O - \bigcirc - \bigcirc - CN$$

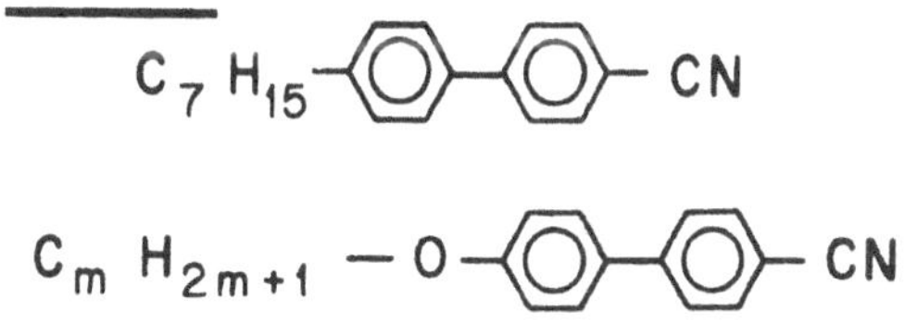

m = 5, 7, 8

We have described a magneto-capacitive technique for measurements of alignment tilt angle in twisted nematic devices. Results of these measurements as a function of angle of oblique evaporation and liquid crystal material have been given. We conclude that Domain C substrates are suitable for application in twisted nematic displays for electronic watches because their finite tilt angle eliminates problems with twist and tilt degeneracy.

## REFERENCES

1    J. L. Janning, Appl. Phys. Lett. 4 (1972) 173
2    E. Guyon, P. Pieranski and M. Boix, Appl. and Eng. Sc. 1 (1973) 19

## DISCUSSION

R. A. Burmeister (Hewlett-Packard)
What was the thickness of the cells used for magneto-capacitance measurements?

K. Toriyama
The cell spacing was 10 $\mu$m.

E. Guyon (Université de Paris-Sud)
The possible change of tilt angle between 22° and 28° for various LC materials can be due either to different anisotropies of elastic constants, which was not considered in our simple analogical solution, or more likely to the chemical mechanisms involved in the adsorption of the first few LC layers.

A. R. Kmetz (Brown Boveri)*
The tilt angle in a twisted nematic cell varies through the layer thickness in a trade-off between tilt and twist energies. Hence the zero field capacitance depends on elastic constant ratios of the material as well as on surface tilt and dielectric constants. The configuration at 9 kGauss is also nonuniform because of the persistance of boundary layers whose magnetic coherence lengths total some 30% of the thickness. The field direction for which the capacitances of these two different nonuniform configurations are equal is an unreliable measure of surface tilt angle. This measurement technique yields unequivocal results only for parallel cells where both configurations are uniform and identical.

*) Comment subsequent to meeting.

# PRINCIPLES OF ELECTROCHROMISM AS RELATED TO

# DISPLAY APPLICATIONS

H. R. ZELLER
Brown Boveri Research Center, Baden, Switzerland

In electrochromic materials an electric current induces a persistent and reversible change in optical properties. The elementary process of coloration consists either in a valency change of one of the constituent ions or in the formation of a color center associated with a lattice defect. Both can be regarded as electrochemical processes and Smakula's equation directly predicts the electrical charge required to achieve a given contrast. For charge neutrality reasons two species of charge carriers (electrons and ions) are needed to induce coloration. In order to make bulk electrochromism feasible, the material has to be both electronically and ionically conducting. The underlying general physical principles provide clear guidelines in materials research and cell design and will be discussed with emphasis on $WO_3$ and solid state technology.

## 1. INTRODUCTION

This paper is intended to give an introduction into the basic principles of electrochromism (EC) for nonspecialists. The emphasis is on the simple physical and electrochemical concepts which apply to the wide variety of hitherto suggested electrochromic materials and devices. More specific materials properties, display and systems aspects will be amply covered by I. F. Chang and J. Bruinink later in this volume.

The best understood example of electrochromism, although technically useless, is the electrochemical formation of F centers in alkali halides.[1] If an

alkali halide single crystal such as KCl is placed at elevated temperature between an anode and a pointed cathode then under the action of an applied
field of ~100 V/cm $Cl^-$ ions migrate towards the anode. For charge neutrality reasons an equal number of electrons is injected at the cathode. At low
concentrations the electrons are located at $Cl^-$ vacancies whereas at high
concentrations aggregates of metallic K are formed. In both cases a strong
optical absorption is induced. Electrochemically it makes little difference
whether the injected electron forms an F center, other defect center or metallic potassium. An F center corresponds chemically to an excess metal
atom and in fact it is possible to create F centers by introducing extra metal
atoms. Thus we define an electrochromic reaction as a reaction in which the
valency of one of the constituent ions is changed by an electrochemical reaction[2] which in turn implies a strong color change. The best known examples as far as display applications go are $WO_3$[3-6] and the viologen salts.[7,8]

It has also been suggested that electrochromism may be due to the direct
action of an electric field without any electrochemical processes.[3-5] At the
low fields ($E \leqslant 10^5$ V/cm) used this possibility can be readily discarded.
Since Ea $\ll$ kT/e (a: lattice constant, e: electronic charge, k: Boltzmann factor), direct field effects such as dissociation, electron-hole generation, etc. in
the bulk are excluded. Also the field is too small to induce a significant space
charge.

## 2. DEVICE PERFORMANCE

Without knowing the details of the coloration process, the mere fact that we
are dealing with an electrochemical reaction allows us to make several predictions on device performance. In its simplest form the EC reaction can be
schematically written as:

$$M^n + e^- + A^+ \rightleftharpoons A^+ M^{n-1}$$

colorless                                    colored

where M denotes an ion of the EC material which can exist in different
valency states and A a mobile cation such as $H^+$ or an alkali. An example is
the formation of tungsten bronze[6]

$$WO_3 + (H^+ + e^-)_x \rightleftharpoons H_x WO_3 \;.$$

transparent                                  blue

An EC display is thus simply a battery with a visible state of charge. Unless
side reactions are involved, the open circuited display retains its charge and
hence its color. EC devices thus in general have memory. The voltage required to drive the cell is of the order of a typical galvanic cell voltage
(~1 V) and the optical density D ($\log_{10}$ of induced absorption) is proportional to the accumulated charge.

$$D_{max} = C \int_o^t I \, dt$$

$D_{max}$ is the optical density at the peak of the absorption band, I the current density across the device and C is a figure of merit. Obviously it is desirable to have C as large as possible but it is easy to show, based on the work of Smakula[9], the C has an upper limit

$$C = 48 \frac{f}{p} \frac{(n^2+2)^2}{n} \frac{1}{W}$$

with p: valency change, f: oscillator strength, n: refractive index and W: half width of the absorption band (in eV), we get C in units of $cm^2/A$ s. Assuming a favorable but still relalistic case (p = 1, f = 1, n = 2, W = 1 eV) we need a charge density of about 1 mA s/$cm^2$ to induce a $D_{max}$ = 10. In practice the current consumption of experimental devices is higher and this number has to be considered as a lower limit. The best performance is obtained by organic dye systems due to the fact that they have a strong absorption band in the visible. $WO_3$ has a comparable oscillator strength but a great part of it is located in the near infrared and is thus physiologically useless. The current consumption of $WO_3$ based devices is about one order of magnitude above the theoretical limit.

An EC device with 10 $mm^2$ active area and a switching rate of 1 $sec^{-1}$ has a power consumption of $\gtrsim$ 100 $\mu$W compared to $\lesssim$ 1 $\mu$W for a twisted nematic liquid crystal display. In applications such as wrist watches where power consumption is very critical, an EC display is only competitive at low switch rates, for instance by displaying hours and minutes. Not only the power consumption but also the inverse lifetime is about proportional to the switching rate. One may expect that the first technical applications of an EC display will be in an area with relatively low switching rate or infrequent change of displayed information.

Speed at a given driving voltage and temperature range depends on the specific materials and systems and shall not be discussed in this context except to say that switching times in the ms region are easily possible and that many systems have a wide operating temperature range.

Electrochromism has no inherent fundamental lifetime limitation. Except for secondary effects such as mechanical damage, corrosion etc., failure of the device is caused by unwanted electrochemical side reactions. For instance in an aqueous electrolyte the cell voltage has to be kept below the dissociation voltage of $H_2O$ in order to avoid strong corrosion and gas bubbles. In some experimental devices proper charging and discharging of the cell is ensured by inserting a reference electrode[7,8] connected to an electronic control circuit. In practice cycle life in excess of $10^7$ cycles is required. $10^7$ cycles correspond to a lifetime of about 4 months at a switching rate of

$1 \text{ s}^{-1}$ or about 20 years at a switching rate of $1 \text{ min}^{-1}$. EC systems with life-times in excess of $10^7$ cycles have been reported.[7]

## 3. ALL-SOLID-STATE DEVICES

The concept of an all solid-state cell is very attractive from several points of view. Manufacturing can potentially be easy and cheap. Virtually all production steps required could be adapted from the highly developed fields of thin film and semiconductor technology. No sealing or containment of liquid EC materials or electrolytes is required. Furthermore electrochemical reactions in solids are less prone to side reactions than in liquid electrolytes. This is due to the fact that in solids in general only one kind of ion is mobile. Charging as well as discharging of many solid-state batteries is self limiting. For a solid-state EC display this leads to a considerable simplification of the electronic driving circuitry. Solid-state EC displays have been reported[10] but are far from a practical realization. In this section we discuss material aspects related to the development of a successful all-solid-state cell.

First we note that an electrochemical reaction such as the EC reaction involves at least two different types of mobile charge carriers (for instance ions and electrons). Secondly for lifetime reasons it is highly preferable to place the EC reaction not at the interface between electrode and EC material but in the bulk of the EC material or, in the language of electrochemistry, all interfaces should be nonpolarizable and reversible. This implies that the EC material has to be conducting for both types of carrier involved, in general electrons and ions (mixed conductor). To ensure nonpolarizability and maximum coloration yield, the EC material is sandwiched between an electronically and an ionically conducting electrode, at least one of which has to be transparent. Contact to the outside circuitry is made by electronic conductors and hence, in order to avoid blocking effects at the ionic conductor, a second mixed conductor has to be introduced between the ionic electrode and the outside circuitry for charge carrier matching purposes. Thus the display will schematically look as shown in Fig. 1. It may be advantageous to make the cell symmetrical, i.e. to make the second mixed conductor also of EC material.

In practice transparent electronic electrodes usually consist of thin films of doped $SnO_2$ or $In_2O_3$. Mixed conducting EC materials are $WO_3$ or other metal oxides. Electrodes with good ionic conductivity for various ions can be found in the literature[11] on solid-state batteries.

As yet a practical all-solid-state cell does not exist, but the simple physical and electrochemical concepts outlined above provide clear and straightforward guidelines in the search for suitable materials. From the above it also becomes evident that the problem of finding proper solid electrode materials is nontrivial.

## 4. THE COLORATION MECHANISM

As stated in Section 2 the figure of merit C is directly proportional to the oscillator strength f. For allowed electric dipole transitions f is of order

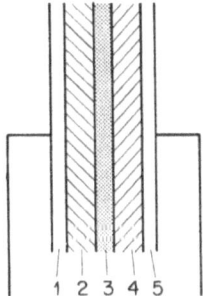

1   Transparent Electrode (electronic conductor)

2   EC Material (electronic and ionic)

3   Solid Electrolyte (ionic)

4   Mixed conductor (electronic and ionic)

5   Counterelectrode (electronic)

Fig. 1. Schematic layer configuration of an all-solid-state electrochromic cell with nonpolarizable interfaces.

.1 to 1 and for all other transitions $f \ll 1$. Strong transitions with $f \sim 1$ can be found in organic dyes, in charge transfer transitions, defect centers[12] etc.

Simple quantummechanical considerations show that if the ion with changed valency in a solid EC material (for instance $W^v$ in $H_x WO_3$) has the same or nearly the same surroundings as the unperturbed ion, then strong optical charge transfer absorption bands necessarily exist.[13] Thus in all solids where a valency change is easy to perform (e.g. the candidates for electrochromism) a strong absorption is in general automatically present.

Physiologically it is not $D_{max}$ which is important for good contrast but the full wavelength dependence of D or C. Optimum contrast at a fixed current is achieved for an absorption in the visible with a bandwidth of $\sim 1$ eV. Many transition metal oxides in which mixed valency can be induced have a large fraction of their oscillator strength in the infrared; some even become metallic. This leads to a marked decrease in current efficiency.

From a charge vs. coloration point of view it is rather unimportant whether the optical absorption occurs at an ion with different valency at a regular lattice site, whether the ion is associated with a lattice defect or whether absorbing lattice defects such as F centers are formed. For the kinetics of the coloration, however, such questions are of utmost importance.

If crystal defects are involved then obviously such defects should be generated while depositing the EC film in sufficient concentration to allow a fast and deep coloration. If the mechanism is intrinsic, a nearly perfect crystal is preferable.

## ACKNOWLEDGEMENTS

It is a pleasure to thank H. U. Beyeler, P. Brüesch and in particular C. Schüler for many enlightening discussions.

## REFERENCES

1   F. Seitz, Rev. Modern Phys. 26 (1954) 7
2   I. F. Chang, B. L. Gilbert and T. I. Sun (J. Electrochem. Soc. 122 (1975) 955) have coined the work "electrochemichromism" which is more exact than the historical "electrochromism"
3   S. K. Deb, Phil. Mag. 27 (1973) 801
4   S. K. Deb and J. A. Chopvorian, J. Appl. Phys. 37 (1966) 4818
5   S. K. Deb, Appl. Optics Suppl. 3 (1969) 192
6   B. W. Fanghnan, R. S. Crandall and P. M. Heyman, RCA Review 36 (March 1975) 177
7   C. J. Schoot, J. J. Ponjee, H. T. van Dam, R. A. van Doom and P. R. Bolwjis, Appl. Phys. Lett. 23 (1973) 64
8   H. T. van Dam and J. J. Ponjee, J. Electrochem. Soc. 121 (1974) 1555
9   A. Smakula, Z. Phys. 59 (1930) 603
10  S. K. Deb and R. F. Shaw, US Patent 3 521 941
11  W. van Gool (ed.), 'Fast Ion Transport in Solids, Solid State Batteries and Devices', North Holland, Amsterdam (1973)
12  D. L. Dexter, in F. Seitz and D. Turnbull (eds.), 'Solid State Physics' 6 (1958) 353
13  M. B. Robin and P. Day, Adv. Inorg. Chem. and Radiochem. 10 (1967) 247

# ELECTROCHROMIC AND ELECTROCHEMICHROMIC

## MATERIALS AND PHENOMENA

I. F. CHANG

IBM Thomas J. Watson Research Center, Yorktown Heights, New York, USA

## SUMMARY

Electrochromism (EC) is broadly defined as a reversible color change induced in a material by an applied electric field or current. A great variety of physical phenomena which bring about such a reversible color change in certain solids or liquids, organic or inorganic materials, are covered by this definition. The physical mechanisms that are responsible for these phenomena are quite different but for ease in discussion may be grouped into two main categories, one being electronic and the other electrochemical in nature. For the latter class of phenomena, a more descriptive nomenclature is electrochemichromism (ECC). An attempt is made in this paper to review various EC and ECC materials and phenomena. A description of each phenomenon is given and the materials which exhibit such phenomena are examined in detail. Emphasis is placed on EC and ECC material properties, especially display performance characteristics and problem areas. Some existing EC and ECC display devices, as well as those being proposed, are evaluated in terms of their applicability and feasibility respectively. Comparisons are made with other non-emissive electrooptic display technologies which already have products available in the market place. The EC and ECC displays are, of course, at a relatively early stage of development. However, they possess some unique features which may ultimately lead to a position of preference for certain applications.

## 1. INTRODUCTION

The electrochromic display is a new non-emissive display technology that, like other exploratory displays, is based on phenomena and materials whose properties are not fully characterized or completely understood. Electrochromism as a physical effect is broadly defined as a reversible optical absorption change induced in a material by an applied electric field or current. As the details of the EC effect are examined, one finds that it may be brought about by a great variety of different physical mechanisms in different materials. These mechanisms may occur in solids or liquids and in organic or inorganic materials. For convenience these EC phenomena and materials are grouped into two main categories, one being electronic and the other electrochemical in nature. For the latter class of materials and phenomena, we adopt the nomenclature electrochemichromism (ECC), in distinction with electronic electrochromism (EC). The primary purpose of this paper is to review the various phenomena and materials and to sort out those that are relevant to display applications.

Table 1 lists all the EC and ECC phenomena that will be discussed in the later sections. There are four different physical mechanisms that produce an

### TABLE 1

#### Electrochromic and Electrochemichromic Phenomena

| Electrochromic | Electrochemichromic |
|---|---|
| (a) Color-center creation | (e) Simple electroredox reaction |
| (b) Charge transfer between impurity centers | (f) Electroplating |
| (c) Franz-Keldysh effect and absorption edge shift | (g) Coloration induced by pH variation |
| (d) Dipole moment change and Stark effect | (h) Electroredox reaction coupled with chemical reaction |
| | (i) Electrochemical reaction involving a solid film |

EC effect. In category a, color centers are created by electron or hole injection into a defect center in a solid material. A new absorption band due to the color centers is thus created whereas in category b, due to charge transfer, an original absorption band is annihilated and a new absorption is produced. In categories c and d, the absorption band is shifted due to an applied field: the Franz-Keldysh shift is an absorption edge shift to longer wavelengths caused by a field-induced tunnelling effect; dipole absorption shifts due to dipole moment change and energy level splitting (Stark effect) are usually superposed and are thus treated as one phenomenon. In the ECC category the physical mechanisms involved are all basically related to an electrochemical redox reaction. However, we separate them into five different subgroups as shown in Table 1 to facilitate discussion.

The tabulated EC and ECC effects can all be triggered by an applied electric field or current, but the device configuration or electrode arrangement may be quite different for optimal operation when they are incorporated in display devices. Figure 1 shows several device configurations for EC and ECC displays, in schematic form. For the EC mechanism, the coplanar electrode

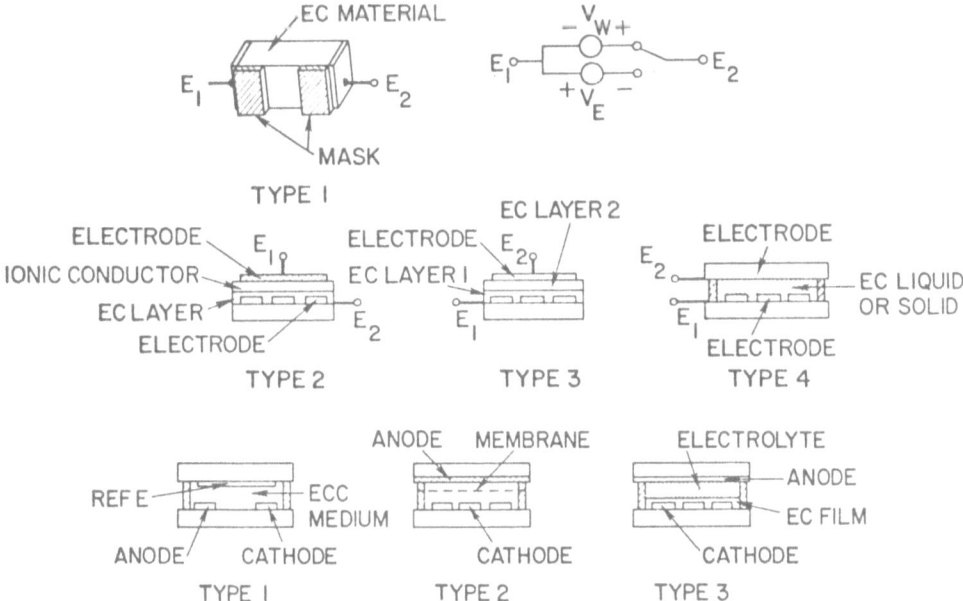

Fig. 1. Typical device configurations for EC (top) and ECC (bottom) displays.

configuration (cathode and anode in the same plane) is usually not suitable due to the requirement of a uniform high field; thus they all have parallel electrode geometries. However, the ECC devices may have coplanar as well as parallel electrode configurations. In addition, a reference electrode may be incorporated to facilitate a three-terminal controlled voltage drive. Furthermore, a membrane may be inserted in the device to separate the cathode and anode compartments, thus isolating cathodic and anodic reactants.

Sections 2 and 3 deal with the EC and ECC phenomena and materials, respectively. The physical principle of each phenomenon and the electrooptic properties of various materials are the main concern. The scientific literature dealing with these phenomena, materials and device applications is given primary emphasis in this review; the patent literature is surveyed in appendices.

Tungsten trioxide, being the most significant EC and ECC material in terms of its potential for display application, is discussed more extensively in Section 4 with the aim of establishing an acceptable model for its EC and ECC behavior. Section 5 summarizes and compares the features and limitations of EC and ECC displays and in turn compares them with other non-emissive displays. It should be emphasized here that our approach is to give a comprehensive review of the subject, since no such review is available. However, adequate references are given for each phenomenon or material dealt with, so that readers may refer to them for more detailed treatment.

## 2. ELECTROCHROMIC MATERIALS AND PHENOMENA

The electrochromic effect was first demonstrated in alkali halide materials[1,2] in connection with the well known phenomenon of color centers in ionic crystals. In contrast to the two related color-center phenomena, photochromism and cathodochromism,[3-5] electrochromism in solids has not yet been thoroughly explored. Some photochromic materials[6] involving charge transfer between impurity centers have also been mentioned as electrochromic materials, however, practical EC devices based on these materials have not yet been demonstrated. The most significant but preliminary effort was a solid state EC imaging device using tungsten trioxide.[7] The device work on $WO_3$ will be briefly reviewed in this section, however we shall defer the discussion of its EC or ECC mechanism to Section 4.

### 2.1. Color Center Effect and Electrolytic Coloration

Color centers in solids are well known phenomena with a vast amount of work published. Several textbooks[8,9] treat this subject in detail. Color centers can be generated in a solid by a variety of techniques. It is the electrolytic coloration that is often referred to as the electrochromic phenomenon observed in many solids. Pohl[1] in 1932 was first to show that color centers can be created in NaCl or KCl crystals at an elevated temperature when a high electric field is applied via a sharp pointed cathode. In the same year, Stasiw[10] showed that a partly additively colored alkali halide crystal under an applied electric field exhibits color cloud migration from cathode to anode at an elevated temperature. The migration is reversed in direction when the voltage polarity is reversed.

A variety of color-center types have been identified in alkali halides. The simplest are the F center, which consists of an electron trapped at an anion vacancy, and the V center, which is a hole trapped at a cation vacancy. Since vacancy defects are created by purely thermodynamic causes,[11] the thermal equilibrium concentrations of color centers in a crystal are governed by thermodynamic equilibrium conditions. The color-center concentrations[12] in alkali halides at room temperature are generally very small, hence the absorption coefficient is low. Consequently the electrolytic coloration in alkali halides is appreciable only at high temperatures, which limits its possibilities for display application.

The color-center effect has also been observed in materials[13-19] other than alkali halides. Deb studied color centers in molybdenum trioxide[13] and tungsten trioxide.[14] (We shall defer our discussions of these materials to Section 4.) Van Raalte[15] observed electrolytic coloration in rutile at 160° C when an electric field was applied along the c-axis of the crystal. He attributed the color emerging from the cathode (absorption centered around 7000 Å) to injected electrons trapped at the anion vacancies and the anode color centers (absorption centered around 6000 Å) to injected holes trapped at impurity ($Fe^{3+}$) sites. An impurity color center may also be viewed as an oxidation, for instance, $Fe^{3+}$ + hole → $Fe^{4+}$ (color absorbing). On the other hand, an impurity ion may be reduced to become color absorbing. It has been shown[16,17] that $Pb^{2+}$ in KCl crystals can be transformed to $Pb^-$ by trapping three electrons through electrolytic coloration.

Electrolytic coloration was also observed in cuprous chloride by Regis.[18] He showed that, when a cuprous chloride crystal between two flat electrodes is submitted to an electric field (~ 1 - 10 V/cm) and heated to 200 - 300° C in an inert or reducing atmosphere, the transparent crystal becomes light green. Regis proposed that there are mobile cations and Frankel defects in the crystal. Under an applied field, interstitial $Cu^+$ ions migrate to the cathode, where they receive an electron and become neutral atoms. On the anode side, the $Cu^+$ ion is oxidized, becoming $Cu^{2+}$ by receiving an injected hole which is assumed to be responsible for the green-blue coloration. As the field is reduced to zero, a reverse current is observed and can be interpreted as due to the accumulated positive space charge. The holes return to the anode and with them the blue color due to $Cu^{2+}$. However, there is still some yellow color remaining which is possibly due to V centers (holes trapped or stabilized at cation vacancies).

Another material, calcium orthovanadate,[19] may be mentioned as exhibiting electrolytical coloration at high temperatures (> 700° C). A black color starts from the cathode when a field of 30 - 40 V/cm is applied. The mechanism involved is believed to be that oxygen vacancies migrate to the negative electrode and capture an injected electron. Local charge neutrality requires two electrons for each vacancy site which thus causes an electron transfer from the vanadium ion (i.e. $V^{5+}$ - e → $V^{4+}$). On the anode side, oxygen ions migrate toward the anode as halogen ions do in alkali halides.

Applications and Devices. The first application of color centers for display is the skiatron tube,[20,21] a cathodochromic CRT in which the target is an evaporated KCl film. Applications of this device have been in radar and facsimile systems. The use of color center material in electrochromic devices has not yet been extensively explored. One preliminary effort[7] was an experimental photoimaging device consisting of a photoconducting CdS layer and an electrochromic $WO_3$ layer sandwiched between two transparent electrodes. Deb[7] claims that the image is formed by the color centers created in the $WO_3$ film, producing a broad absorption band peaked about 9000 Å. The image may be erased by applying a reverse potential. Unfortunately the electrochromic coloration in $WO_3$–CdS is not polarity sensitive; thus simultaneous recoloration may occur while bleaching is taking place.

In a US patent, Deb and Shaw[A1] (see Appendix A) proposed an EC display device with a charge-carrier-permeable insulator adjacent to the EC layer. They claim that the insulator renders the EC device polarity sensitive, i.e. the application of a reverse field will erase the coloration without simultaneous recoloration. Solid-state insulators such as silicon oxide, calcium fluoride and magnesium fluoride were suggested to form an all-solid-state device. However no performance data on this type of device have yet been reported. An attempt to utilize the $WO_3$-EC film in conjunction with a SEBIC layer[A10] (susttained electron bombardment induced conductivity) for a projection display was made by Robertson[25] a few years ago. The idea was to take advantage of the conductivity persistence in SEBIC to compensate for the slow response time of the $WO_3$ film. However he encountered a variety of fabrication problems which prevented a realistic evaluation of the SEBIC-EC concept.

Recently Chang and Howard[26] investigated the EC effect in $WO_3$ films and evaluated its potential for display application. They compared the $WO_3$ film operated in a solid-state device with that operated in an electrochemical cell. The $WO_3$-ECC device was found to be superior in terms of maximum contrast, coloration speed, and reversibility. This sensitivity to the mode of activation is the reason that workers in this field are currently concerned with the mechanisms of electrocoloration in $WO_3$ films. Is the color center model applicable at all? Or is it entirely an electrochemical reaction? These questions of current interest are discussed in Section 4.

One other exploratory effort on color-center electrochromic devices should be mentioned here. This is a multilayer EC panel consisting in sequence of Al-Alkali halide (or oxide such as $WO_3$) -- $Bi_2O_3$ (or $Sb_2O_3$) -- Bi (or Sb) as described by Robillard.[27] The electrocoloration is said to be achieved by electron injection from the $Bi_2O_2$ or $Sb_2O_3$ junction to the EC layer, forming color centers when +35 V is applied to the Al-layer. When the voltage polarity is reversed, the coloration is erased. However, it is not clear how such an erasure process would not cause simultaneous recoloration. There has been no further work reported on this device.

There are a few noteworthy patents[A1-A14] related to color- center phenomena. Dreyer[A5] describes an electric light valve which uses the phenomenon of F color-center migration under the influence of an electric field in EC materials such as alkali halides, oxides, and sodalites. The electric field moves the color-center cloud in and out of the light-transmission path as in the arrangement shown in Fig. 1a (Type 1).

Another electrochromic device was proposed by Robillard[A6,A7] for electrophotographic applications. The device consists of a photoconductor, a catalyst and a semiconductor layer sandwiched between electrodes. The principle of the device is that an image projected on the photoconductor causes catalytic ions (such as $Ag^+$) to be injected into a semiconducting oxide (such as $CeO_2$, $TiO_2$, $Ta_2O_5$, $SnO_2$ etc.) with concentrations corresponding to the optical density of the input image. The catalytic ions then reduce the cation in the oxide to a light-absorbing lower state, forming a color image.

A device having a configuration similar to Type 2 of Fig. 1a is described in a patent issued to McIntyre and Harrison.[A9] The EC layer is a film of tungstic acid ($H_2WO_4$) or molybdic acid ($H_2MoO_4$) or a mixture of both, and the ionic conductor is a thin film of silver chloride containing a trace amount of silver. It is claimed that such panels may be addressed and erased with voltages from about 3 to 10 V by potential reversal.

Letter[A11,A12] describes an EC glass material in which an ion $Pb^{+2}$ is reduced to Pb near the cathode and is oxidized to $Pb^{+4}$ near the anode causing color absorption under an applied voltage of 2.5 - 5 V. This is similar to the EC effect observed in CuCl crystals discussed above. One application for this material is EC sunglasses, however the induced density and coloration speed are both low at room temperature.

In summary, the color-center materials discussed above (except $WO_3$) generally have very low coloration speed at room temperature if they show color at all. The optical density achievable is also low for reasonable thicknesses due to small absorption coefficients. Therefore display devices based on color-center effects (electrolytic coloration) may not be very practical.

## 2.2. Charge Transfer between Impurity Ions

A charge-transfer process between impurity ions certainly can be viewed as an oxidation-reduction process changing the charge state of the impurity ions. The term charge transfer is used to describe a special class of solid materials. The charge transfer process in these solids can be excited by high energy photons and has been observed as a photochromic phenomenon.[28-31] Examples of materials that belong to this class are the titanates doped with transition and refractory metal ions. This type of electrochromic effect was first observed by Kosman and Bursian[32] in $BaTiO_3$ at $300°$ C and at room temperature with an electric field $< 10^4$ V/cm. The most extensively studied materials are $SrTiO_3$: $Mo^{+6}$, $Ni^{+2}$ and $SrTiO_3$: $Mo^{+6}$, $Fe^{+3}$. Faughnan and Kiss,[28-31] who first investigated their photochromic properties, have established the charge-transfer mechanism by optical as well as by electroparamagnetic resonance studies.[31] They, along with others,[33-36] identified that transition metals on Ti sites can exist in several formal oxidation states and that such ions can be changed from one charge state to another by oxidation and reduction heat treatment. Neither $Fe^{+3}$, $Ni^{+2}$ nor $Mo^{+6}$ have significant absorption in the visible. However, when $Mo^{+6}$ is reduced and $Fe^{+3}$ or $Ni^{+2}$ is oxidized, they become color absorbing.

The electrochromic effect in transition metal-doped $SrTiO_3$ shows an absorption spectrum similar to the photochromic spectrum; however the excitation mechanism responsible for the EC effect is different. The EC phenomenon in $SrTiO_3$: Mo, Ni, Al has been investigated by Blanc and Staebler.[6] Aluminum is used for compensation if Mo and Ni are kept at the same concentration. When an electric field ($\sim 50 - 100$ V/cm) is applied to a sample at a temperature above $130°$ C in air, vacuum or inert atmosphere, the negative electrode region develops a blue color characteristic of the reduced form of refractory molybdenum ($Mo^{+6} \rightarrow Mo^{+5}$). A brown color characteristic of oxidized transition element iron ($Fe^{+3} \rightarrow Fe^{+4}$) is developed adjacent to the positive electrode. Reversal of the field polarity leads to a retreat of the colored zones from each other, leaving a well defined colorless region in between. If the reverse process is allowed to continue, a reverse coloration will begin at the electrodes. As electro-coloration is prolonged sufficiently, the two color fronts meet and eventually the brown oxidized region enlarges and the reduced region becomes green. Figure 2 shows the absorption spectrum

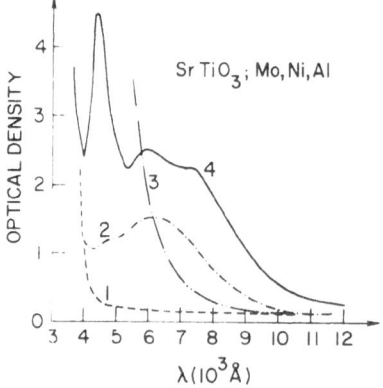

Fig. 2.    Absorption spectrum of $SrTiO_3$: Mo, Ni, Al (all $\sim 4$ - $5 \times 10^{18}$ cm$^{-3}$). 1) clear state, 2) blue coloration near cathode (electro-reduction), 3) brown coloration near anode (electro-oxidation) and 4) highly reduced sample (from Blanc and Staebler[6])

of these colored regions before they meet and after they mix. In the heavily reduced region two strong peaks at $\sim 4500$ Å and 5900 Å appear. An absorption coefficient approaching $10^4$ cm$^{-1}$, two orders of magnitude higher than that obtainable by the photochromic process, is claimed.[37]

The mechanism of the EC phenomenon is believed to be initiated by thermally activated oxygen vacancies drifting in the applied field. Charge compensation by hole and electron injection at anode and cathode respectively produces colored oxidized and reduced regions. No practical device has been made to date based on this EC effect. However, a recent patent issued to Kiss[A13] describes a device consisting of two solid-state layers with one layer having reducible ions. A color absorption is induced by at least one of the ions undergoing a valence-state change.

In summary, this class of materials has not been explored extensively. Studies were limited to single crystal materials. It is conceivable that the charge transfer effect may take place in polycrystalline thin films and may even exist in amorphous or glassy materials. However, with crystalline titanates   the limitations on the doping levels of impurities (limited to $10^{18}$ - $10^{19}$ cm$^{-3}$) and the mobility of ionic carriers ($10^{-7}$ - $10^{-5}$ cm$^2$/Vs) imply only a slow and insensitive EC effect.

## 2.3. Franz-Keldish Effect and Absorption Edge Shift

The effect of a strong electric field on the absorption of light is often referred to as an electrochromic effect. The shift of a fundamental absorption edge toward longer wavelengths induced by high electric fields ($> 10^5$ V/cm) was first discussed by Franz[38] and Keldish.[39] One example of such a shift ($\sim 70$ Å) was observed in CdS.[40]

Another electric field induced absorption edge shift was observed in antimony sulfo-iodide SbSI.[41,42] In contrast to the Franz-Keldish effect, the absorption edge is shifted to a shorter wavelength by about 80 Å for an applied field about 5000 V/cm along the c-axis at room temperature. The shift is roughly proportional to the square of the field. Since accompanying the absorption edge shift there is an elongation of the crystal along the c-axis of a few tenths of a per cent, the EC effect is very likely an indirect result of the dielectric (electrostrictive) deformations. Similar effects have also been observed in BaTiO$_3$.[43]

In summary, since the above described EC effects merely cause a small absorption edge shift, the integrated optical spectral density change is usually too small to show any color variations to human eyes. Thus this EC effect does not appear to have direct use in display applications, although it may have some applications in monochromatic electrooptical devices.

## 2.4. Dipole Moment Change and Stark Effect

Changes in the absorption spectra of dye molecules in solutions, solid matrices and crystalline forms, observed under an applied electric field, have also been given the name electrochromism.[44,45] A few review articles[45,46] have been published on this subject. Liptay[47] has given a theoretical treatment of the effect of an external electric field on the optical absorption of molecules in solution. He has obtained an expression for the frequency shift between a ground state and an excited state in which three terms are independent of the external field and give rise to a solvatochromism (the solvent dependence of the optical absorption),[48-50] while three other terms represent a shift which is dependent on the external field. The field dependence of the optical absorption has been described by Blumenfeld et al.[46] in terms of possible mechanisms. The optical density change $\Delta D$ at a given wavelength in the absorption band can be shown to consist of three major contributions: an orientational term, a mixed term and a molecular term. It can be shown [46,51] that the orientational effect (Kühn effect) cannot be observed if the angle between the electric vector of the light wave and the direction of the applied electric field ist $45° 44'$; in this case, only the mixed and the molecular terms contribute to electrochromism. The Kühn effect can only be observed alone for molecules incapable of polarization with a dipole moment independent of excitation. Usually it is accompanied by a shift of the absorption band due to the "mixed" and molecular effects. The "mixed" and orientational effects cannot be observed in solid media where the motions of molecules are completely hindered; thus solids have only the molecular contributions, which are usually very small.

The expectation of a large electrochromic effect originated with Platt,[44] who first proposed the hypothesis that in strong electric fields of $10^6$ - $10^7$ V/cm, the absorption spectra of some organic dyes containing long chains of conjugated bands could exhibit shifts of thousands of angstroms. This hypothesis was based on the work of Brooker and her coworkers[52-54] and on her own earlier work.[55] Kuwamoto et al.[56,57] attempted to detect such an electrochromic shift in methyl red and phenol blue in polystyrene. However, in both cases only very small spectral changes (~ 10 Å) were detected. Later Chernyakovskii et al.[58] developed a low frequency ac technique to investigate the small shift in methyl red and other azo dyes in polystyrene. With an ac applied field they were able to obtain directly the incremental optical density, which they term the EC spectrum. The EC spectrum exhibits a zero point for each absorption maximum or minimum that is shifted under an applied electric field. This technique is a useful analytical tool for investigating the motion of polymer molecules[59-62] and the deorientation of molecules, or the relaxation of rotational motions.[59]

The field-induced molecular absorption has been observed not only in solutions and solid matrices as discussed above but also in molecular crystals.[63-65] Stevenson and coworkers[63] observed such electro-absorption effects in metanitroaniline (MNA) single crystals. An electric field applied along the c-axis of the crystal causes a change in the optical transmission near the band edge (between 505 and 540 nm). The incremental change varies linearly with applied field and also reverses with field polarity. They described the phenomenon as a linear electrochromic effect. MNA also exhibits a linear electrooptic effect.[64]

In summary, the phenomena and materials discussed here, while seemingly useful have very little practical value for display applications. However, the EC effect may be a very useful tool for studying polarization of electronic transitions in molecules and their dipole moments in excited states. Also it is difficult to exclude the possibility that Platt's prediction of a large EC effect may yet be fulfilled in some new class of materials.

## 3. ELECTROCHEMICHROMIC MATERIALS AND PHENOMENA

In this category, the electro-coloration phenomenon by definition involves an electrochemical reaction. Strictly speaking, the oxidation and reduction of impurity ions in a solid discussed in Section 2 may also be considered as electrochemical in nature. However, it is more convenient to describe such effects in solids in terms of electron and hole injection and solid-state theory, whereas the redox reactions in liquid solutions are customarily described in terms of chemical and electrochemical reactions. Thus the electrochromic effects which involve a liquid medium are included in this section and are termed electrochemichromic effects. These ECC phenomena and materials are subdivided into several types depending on whether, and which, chemical reactions are involved.

## 3.1. Electro-redox Reaction

A redox reaction can be described as A ± ne → B (colored species), i.e. ions or molecular complexes in solution are either reduced at the cathode by gaining an electron or oxidized at the anode by losing an electron. When the reaction product turns out to be color absorbing or to have a different color absorption from the original species, the electro-redox reaction exhibits an electrochemichromic behavior. Typically the colored form has much lower solubility in an aqueous solution than the colorless form so it is possible to have local precipitation and color persistence. Numerous materials which exhibit an ECC phenomenon are reported in the scientific and patent literature (see Appendices B-E). In this subsection, we briefly review several ECC materials which have been investigated recently for applications in displays.[66,26]

Viologen family (4,4'-depyridinium compounds). The viologen family has the general chemical formula

$$\left[ R - N\bigcirc\!\!\!-\!\!\!\bigcirc N - R \right] X_2$$

where R may be an alkyl, cyclo-alkyl or other substitute and X is a halogen. The name viologen was coined by Michaelis and Hills[67] for the 4,4'-dipyridinium compounds because they become deeply blue-purple on reduction. The viologen ion,

$$\left[ R - N\bigcirc\!\!\!-\!\!\!\bigcirc N - R \right]^{+2}$$

can have a two step reduction,[67,68] i.e. one-electron or two-electron reduction. The most extensively studied material is methyl viologen[67,69-77] (MV). Elofson and Edsberg[69] reported the first-electron reduction of MV at −0.68 V vs. a saturated calomel electrode SCE. It is reversible (the coplanarity of the two heterocyclic nuclei facilitates the reversible reduction) and independent of pH. The second-electron reduction is not electrochemically reversible, but MV can be reoxidized by air. The second reduction was reported at −1.038 V vs. SCE in the range of pH = 5 to 13 and slightly pH-dependent below pH = 5. A polarograph[66] of methyl viologen dichloride in an aqueous solution with pH = 5.3 clearly exhibits a two-step reduction with two half-wave potentials located at −0.68 V and −1.07 V vs. SCE, respectively, confirming the results of Elofson and Edsberg[69] and the cyclic voltammetry results.[74,76]

The absorption spectrum of $MV^+$ has two main absorption peaks[68,71,73] at 3900 Å and 6050 Å. Kosower and Cotter[71] have reported an absorption coefficient of 10700 cm$^{-1}$M$^{-1}$ℓ for the 6050 Å peak, citing Schwarz's work.[78] However, it appears to be a misquote since Schwarz only published the absorption spectrum of diethyl dipyridinium radical. It is important to point out that Strojek et al.[73] have monitored the absorbance at the two absorption peaks during electrochemical reduction as a function of time. They observed that the absorbance is proportional to $t^{1/2}$ at either the one- or two-electron reduction potential. This behavior has also been observed in other electrochemichromic systems[26] and is direct evidence of a diffusion-rate-controlled electro-redox reaction. In fact, assuming such a diffusion-limited electrochemical reaction, an analytical model[66] has been derived for the ECC effect. The expressions for the current and the absorbance can be used to estimate[26] the speed and contrast obtainable in a given ECC material system if its absorptivity and diffusion constant are known. Conversely, the diffusion constant may be estimated by measuring the absorbance as a function of time. The diffusion constant for MV is estimated to be $\sim 9.5 \times 10^{-6}$ $cm^2$/s in agreement with the data of Osa and Kuwana.[74]

Another interesting observation on the absorbance vs. $t^{1/2}$ plots is that the two absorption bands (3900 Å and 6050 Å) of reduced MV appear during both one- and two-electron reductions. Therefore they are characteristic of the $MV^+$ free radical. Furthermore, the slopes of the absorbance vs. $t^{1/2}$ plots for both absorption peaks at the second-electron reduction potential are double those at the first-electron reduction. This suggests that there must be a disproportionation-type reaction such as

$$MV^{+2} + e \rightarrow MV^+ \text{ at } E_{1/2}^1$$

$$MV^+ + e \rightarrow MV \text{ at } E_{1/2}^2$$

$$MV + MV^{+2} \rightleftharpoons 2 MV^+$$

The equilibrium of the disproportionation reaction must lie far to the right in the solution to account for the larger slope at $E_{1/2}^2$ mentioned above.

O-Tolidine. O-Tolidine, $C_{14}H_{16}N_2$ (4,4'-diamino-3,3'-dimethylbiphenyl), is an ECC material that becomes color absorbing upon oxidation.[79-82] It has two oxidation steps with the first oxidation potential at +0.58 V vs. SCE.[66] The intermediate free radical (one-electron oxidation) tends to form a dimer[82] which has absorbance maxima at 3650 Å and 6300 Å with absorptivity $4.8 \times 10^4$ and $2.8 \times 10^4$ $cm^{-1}M^{-1}\ell$ respectively. The two-electron product[81,82] has only one absorption peak at 4370 Å, with molar absorptivity $6.1 \times 10^4$ $cm^{-1}M^{-1}\ell$. The induced absorbance during the electroredox reaction is found[81] to be proportional to $t^{1/2}$. From this data, we estimate the diffusion constant to be $6.5 \times 10^{-6}$ $cm^2$/s.

Polytungsten anions (PTA). Besides the organic materials mentioned above, there are some inorganic materials that also exhibit ECC phenomenon via a simple redox reaction. The polytungsten anion (PTA) family is an example[26,66] of such a reaction. PTA's exhibit deep blue color upon reduction. Some of these anions are prepared from tungsten trioxide, tungstic acid and tungstates in various acids, thus the redox properties of these PTA's may throw some light on the $WO_3$ electrochemichromic behavior. Isopolymolybdenum anions[105-109] have drawn the attention of chemists and electrochemists for quite a long time. The color of these anions and its persistence even in very dilute solutions suggest the possibility of novel structure and binding features. Thus they are considered as unique candidates for studies on structures and bondings in complex ions.

There are several heteropolytungsten or -molybdenum anions that exhibit a deep coloration upon reduction, for instance, the twelve tungsten HPTA,[102] silicomolybdate,[105] and phosphotungstate,[104] whereas the family of isopolytungsten anions are all reducible to blue form. We have studied one PTA system in detail,[26,66] which consists of sodium tungstate and chloroacetic acid in aqueous solution. Our polarographic and voltammetric measurements suggest a two-step reduction; however, the nature of the anion has not been identified. From measurement of current as a function of reduction time under constant voltage, we have estimated the diffusion coefficient to be less than $10^{-6}$ cm$^2$/s. This suggests that the anion may be fairly large, incorporating more than 12 tungsten atoms, since a 12 W-PTA has a diffusion coefficient[96,104] of about 2 - 6×$10^{-6}$ cm$^2$/s. The absorption spectrum measured after reduction is a broad band peaked around 7500 - 8000 Å and somewhat resembles the spectra for reduced silicotungstate and phosphotungstate anions measured by Stonehart et al.[104] The reduction process of most PTA's[66,96,104] can also be described by an analytical model.[66]

Applications and devices. Although there are quite a few applications based on electro-redox systems proposed in the patent literature (Appendix B), there is little display device work published in the scientific literature. Treier and Shapiro[110] were the first to report a low-voltage indicator display device utilizing the electrochemical redox reaction of methyl viologen. Their device basically consists of an inert electrode, a silver/silver chloride electrode and an aqueous electrolyte with 0.05M dimethyldipyridinium dichloride and 1M KCL. As with many other redox reaction systems with soluble colored species, this device suffers from a color smearing problem.[26]

In the patent literature the electrochromic redox reaction has often been proposed for light filters.[B1,B5,B6] Jones et al.[B1] describe a light filter based on the reduction of molybdates, tungstates and vanadates. Hall et al.[B5] specifically mention the methyl viologen system for an electrically responsive light-filtering device, whereas Rogers[B6] proposed the 4,4'-dipyridylium, diazapyrenium and pyrazidinium compounds for such applications. Manos[B3] and Kissa[B4] also described electrochromic display devices based on redox systems. In particular, Kissa et al. suggested the use of a membrane in the device as illustrated in Fig. 1b (Type 2). The concept of using a membrane is significant in that it is more likely that one can achieve totally reversible reactions in such a two-compartment membrane-separated system.

In summary, the electro-redox reaction does have possibilities for display-related applications; however, there are certain problems such as color drifting and short persistence (no memory) which may not be tolerable in certain display applications. In Section 3.4 we discuss some solutions to these problems.

### 3.2. Electroplating

An electro-redox reaction wherein a metal ion is deposited on the electrode, forming a solid metallic layer, is generally referred to as electroplating.[111]

Electroplating has been proposed as the basis for a light modulating device. The first such proposal[C1] actually used nonmetal ions such as iodine. Thus, in this section, we take the broad view of electroplating to include both metallic and non-metallic ions as long as they can be plated out as solid atomic or molecular layers with sufficient light absorbing power for light modulation and display applications.

Zaromb[112,113] has analyzed the design principle of a reversible electroplating (metal ions) light modulator in the coplanar electrode configuration as well as the parallel electrode scheme (Fig. 1). He has made estimates on the coloration efficiency which is defined as induced optical density change per applied charge density ($C^{-1} cm^2$). Table 2 lists the coloration efficiencies for a number of elements and their characteristic absorption wavelengths. The efficiency can be shown to be

$$\eta = 0.43 \ \alpha W/Fzd \tag{1}$$

where $\alpha$, W, F, z and d are the absorption coefficient, atomic weight, Faraday constant, redox state and density respectively. One notes from Table 2 that the best plating metal element is Ag which has a coloration efficiency $\sim 30 - 40 \ C^{-1} cm^2$. This is somewhat smaller than that for some of the organic ECC systems.[26] Zaromb[112,114] has shown that it is possible to supplement a reduction with an oxidation plating in the same electrochemical cell for a light valve application. For example, the AgI solution (with NaI electrolyte) could give silver plating on the cathode and $I_3^-$ (or $I_2$) on the anode. The $I_3^-$ can supplement the absorption in the blue-UV region, thus slightly improving the coloration efficiency.

There are quite a few light valve devices described in the patent literature. Smith[C1] in 1929 described a light valve for use in conjunction with automobile head lamps. An electrolyte solution of potassium iodide and sodium hydroxide was used. A brown coloration which originated at the anode and diffused throughout the electrolyte is believed to be due to oxidized iodine. Flanagan et al.[C2] have described a light filter for use in optical systems based on the plating of metallic ions such as $Cu^+$ in a 10% $HClO_4$ solution. It has been claimed that when gelatin or polyacrylamide is added to the electrolyte, the uniformity of the metal deposition is improved. The use of a salt bridge layer (for instance, an anion-saturated gelatin layer) between a double electrolyte sandwich is a preferred form (Fig. 1b, Type 2) to allow more reversible operations. This is because the ions in one of the electrolyte layers undergoes a redox reaction but requires no optical density change. Zaromb[C3] proposed using a so-called oxidizer in the electroplating system to chemically deplate or erase the plated metal rather than to deplate electrically, which often is not satisfactory in terms of speed and uniformity. An example of such a system is $LiBrO_3$ in a silver bromide salt-silver plating solution. Another light filtering and polarizing apparatus has been proposed by Land and Roger.[C4,C5] The concept of this light polarizer is based on the process of plating (they refer to it as staining) a molecularly oriented plastic

## TABLE 2

Estimated Coloration Efficiency for Various Elements and Wavelengths

| Element | $d$ (g cm$^{-3}$) | $W$ (g) | $\lambda$ (nm) | $\alpha$ ($10^6$cm$^{-1}$) | $\eta$ (C$^{-1}$cm$^2$) |
|---|---|---|---|---|---|
| Cd$^{+2\to0}$ | 8.64 | 112.4 | 589 | 1.07 | 31.3 |
| Cr$^{+3\to0}$ | 7.2 | 52.0 | 579 | 1.05 | 11.4 |
| Co$^{+2\to0}$ | 8.9 | 58.9 | 500 | .93 | 14.0 |
| | | | 650 | .85 | 12.7 |
| Cu$^{+2\to0}$ | 8.92 | 63.6 | 347 | .53 | 8.5 |
| | | | 500 | .59 | 9.4 |
| | | | 650 | .63 | 10.1 |
| | | | 870 | .56 | 8.9 |
| Au$^{+1\to0}$ | 19.3 | 197 | 441 | .53 | 24.2 |
| | | | 589 | .60 | 27.6 |
| I$^{-1\to0}$ | 4.93 | 126.9 | 589 | .12 | 14.1 |
| Ir$^{+4\to0}$ | 22.4 | 193.1 | 579 | 1.06 | 10.2 |
| Pb$^{+2\to0}$ | 11.3 | 207.2 | 589 | .74 | 30.6 |
| Mn$^{+2\to0}$ | 7.2 | 54.9 | 579 | .84 | 14.5 |
| Ni$^{+2}$ | 8.9 | 58.7 | 420 | .76 | 11.2 |
| | | | 589 | .71 | 10.5 |
| | | | 750 | .73 | 10.8 |
| Pt$^{+2\to0}$ | 21.5 | 195.2 | 441 | .90 | 18.4 |
| | | | 589 | .76 | 15.4 |
| | | | 668 | .69 | 14.1 |
| Rh$^{+3\to0}$ | 12.4 | 102.9 | 579 | 1.01 | 12.6 |
| Se$^{-2\to0}$ | 4.8 | 79.0 | 400 | 2.13 | 79.0 |
| | | | 490 | 1.19 | 44.1 |
| | | | 589 | .28 | 10.4 |
| | | | 760 | .03 | 0.95 |
| | | | 332 | .24 | 11.4 |
| Ag$^{+1\to0}$ | 10.5 | 107.9 | 395 | .61 | 28.1 |
| | | | 500 | .74 | 34.2 |
| | | | 589 | .78 | 35.9 |
| | | | 750 | .86 | 40.1 |
| Sn$^{+2\to0}$ | 7.28 | 118.7 | 589 | 1.12 | 41.1 |
| Zn$^{+2}$ | 7.14 | 65.4 | 441 | .91 | 18.7 |
| | | | 589 | .99 | 20.5 |
| | | | 668 | .96 | 19.7 |

such as polyvinylalcohol with a staining material such as iodine. Land[C10] has claimed that staining by iodine renders the plastic material dichroic.

There are also a few display devices[C6-C9] based on electroplating. One such device is an electrically addressable moving indicator, such as a clock display, proposed by de Koster.[C7] The device employs a silver nitrate electrolyte with a dark platinized platinum cathode and has been cited as an attractive system. The redox potential for plating Ag from a silver nitrate solution is about 0.8 V. The silver is plated and deplated successively on an array of electrodes controlled by an electronic switching circuit or a mechanical stepping motor, thus appearing as a moving indicator in a reflective mode. Another reflective display device[C9] is based on an organic electroplating system such as acetylcholine iodide $[N(CH_3)_3CH_2CH_2OCOCH_3]^+I^-$ in aqueous solution. In this system the iodine ion is believed to be oxidized and plated out as iodine molecules $(I_2)$ in film form. Because of the wetting nature of the pentavalent nitrogen organic solution, the iodine film has a strong covering power and persists for over 1 minute on noble metal electrodes. However, there is a problem in operating life with this system, i.e. a fraction of the iodine film migrates to the solution as free iodine and cannot be converted back to ionic iodine.

In summary, the electroplating ECC systems have generally the same characteristics as the simple redox systems. However, one problem in metal electroplating particularly worth mentioning is the plating uniformity. Since the plated material is very conductive, except in the case of an iodine film, any nonuniformity in conductance of the electrode will cause severe nonuniformity of plating. In the case of nonconducting species such as $I_2$, the nonuniformity is self-correcting during the plating process. A significant improvement with respect to problems of this nature is achieved in an electroredox system coupled with a chemical reaction.

### 3.3. pH Variation Induced Electrocoloration

pH indicators have been known to chemists for almost a century. Kolthoff[115] has given an excellent review on various pH indicators. There are two classes of pH indicators; one is a fluorescence indicator and one is a color indicator. Alburger[116] first proposed both types of indicators for display and data storage application.[D2,D3] However, we are concerned here mainly with the non-emissive color indicators. The color indicator works on the principle that a dye ion assumes different structures when it is placed in solutions of different pH values. A color change results from the change of absorption spectrum associated with different chemical structure.

The applications of pH variation-induced color change for optical and display devices are mostly contained in the patent literature.[D1-D5] A few important ones are discussed here. Sziklai[D1] in 1949 described an electrochemical color filter in which the color of an indicator solution is varied by changing the applied potential to the electrodes in the solution. The indicator, such as phenylphthalein, changes color from colorless to red violet as the pH is reduced below the range of 8.0 - 9.8. Alburger[D3] discloses a long list of color indicators as well as fluorescence indicators most of which have been discussed in detail by Kolthoff in his book. Another application proposed by Donnelly and Cooper[D4] is an optically variable mirror for use in vehicles as a rear view mirror. It is our opinion that the pH effect and other ECC systems may not be practical for such an application because of its rough operating environment (wide temperature range, vibration etc.).

A few general comments may be made about the applicability of this effect for displays. It certainly has the general characteristics of an electrochemical redox reaction (essentially a redox reaction for $H^+$ and/or $OH^-$ ions). In addition, the pH effect is limited to aqueous solutions, which limits the applied potential to below 1.4 V; otherwise, electrolysis of water sets in. Unlike electroplating, the pH change does not have memory, i.e. the $H^+$ and $OH^-$ ions accumulated near the electrodes tend to diffuse out and recombine. This also produces color diffusion (fuziness). The speed of coloration is, as with other redox systems, limited by an ion-diffusion process and the maximum optical density achievable in a pH indicator system is limited by the solubility of the indicator and by its dissociation constant. However, since most indicators are only slightly soluble,[115] the maximum optical density obtainable is usually low. Therefore the pH color indicator effect does not appear to be a very useful effect for display applications.

### 3.4. Electrochemical Reaction Coupled with a Chemical Reaction

This category can be separated into two groups. In one group the chemical reaction is responsible for creating a colored-insoluble species and in the second group the chemical reaction annihilates the colored species.

Insoluble film. The first group is exemplified by the heptyl viologen bromide system described by Schoot et al.[117,118] The chemical reaction following the electrochemical reduction forms an insoluble purple compound, heptyl viologen bromide. Because of its insolubility, it adds memory capability in a display device, provided that the external circuit does not have a low-impedance discharge path. Van Dam and Ponjee[119] have shown that there exists a correlation between the solubility of the reduced radical compound and the sizes of both the hydrocarbon substitute R in the viologen and the anion X available in the solution. The larger the radius of $X^-$, the greater the stability of the salt. The viologen bromide induced by the electrochemical reaction followed by a chemical reaction has sufficiently low solubility to be a permanent film if the substitute R is larger than five ($CH_2$) units. This chemical reaction is preferred in water solution although nonaqueous solvents such as acetonitrile, propylene carbonate and nitromethane are suggested in a Dutch patent application.[E1] Unfortunately the electrochemically reduced product in these organic solvents does not adhere to the cathode as a film. Recently Kawata et al.[120] investigated a few such systems. A second example in this category is the oxidation of Cu-ethlenediamine[121] which results in a thin layer of solid deposit on the anode. However, no electrooptical properties are available for display consideration.

It is interesting to note that Simon[E2] has described a class of dipyridyl polymers as being electrochromic and has disclosed methods of preparing them. This class may be expressed as

$$\left[ -{}^+N\!\!\bigcirc\!\!-\!\!\bigcirc\!\!N^+ - R_2 - \right]_n$$

where R and $R_1$ are hydrogen or methyl, $R_2$ is alkylene or $R_3$-phenylene-$R_4$ wherein $R_3$ and $R_4$ both are alkylene, n is at least 2 and $A^-$ is an anion such as $Cl^-$, $Br^-$, $I^-$, $BF_4^-$, $SO_4^-$, $ClO_4^-$ or tosyl. Rogers[E3,E4] proposed the use of this class of materials as well as the 4,4'-dipyridinium compounds

$$\left[ R_1 - N\!\!\bigcirc\!\!-\!\!\bigcirc\!\!N - R_2 \right]_2 A_2$$

the diazapyrenium compounds and pyrazindinium compounds for a light filtering device application. However the relative merits of these materials have not been compared. Another specific dipyridyl compound has been described by Kenworthy[E5] in a variable light transmission device. It is claimed that the N,N'-di-(p-cyanophenyl)-4,4'-bipyridinium dichloride in aqueous sulfuric solution at a potential of −0.2 V (vs. Ag/AgCl reference electrode) can be reduced to an insoluble green radical cation as an even layer on the cathode. A low reduction potential is generally desirable in aqueous solution since the electrolysis of water at 1.4 V limits the applied potential. However, in this case, −0.4 V is the limiting reduction potential because a two-electron irreversible reduction sets in at that voltage. Any irreversible reaction in an ECC device is obviously undesirable, since it limits the device's cycle life-time.

In summary, the redox materials in this category have the unique feature that the radical ions formed after the electro-redox reaction can undergo a chemical reaction to form insoluble films, which provide memory. This is very useful in certain display applications and may affect the slow writing speed which generally prevails for ECC devices. The above mentioned dipyridinium compounds all possess this unique feature but apparently only in aqueous solutions. Of course, there may be other materials having this feature in a nonaqueous solution, thus eliminating the problem of water electrolysis. A serious exploratory effort is needed in this direction. One likely candidate which may be mentioned here is tetramethoxy biphenylene in dry acetonitrile $(10^{-3}$ M).[122] It is said to be electrochemically oxidized and precipitated as a dark blue solid.[122] No details are available for display device consideration.

Regenerative process. In this category the electro-redox reaction is coupled with a chemical reaction which controls the rate of reversal of the colored species back to its uncolored state, thus providing a variable color persistence. The overall electrochemical reaction may be expressed as

$$A + e \rightarrow B$$
$$B + Z \rightarrow A + Z'$$
$$Z' - e \rightarrow Z$$

where A is the original redox species, B is the colored species and Z is an added chemical providing the regenerative reaction. Chang and Howard[26] have studied one system where A is PTA and Z is $H_2O_2$. This system eliminates the color-drift problem typically associated with soluble colored species produced at the electrode at the expense of memory (the information must be refreshed periodically). Furthermore, this feature in some ways facilitates the use of matrix addressing (multiplexing) in a display system. However one should be aware that, by including a regenerative mechanism, the overall electro-coloration efficiency is reduced. Experimental data related to the reduction of efficiency has been shown.[26] Kuwana and coworkers[76] have studied a regenerative system involving methyl viologen. Although they were not interested in the electrooptic properties for display applications, their discussion of electrochemical reaction kinetics is very useful. An analytical model for this type of regenerative reaction has been derived.[66]

In summary, very little effort has been directed to this type of system, except for a few exploratory experiments.[E6,E7] The ECC effect in this type of system is more sensitive to reaction kinetics and reversibility than other ECC systems. However the characteristic of variable persistence is a desirable feature in certain display applications such as television displays, so it may merit some research effort.

### 3.5. Electrochemical Reaction Involving a Solid Film

In this category, a solid film deposited on an electrode undergoes an electro-redox reaction and exhibits intense color. The most significant system discovered thus far is the tungsten trioxide film in an acidic solution.[F1-F15] As a negative potential is applied to a $WO_3$-coated electrode in the configuration of Type 3 in Fig. 1b, a blue color is created in the $WO_3$ film. Recently the electrochemical and electrooptical properties of this type of electrochromic device have been investigated in detail.[26,66] It was concluded that the hydrogen ions in the electrolyte and their reduction play an important role in the electro-coloration. In fact, the rate of electro-coloration is determined by the diffusion of hydrogen ions.[66] Therefore the device should be classified as an electrochemichromic device. The mechanism is definitely not a simple color-center effect as proposed by Deb[7] in his discussion of $WO_3$ devices. The detailed mechanism of the ECC $WO_3$ device is of great interest since this technology may be a competitive candidate for certain applications such as watch and calculator displays.

Talmey[F1] seems to be the first to suggest the electrolytic reduction of tungsten and molybdenum compounds, including oxides, but the major work is contained in a series of patents.[F2-F9,F11,F14,F15] The most significant teachings in these patents are in the area of achieving high electrolyte conductivity in gel form, for example by using polyvinyl alcohol with high concentration of sulfuric acid[F2-F5] or lithium stearate containing propylene carbonate and p-toluene sulfonic acid together with Li and Na salts.[F6] Claims have also been made relative to more efficient and longer-life counterelectrodes (such as hot-pressed graphite with $WO_3$)[F4] and other improvements of device lifetime.[F8-F10] A few specific applications like watch or time-indicator displays have been discussed.[F11] It has also been suggested that by using a $WO_3$ layer as a counter-electrode,[F12,F15] better reversibility can be obtained. Recently, Giglia[123] gave a short account of the performance properties of this type of device in a better packaged form. Several million write-erase cycles have been achieved.

In summary, the patent literature may represent some of the development efforts in this technology; however the teachings are very empirical and sometimes speculative. Therefore it is essential to gain a better understanding of the basic mechanisms involved in the $WO_3$ ECC effect in order to evaluate the past effort and possibly develop a reliable technology. Furthermore, it is also likely that one could find other solid films, perhaps even organic films, that can operate in a manner similar to $WO_3$. One such example is rare-earth diphthalocyanine ($RE-PC_2$).[124-128] Moskalev and Kirin have shown that a lutetium-diphthalocyanine thin-film-coated electrode in an aqueous KCl solution can undergo a color change if a potential is applied to the electrode. Figure 3 shows the absorption spectrum change as a function of applied po-

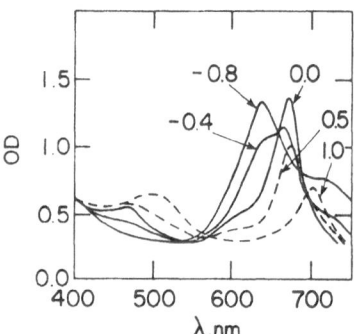

Fig. 3. Absorption spectrum of $Lu-PC_2$ film deposited on tin oxide electrode and biased at voltage indicated (from Moskalev and Kirin[124]).

tential.[124] A large change in amplitude and a large shift in wavelength is observed within a range of applied voltage between −.8 and +1 V. For positive potentials the absorption is attributed to the formation of molecular complexes with an electron acceptor such as $O_2$ or $H_2O$, whereas for negative potentials the change in spectrum is due to ionization of the $RE-PC_2$ molecule. It is conceivable that materials of this type may find applications in multicolor or gray-scale displays.

## 4. PROPERTIES, MECHANISMS AND MODELLING OF EC AND ECC $WO_3$

As we pointed out, the electro-coloration in $WO_3$ was recognized and suggested for recording and display applications as early as in 1939, but the understanding of this phenomenon is by no means clear and complete. Deb[14] first interpreted the electro-coloration effect in $WO_3$ in terms of a color-center model. It was later shown by Chang and Howard[26] that the electro-coloration of $WO_3$ is more efficient if operated in an electrochemical mode, i.e. a higher ultimate contrast can be achieved. They further illustrated the important role of hydrogen ions and water in the ECC effect by cyclic voltammetry measurements.[66] Later, Hurditch[129] correlated the ease of electro-coloration with the presence of water by absorption measurements. McGee et al.[130] showed that $WO_3$ can be reduced in a solution of $LiClO_4$ in propylene carbonate to form lithium tungsten bronze, which has a similar absorption spectrum to that of the $WO_3$ reduced in an aqueous solution. In the latter case, the result is most likely to be a hydrogen tungsten bronze. All this new evidence points to the fact that a simple color-center model is not adequate for interpreting the $WO_3$ ECC effect. A more detailed model taking into account the electrochemical reactions has to be developed. Furthermore, there are still questions about how to interpret the behavior of a totally solid-state $WO_3$ EC device. In this section, we attempt a thorough review of the physical properties of $WO_3$ and its reduction chemistry, with the purpose of deducing a reasonable model to account for the $WO_3$ device characteristics.

### 4.1. Properties of $WO_3$

Phase transitions and substoichiometry. $WO_3$ is one of the transition-metal oxides which has a number of interesting physical properties. The structural [131-133] properties of $WO_3$ have been studied a great deal in recent years because of its multiple-phase transitions[132,134] and its ferroelectric properties.[135,136] An electric field-induced absorption change near the fundamental edge has also been reported by Salje.[137] Furthermore, it was observed that $WO_3$ shows pleochroism[138,139] at room temperature which may be attributed to phase transitions or to lattice imperfections of anisotropic character such as oxygen vacancies.[139] Single-crystal $WO_3$ at room temperature has a monoclinic lattice[131,132] (a = 7.3 Å, b = 7.53 Å, c = 7.68 Å, $\beta$ = $90°54'$) and possibly exists also in a triclinic phase.[134] It shows several phase transitions from monoclinic to triclinic near $17°$ C, back to monoclinic near $-50°$ C, to orthohombic at $350°$ C and to tetragonal at $740°$ C, with two other tetragonal phases above $900°$ C.[132,139] Another property of $WO_3$ is that there exist numerous substoichiometric crystals.[140-145] These crystals may be easily obtained by reducing $WO_3$ at high temperatures under vacuum. The basic defect unit in the stoichiometric crystals is a shear plane.[140,141]

Electrical properties of $WO_3$. Due to the phase transitions, $WO_3$ exhibits resistivity jumps[132,144-146] as a function of temperature. At room temperature, $WO_3$ is n-type, having a resistivity of 0.3 to 10 $\Omega$ cm[132,146-148] and a carrier concentration[148] $n = 10^{17}$ to $2 \times 10^{18}$ cm$^{-3}$. As oxygen vacancies are introduced, there is a parallel increase in carrier concentration, but the resistivity behavior is governed primarily by carrier mobility which decreases with oxygen deficiency. Sienko and coworkers[145,148] reported that $WO_3$ possesses a normally empty conduction band to which electrons can readily be excited from shallow (.03 - .05 eV) electron donors, such as oxygen vacancies or the W-W interaction pairs that result from the collapse of a shared octahedron corner. In the case of tungsten bronzes, the metal atoms [149] or protons[150] are introduced to the $WO_3$ host lattice interstitially, and they donate electrons to the wide conduction band formed by the overlap of $W_{5d}$ and $O_{2p}$ orbitals.

It has been reported that $WO_3$ can readily be reduced by hydrogen to form hydrogen tungsten bronze,[150-154] whereas tungstic acid $WO_3 \cdot H_2O$ can be reduced by hydrogen to form reduced tungsten oxide hydroxide $WO_{3-x}(OH)_x$ which is a closely related compound.[155] The crystals of $H_x WO_3$ are usually colored blue to purple and show metallic conductivity.

Amorphous $WO_3$ film. It is the amorphous state of $WO_3$ which is interesting from a display device point of view. This is because the amorphous $WO_3$ film prepared by vacuum evaporation at a pressure of about $10^{-5}$ to $10^{-6}$ torr can readily be colored by UV irradiation, electric field or chemical reduction. The evaporated film is usually slightly substoichiometric with a density of 5.5 - 6.5 g/cm$^3$, which is much less than the density of crystalline $WO_3$ ($\sim 7.16$ g/cm$^3$).[156] The low density may be an indication of a porous structure. Deb[14] has investigated the optical properties of amorphous $WO_3$ which turns out to be slightly less absorbing when compared to the crystalline $WO_3$ but has a similar absorption spectrum except for a small absorption edge shift toward higher energy. This suggests that the essential features of the energy band structure of crystalline $WO_3$ may be retained for the amorphous phase. The energy gap can be estimated from the absorption spectrum to be $\sim 3.27$ eV for amorphous films. The blue coloration in a $WO_3$ film induced by UV irradiation ($> 3.5$ eV) results from an absorption spectrum similar to that induced by electro-coloration. The color formation by both methods is very sensitive to ambient conditions, especially moisture. Figure 4 shows a comparison of the absorption spectrum induced by electrochemical reduction to that induced by UV excitation. The insert in the figure shows that the induced optical density at the wavelength of peak absorption is linearly proportional to the charge density applied. The resistivity of an amorphous $WO_3$ film is very sensitive to ambient conditions. The temperature dependence of resistivity also exhibits a jump around 470° K. This has been attributed to phase changes taking place in the film in analogy with the jumps observed in crystalline samples.[132,144-146]

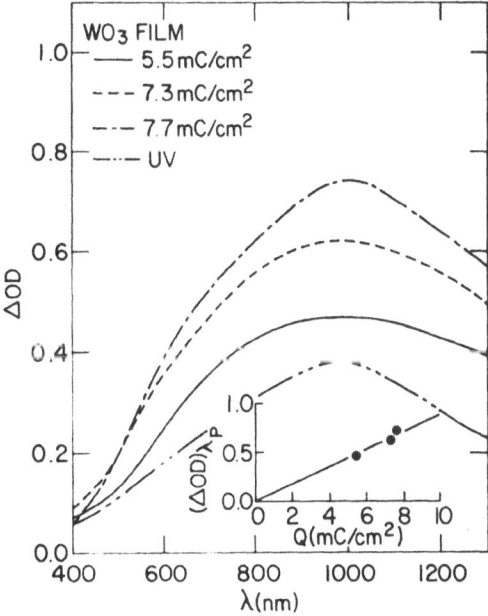

Fig. 4. Absorption of $WO_3$ film induced by UV excitation ($- \cdot \cdot -$) and electro-chemical reaction ($\longrightarrow$ 5.5 mC/cm$^2$; $- - -$ 7.3 mC/cm$^2$; $- \cdot -$ 7.7 mC/cm$^2$). The insert shows that the optical density at peak absorption $(\Delta OD)_{\lambda_p}$ is a linear function of applied charge density.

## 4.2. Reduction of $WO_3$

It is well known that $WO_3$ can be chemically reduced to form a blue com-pound.[150,157-164] As early as 1815, Berzelius[157] reported that when hydrogen is passed over gently warmed tungsten trioxide, a blue oxide is formed. Austin[160] showed that hydrogen molecules accomplish the reduc-tion of $WO_3$ at 200° C but not at room temperature. Later Khoobiar[161] re-ported that when platinum is mixed with $WO_3$ and $Al_2O_3$, the reduction be-comes more effective. He postulated that Pt, being a catalyst, dissociates the hydrogen molecule to atoms, which migrate to $WO_3$ and reduce it. This ex-periment was later repeated by Kohn and Boudart,[162] who found that the hydrogen reduction of $WO_3$ occurs only when the mixtures are exposed to air. They concluded that the adsorbed oxygen is essential to the release of hydrogen atoms by the Pt catalyst, as proposed by Taylor.[165] Later Benson et al.[163] again investigated this chemical reduction. They confirmed that it is actually water that is enhancing the reduction at room temperature to form hydrogen tungsten bronze as the end product. This has been verified by x-ray powder pattern measurement. The enhancement is attributed to the fact that the water molecules absorbed on $WO_3$ allow rapid transport of hydrogen atoms. In a different experiment of reducing $WO_3$ crystals by Zn and HCl, Dickens and Hurditch[150,166] also obtained hydrogen tungsten bronze as verified by x-ray diffraction studies.

The electrochemical reduction of $WO_3$ has also been of interest to chemists and electrochemists for a long time.[167-169,151-154] In 1930, Kobosew and Nekrassow,[167] in an experiment of electrochemically reducing $WO_3$ in acidic solution, observed that $WO_3$ can be reduced to a blue compound. The blue color is induced uniformly in the separate grains of the powder substance, not merely in those in contact with the metal cathode but also in those at some distance. Therefore they proposed that atomic hydrogen is responsible for the reduction. They claimed that the hydrogen ions are reduced at the electrode, becoming hydrogen atoms which can diffuse away and reduce the $WO_3$ grains. Later Bagotsky and Jofa,[168] repeating such an experiment, observed that an individual grain of $WO_3$ is reduced only when in contact with the surface of the electrode at least at one point, whereas the other particles which were not in contact with the electrode nor in contact with those grains touching the electrode surface did not change color at all. It was also observed that the blue coloration can spread from the contact point to the entire surface of the grain. This spreading continues even when the voltage is switched off, provided the grains are moist. The coloration can spread from particle to particle if they are in contact. The coloration is not confined to the surface of the microcrystals, which has been proven[168] by grinding them. It is interesting to note that Albrecht et al.[170] showed that grinding of $WO_3$ in a vibrating mill results in reduction of $WO_3$ to amorphous $WO_{2.98}$. However, we believe that the claim made by Bagotsky and Jofa is valid. Based on their experiment and on molecular kinetic considerations (hydrogen atom evaporation and diffusion rate), Bagotsky and Jofa rejected Kobosew and Nekrassow's concept of active hydrogen atoms. Instead they recognized the electronic conductivity in $WO_3$ and its increase with reduction and proposed an electrochemical reduction scheme in which the $WO_3$ is electronically reduced by an electron supplied by the electrode and neutralized by a hydrogen ion from the solution. They further conclude that the propagation of the reduction along the surface of $WO_3$ is determined by the movement of the electrons inside the particle and the movements of hydrogen ions in the solution. However, this still does not explain the mechanism of reduction of propagation into the bulk of the microcrystals. They speculated on the possibility of oxygen being diffused out to the surface. In our view this is not very probable due to the low diffusion constant of oxygen (see Table 3). On the other hand, there is a possibility of $H^+$ reduction to active hydrogen, taking place inside $WO_3$, if electron injection and $H^+$ diffusion are assumed. Recently, Hobbs and Tseung[151-154] studied the electrochemical reduction of $WO_3$ admixed with Pt. They found that the reduced blue compound is hydrogen tungsten bronze having the same x-ray diffraction data as that produced by a chemical reduction such as $WO_3$ plus Zn plus HCl acid.

The question thus remains: what is the reduction product — lower oxides, tungsten oxide hydroxide or hydrogen tungsten bronze? After care-

## TABLE 3

Diffusion Constants for Various Species Which May Participate
in the $WO_3$ ECC Reaction

| Species | Diffusion Constant $(cm^2/s)$ | Mobility $(cm^2/Vs)$ | Ref. |
|---|---|---|---|
| Hydrogen in $H_xWO_3$ | $7 \times 10^{-6}$ | $(2.7 \times 10^{-4})$ | 177 |
| | $10^{-11}$ to $10^{-10}$ | $(3.9 \times 10^{-10}$ to $3.9 \times 10^{-9})$ | 178 |
| | $8 \times 10^{-7}$ | $(3.1 \times 10^{-5})$ | 179 |
| $Na^+$ in $H_xWO_3$ | $10^{-15}$ | $(3.9 \times 10^{-14})$ | 180 |
| | | | 181 |
| $H^+$ in $WO_3$ (with .09 M to .23 M $H_2O$) | $(5.2 \times 10^{-9}$ to $1.1 \times 10^{-6})$ | $2 \times 10^{-7}$ to $4.2 \times 10^{-5}$ | 129 |
| e in $WO_3$ single crystal | $(.285)$ | 11 | 144 |
| | | | 145 |
| | | | 148 |
| e in $WO_{3-x}$ single crystal $(x < .001)$ | $(.04 - .12)$ | 1.5 to 4.5 | 144 |
| | | | 145 |
| e in amorphous $WO_3$ | $(.259)$ | 10 | 14 |
| | $2.5 \times 10^{-3}$ | $(.1)$ | 182 |
| H in $H_2O$ | $7 \times 10^{-5}$ | $(2.7 \times 10^{-3})$ | 183 |
| $H_2$ in $H_2O$ | $3.4 \times 10^{-5}$ | $(1.3 \times 10^{-3})$ | 184 |
| $H^+$ in $H_2O$ monolayer | $10^{-8}$ | $(3.9 \times 10^{-7})$ | 185 |
| Hydrated e in $H_2O$ | $4.8 \times 10^{-5}$ | $(1.8 \times 10^{-4})$ | 186 |
| $O_2$ in $WO_3$ $(568^\circ C$ to $980^\circ C)$ | 1.5 to $2.5 \times 10^{-9}$ | $(5.8$ to $9.6 \times 10^{-8})$ | 187 |

ful examination of the evidence presented above, one may conclude that the lower oxides can only result from a high temperature ($> 200°$ C) chemical reduction of $WO_3$. This is found to be true also for $MoO_3$.[171] On the other hand, in the presence of water, $WO_3$ can be reduced at low temperatures chemically and electrochemically to hydrogen tungsten bronze. In the case of chemical reductions, this has been confirmed by the work of Benson et al.[163] and Dickens et al.;[150,166] both have made x-ray diffraction studies. In electrochemical reductions, confirmation is provided by Hobb and Tseung,[151-154] whose x-ray diffraction results show no evidence for the presence of lower oxides and agree entirely with the hypothesis of hydrogen tungsten bronze phase. Furthermore a similar reaction also occurs in $MoO_3$.[171] Tungsten oxide hydroxide $WO_{3-x}(OH)_x$ is the reduction product of hydrated tungsten oxide ($WO_3 \cdot nH_2O$). Its structure has been shown to be identical to $H_xWO_3$.[155,164] Therefore there is little doubt that $H_xWO_3$ is the reduction product of $WO_3$ ECC reaction.

## 4.3. Model for $WO_3$ EC and ECC Display Devices

The recent surge of interest in $WO_3$ EC and ECC effects results from its potential for use in low power, low voltage displays. Several workers have investigated these effects and have proposed models for them.[26,66,129, 130,173,174] We have considered both EC and ECC devices.[26,66] In the EC case, i.e. no moisture or electrolyte, the coloration is considered to be due to color centers and has limited optical density change comparable to that induced by UV excitation. In the ECC case, in view of the evidence described in the previous section, the end product is more likely to be hydrogen tungsten bronze than the lower oxides. Accepting $H_xWO_3$ as the end product, then the key issue in modelling the ECC effect is: how is $H_xWO_3$ formed? What is happening to the hydrogen ion in the electro-coloration process?

Faughnan et al.[173] and Hersh et al.[174] both proposed a double injection model and claimed the end product to be $H_xWO_3$. The former group has attributed the color absorption to electrons localized on tungsten sites; i.e. $W^{6+} \rightarrow W^{5+}$ and the protons remain ionized. On the other hand, the latter group found no evidence[174,175] of $W^{5+}$ by x-ray photoelectron spectroscopy. This measurement technique was first used by Rabalais et al.[176] in studying the UV-colored and electrically colored $MoO_3$ amorphous films. They observed an electron distribution corresponding to an increased density of states in the band gap of the colored films and attributed the change to electrons trapped in oxygen vacancies. One should be aware of the fact that this technique only reveals density-of-state information for a shallow surface ($< 50$ Å). Hersh et al.[174] observed a similar energy band in colored $WO_3$ and attributed it to positive ions $H^+$, $Li^+$ or $Na^+$, which they employed in the electrolyte. We too have proposed somewhat similar models.[66] These models assume that it is the active hydrogen that plays a key role in the reduction of $WO_3$ to the tungsten bronze.

One scheme similar to double injection (Scheme 1 in Fig. 5) assumes that the electrons are injected from the cathode and the hydrogen ions are in-

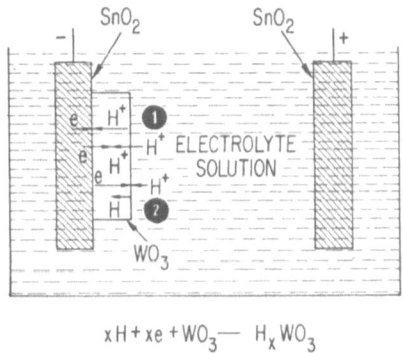

$$xH + xe + WO_3 \longrightarrow H_x WO_3$$

Fig. 5. Model for $WO_3$ ECC Reaction (see text).

jected into $WO_3$ from the solution; the electrons reduce the hydrogen ions to form active atomic hydrogen which in turn reacts with $WO_3$ to form $H_x WO_3$. It may be described by the following electrochemical reactions:

$$H^+ + e \rightarrow H, \ xH + WO_3 \rightarrow H_x WO_3, \ \text{or} \ xH^+ + xe \div WO_3 \rightarrow H_x WO_3 \ (x < 1).$$

Alternatively, the hydrogen ions are reduced at the $WO_3$-solution interface (Scheme 2 in Fig. 5). The hydrogen atoms then diffuse into $WO_3$ to form $H_x WO_3$. Hurditch[129] essentially supports the active hydrogen atom concept, except that he considers a planar configuration with a hydrated $WO_3$ film covering two molybdenum electrodes spaced 1 mm apart. He assumes that the hydrogen ions are dissociated from $H_2 O$ molecules in the film. It should be pointed out that once having assumed electron injection, it is reasonable to expect that $H^+$ can be reduced to active hydrogen atom in the $WO_3$ film. Therefore it is not necessary for the $H^+$ ions to get reduced at the electrode surface by diffusing through the $WO_3$ film as considered an improbable process by Bagotsky and Jofa.[168]

Besides the question of active hydrogen, the above models involve different kinetic mechanisms. Therefore a crucial test of their validity is to examine the rate-limiting process and compare it to the experimentally observed electro-coloration rate which is on the order of $1 \cdot 1000 \ s^{-1}$ for obtaining an optical density OD = 1 in a typical ECC cell. In Table 3 we list all the diffusion coefficients for various species[177-187] that may be involved in an ECC reaction. As one can see, the hydrogen transport rather than the electron transport must be the limiting process for the electro-coloration rate. Which step, then, of the hydrogen transport is limiting the overall reaction rate? This is inseparable from the question of how $H_x WO_3$ is formed. The experimental evidence obtained so far does not offer a satisfactory picture for the overall reaction. Particularly, there is no information concerning the hydrogen transport across the electrolyte-$WO_3$ interface However the current response curves of ECC cells,[26] i.e. $i \propto t^{-1/2}$, imply that the overall electrochemical reaction is limited by hydrogen ion diffusion in liquid solution. In other words, the electrochemical or chemical reaction at the $WO_3$ electrode is very fast (at least for high electrode voltages) compared to $H^+$ diffusion in the solution. This conclusion is supported by the work of Boudart et al.[188,177,163] who have endorsed the active hydrogen concept and suggested that in the presence of $H_2O$, the active hydrogen forms $H_3O$ intermediates and that the large internal surface of $WO_3$ (particularly in a $WO_3$ film) allow rapid surface diffusion of the active reactant.

One might wonder how the above model explains the Li or Na tungsten bronze formation in the experiment of McGee et al.,[130] since those ions have much smaller diffusion constants in $WO_3$ (Table 3). The answer may be that it is the surface rather than bulk diffusion process that applies to the bronze electrochemical reactions. The surface diffusion is enhanced by the water adsorbed along the grain boundaries of $WO_3$ films. For the diffusion lengths involved in a $WO_3$ ECC film, even a surface diffusion constant of $10^{-8}$ $cm^2/s$ is adequate to give a device response time on the order of one second. However, when the ECC reaction is operated at very low voltages, the ion transport across the electrolyte-$WO_3$ interface may become more important. An ion-diffusion-limited (in solution) model may not apply any more. Further research in this area is needed before a complete understanding of the overall electrochemical reaction can be elucidated. Detailed studies on a solid-state electrochemichromic cell such as the $Ag \leftrightarrow Ag_x WO_3$ reaction,[189]

$$Ag \rightarrow Ag^+ + e \qquad\qquad (Ag^+ \text{ in } RbAg_4I_5)$$
$$xAg^+ + xe + WO_3 \rightarrow Ag_x WO_3,$$

may indeed be very rewarding.

Another issue in the $WO_3$ ECC device is its bleaching behavior. Faughnan et al.[179] studied this problem recently. They succeeded in explaining part of the bleaching process by a model assuming that the extraction of protons out of $WO_3$ obeys the condition of a space-charge-limited current flow. The exact nature of the entire bleaching process is not yet clear.

## 5. COMPARISONS AND CONCLUSIONS

### 5.1. Comparison of EC and ECC Effects and Displays

EC and ECC phenomena and materials that have possibilities for display applications satisfy three necessary criteria. The first is that the material or the system of materials must have an optical absorption band which can be created and annihilated by receiving or releasing an electron (or hole). The second condition is that the material must allow electron (or hole) injection or conduction. The third condition is that the material allows ionic conductivity. The electronic conduction provides the means for altering the absorption state, while the ionic conduction guarantees charge neutrality and current continuity during charge injection. In solid EC materials, the third condition is often not satisfied at room temperature because of the limited ionic conductivity in solids. This is why most EC phenomena discussed in Section 2.1 and 2.2 are observable only at high temperatures. On the other hand, this condition is usually not a problem for ECC liquids, except for certain nonaqueous systems that have limited electrolyte solubility. The second condition, however, is different for EC solids and ECC liquids. In solids, the concentration of switchable color centers is usually of the order of $10^{17}$ to $10^{19}$ $cm^{-3}$. This tends to limit the maximum induced optical density. In contrast, the switchable density of color absorbing centers or ions can be as high as $10^{22}$ $cm^{-3}$ in the case of a solid precipitate like the viologen bromide (Section 3.4). Therefore, it is no surprise that ECC systems are capable of very high ultimate contrast. In addition, the charge transfer efficiency in an ECC system is typically 0.5 to 1 absorbing ion per electron whereas in solids it is typically as low as $10^{-2}$ color centers per electron. Another difference is that ECC systems typically have a threshold voltage determined by the electrochemical redox reaction involved, whereas the threshold condition in EC systems is quite sensitive to the history of write-erase cycles. Therefore it appears in general that ECC systems are superior to EC systems.

At first, the solid $WO_3$ film seems to be the only exception. It allows reasonable ionic conductivity (particularly for $H^+$ ions) due to its open lattice structure and has a very high density of color absorbing centers ($\sim 10^{22}$ $cm^{-3}$) due to the possibility of forming $H_x WO_3$, as described in Section 4. However these characteristics have only been obtained in an ECC liquid cell configuration. It has yet to be shown that a totally solid $WO_3$ system, for instance the $Ag$-$RbAg_4I_5 \cdot WO_3$-$SnO_2$ system, can give the same performance.

Since all EC and ECC systems (except the field effect systems) are charge-responding devices, i.e. the induced optical density is proportional to charge per unit area of electrode cross-section, the proportionality constant can be defined as the EC and ECC device efficiency. Figure 6 presents the efficiency data for several ECC systems. The slope passing through each set of data points is the efficiency parameter. For instance it is 135 $C^{-1}cm^2$ at 544 nm for heptyl viologen bromide. So far there is no accurate data taken for EC systems, including the $WO_3$ EC device. This is mainly because $WO_3$ EC device characteristics are not very reproducible due to extreme sensitivity to fabrication and ambient conditions. The efficiency parameter of a device is an important determinant of its speed. The speed for ECC devices, i.e. the time required to achieve unit optical density, ranges from 10 ms to seconds. It is typically slower for EC devices (100 ms to minutes).

While ECC systems have better efficiency and better defined threshold than the EC solids, the former requires more care in device packaging. In

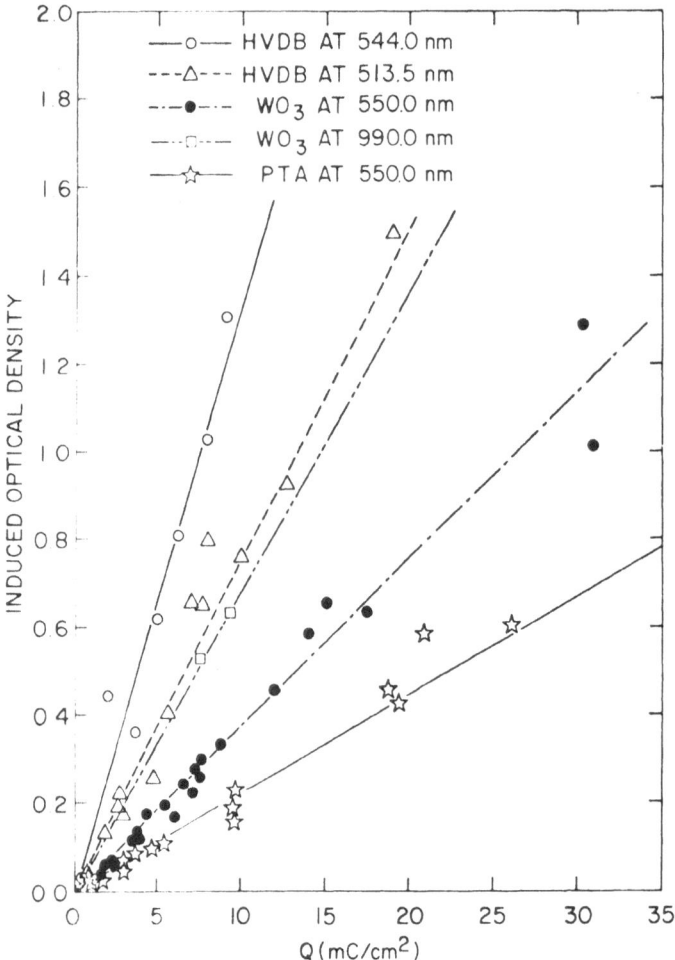

Fig. 6. Comparison of electrochemichromic efficiency. Heptyl viologen dibromide (HVDB): 135 $C^{-1}cm^2$ at 544 nm, 75.4 $C^{-1}cm^2$ at 513 nm; $WO_3$: 70.2 $C^{-1}cm^2$ at 990 nm, 38 $C^{-1}cm^2$ at 550 nm; polytungsten anion (PTA): 22 $C^{-1}cm^2$ at 550 nm.

aqueous systems, the electrolysis of water is an undesirable competing and destructive reaction, whereas in nonaqueous systems the solution conductivity is generally low due to low electrolyte solubility. Furthermore the low threshold voltages, the internal emf and the low impedances of ECC systems also present some problems with respect to large matrix-addressed arrays. Specifically, the emf presents a severe problem in crosstalk, and the low device impedance requires a low impedance driving network which cannot easily be achieved with transparent conducting electrodes. Considering these limitations, the ECC devices seem to be best suited for direct-addressed and small matrix-addressed displays.

## 5.2. Comparison with Other Non-Emissive Displays

The most attractive features of ECC displays are their appearance, high contrast even with a high level of ambient light, their wide viewing angle capability and in some cases their low maintenance power. The memory and variable-persistence capabilities certainly add to their assets. In Table 4 we compare the EC and ECC displays with other non-emissive displays, namely,

TABLE 4

Comparison of Non-Emissive Displays (see Text)
T: transmission, R: reflection, P: projection

|  | EC | ECC | EP | FELC | DSLC | DS |
|---|---|---|---|---|---|---|
| Contrast ratio | 10 | 40 | 40 | 10 | 10 | 10 |
| Viewing angle | wide | wide | wide | narrow | narrow | narrow |
| Optical mode | T, R, P | T, R, P | R | T, R, P | T, R, P | T, R |
| Color | 2 | 2 | 2 | B/W or dye | B/W | B/W |
| Resolution | mostly electrode limited | | | | | ? |
| Operating mode | dc pulse | dc pulse | dc pulse | ac | ac | ac |
| Voltages (V) | .25 - 20 | .2 - 2 | 30 - 80 | 2 - 10 | 10 - 30 | 2 - 30 |
| Power ($\mu$W/cm$^2$) | | | 15 | $<$ .1 | 1 - 10 | 1 - 10 |
| Energy (mJ/cm$^2$) | 10 - 100 | 5 | $6 \times 10^{-4}$ | | | |
| Memory | yes | yes | yes | no/yes | no/yes | no |
| Threshold | poor | yes | no | poor | poor | poor |
| Write time (ms) | 100 - 1000 | 10 - 100 | 60 | 20 | 20 | 20 |
| Erase time (ms) | 100 - 500 | 10 - 50 | 30 | 1'500 | 100 | 30 |
| Operating life ($10^4$ h) | ? | ? | ? | $> 2$ | $>1$ | ? |

the electrophoretic (EP) display,[190] dynamic scattering (DSLC) and field effect (FELC) liquid crystal displays[191,192,193] and the dipole suspension (DS) display.[194,195] Only the first three types have wide viewing angle and memory capability and they are typically driven with dc pulses. The latter three require ac drive and have the advantage of low power consumption. The EP display also has the advantage of low power consumption, but its high voltage and poor threshold (worse than liquid crystal displays) place it in a much less favorable position. The DS display has characteristics similar to the liquid crystals but it faces an uphill fight against a mature technology having a respectable operating life ($> 20$'000 h). It is perhaps relevant to point out here that one type of DS display[A14] actually exhibits an EC property, i.e. an absorption band change is induced by the applied field. However the sensitivity or efficiency is too low to be really useful.

Except for ECC displays, the others all have rather poor threshold behavior or low contrast ratio (low discrimination). This certainly limits their applicability in matrix-addressed displays. Even in the case of ECC displays (which have reasonable threshold characteristics), the response is too slow to allow any significant size of matrix addressing. Slow response is a problem common to all non-emissive displays and limits their usefulness to simple displays with limited multiplexing capability.

## 5.3. Potentials and Problem Areas for EC and ECC Displays

Considering the characteristics of EC and ECC effects in their present state, one may conclude that only ECC systems are suitable for display application in the near future. The obvious applications for ECC's are direct-addressed displays. However, there are definitely other possibilities for ECC's and even perhaps for EC's. One example is a TFT (thin film transistor)-ECC (or EC) display system[196] whereby a complex array of ECC or EC cells is controlled by a matching array of matrix addressed TFT's. Since the EC or ECC systems have switching voltages compatible with TFT's, it is conceivable that an ECC panel may be constructed with integrated thin film drivers as has been done for liquid crystals.[196] Another potential device is a laser beam addressed system in which one could either incorporate a photoconductive material to form an EC (or ECC)-PC sandwich[B2,E7] or introduce some photosensitive materials[E8,E9] directly into an ECC liquid. For instance, if an electron donor material is added to a viologen, a viologen-donor complex is formed.[E8] As a consequence, the viologen becomes electrochemichromic at much lower voltages ($\sim 0.2$ V). Such low voltages are within the range of photovoltaic effects in certain organic materials[E8] induced by radiation. Thus it is conceivable that by combining these organic materials one could achieve a light-addressed passive imaging device. It is worthwhile pointing out that some EC or ECC materials (tungstic acid related) are also photochromic.[197-201] In fact, photographic systems have been developed based on some of these materials.[202,203] Obviously, one great advantage in a laser beam-addressed scheme would be that the definition of electrodes and the isolation of individual cells might turn out to be unnecessary.

As far as problem areas are concerned, the EC and ECC displays, like most other developing display technologies, have quite a few serious problems such as reversibility, side reactions, packaging problems, operating lifetime, etc. Since the materials presently known are by no means completely satisfactory, there is certainly a need for continued efforts to find new materials as well as a need for improvements of existing systems. For instance, most ECC systems known are aqueous systems which always face the problem of water electrolysis. Thus it is desirable to search for non-aqueous ECC's. A criterion to be applied in seeking out efficient ECC's has been schown[66] to be that of maximizing the parameter $p = a\, C^{\circ}\, D_{0}^{1/2}$, where a is the absorptivity of the ECC ion, $C^{\circ}$ is the maximum solubility of the ECC species and $D_{0}$ is its diffusion constant. One should realize that, in general, the larger the ion size the greater the absorptivity but the smaller the diffusion constant. Therefore the search for an optimum molecular structure is an interesting problem for synthesis chemists.

## 5.4. Concluding Remarks

A comprehensive review of various EC and ECC phenomena and materials has been attempted. These phenomena involve many different physical mechanisms in various materials. Because of this broad coverage, detailed and in-depth discussions of each phenomenon have been avoided; however adequate references are given for further information. Some physical phenomena such as the Franz-Keldish effect and the Stark effect do not have direct use as display devices but most others do have attractive features for display applications. In particular the ECC viologen system and $WO_3$ system have already been shown to be feasible for numeric displays. Their full potential will be known only after further research and development. $WO_3$ is a unique material which exhibits an ECC effect in a completely solid-state system as well as in an electrochemical liquid system. A thorough review has been presented in Section 4 of its physical, chemical and electrochemical properties.

## ACKNOWLEDGMENTS

I would like to thank Dr. W. E. Howard for his encouragement and interest during the course of this work. I would also like to thank my colleagues, B. L. Gilbert, C. Lanza and Dr. P. M. Alt for many helpful discussions. Special gratitude to Isabel Cawley and Ellen Howing of our library staff for their assistance in the literature search is acknowledged.

## REFERENCES

1   R. W. Pohl, Naturwissenschaften 20 (1932) 932
2   R. W. Pohl, Proc. Phys. Soc. 47 (1937) 3
3   R. C. Duncan, Jr., B. W. Faughnan and W. Philips, Appl. Optics 9 (1970) 226
4   I. F. Chang, J. Electrochem. Soc. 121 (1974) 815
5   I. F. Chang and A. Onton, J. Materials for Electronics 2 (1973) 17
6   J. Blanc and D. L. Staebler, Phys. Rev. B. 4 (1971) 3548
7   S. K. Deb, Appl. Optics Suppl. 3 (1969) 192
8   J. H. Schulman and W. D. Compton, 'Color Centers in Solids', Pergamon, New York (1962)
9   J. J. Markham, 'F-Centers in Alkali Halides', Academic Press, New York (1966)
10  O. Stasiw, Nachr. Akad. Wiss. Göttingen (1932) 261 and (1933) 387
11  N. F. Mott and R. W. Gurney, 'Electronic Processes in Ionic Crystals', Oxford (1950)

12   Y. I. Gritsenko, Phys. Status Solidi A7 (1971) K 113

13   S. K. Deb, Proc. Roy. Soc. 304A (1968) 211

14   S. K. Deb. Phil. Mag. 27 (1973) 801

15   J. A. Van Raalte, J. Appl. Phys. 36 (1965) 3365

16   V. Topa and B. Velicescu, Phys. Status Solidi 33 (1969) K 29

17   M. Yuste and W. Bogusz, Phys. Status Solidi B52 (1972) K 133

18   M. Regis, J. Phys. Chem. Solids 33 (1972) 1997

19   M. Sayer, H. Erdogan and C. D. Cox, J. Electrochem. Soc. 119 (1972) 265

20   A. H. Rosenthal, Proc. IRE 28 (1940) 203

21   P. G. R. King, Proc. IEE (London) 93A (1946) 171

22   P. H. Hyeman, I. Gorog and B. Faughnan, IEEE Trans. Electron Devices ED-18
     (1971) 685

23   P. T. Bolwijn and R. A. Van Doorn, J. Phys. D5 (1972) 896

24   Y. Uno and H. Maeda, S.I.D. Int. Symp. Digest 3 (1972) 76

25   G. D. Robertson, Jr , 'Bistable Electrophotographic Display Devices', AD 732301
     (1971)

26   I. F. Chang and W. E. Howard, IBM Report RC 5136, Nov. 1974; IEEE 1974 Display
     Devices and Systems Conference Record, 148 and IEEE Trans. Electron Devices
     ED 22 (1975) 749

27   J. J. Robillard, AGARD Conf. Proc. No. 50, paper 13, 17th Tech. Symp. Tonsberg.
     Norway, September 29-Oct. 3, 1969

28   B. W. Faughnan and Z. J. Kiss, Phys. Rev. Lett. 21 (1968) 1331

29   Z. J. Kiss, IEEE J. Quantum Electron. QE-5 (1969) 12

30   B. W. Faughnan, Phys. Rev. B4 (1971) 3623

31   B. W. Faughnan and Z. J. Kiss, IEEE J. Quantum Electron. QE-5 (1969) 17

32   M. S. Kosman and E. V. Bursian, Sov. Phys. Doklady 2 (1957) 354

33   K. A. Müller, Th. von Waldkirch, W. Berlinger and B. W. Faughnan, Solid State
     Commun. 9 (1971) 1097

34   K. W. Blazey, O. F Schirmer, W. Berlinger and K. A. Müller, Solid State Commun. 16
     (1975) 589

35   O. F. Schirmer, W. Berlinger and K. A. Müller, 'Electron Spin Resonance and Optical
     Identification of $Fe^{4+}$-$V_o$ in $SrTiO_3$', IBM Report RZ 679 (1974)

36   C. T. Luiskutty and P. J. Ouseph, Solid State Commun. 13 (1973) 405

37   D. L. Staebler, J. Solid State Chem. 12 (1975) 177

38   W. Franz, Z. Naturforsch. 13a (1958) 484

39   L. W. Keldysh, Soviet Phys. JETP 7 (1958) 788

40   R. Williams, Phys. Rev. 117 (1960) 1487

41   R. Kern, J Phys. Chem. Solids 23 (1962) 249

42   G. Harbeke, J. Phys. Chem. Solids 24 (1963) 957

43   V. M. Fridkin and K. A. Verkhovskaya, Appl. Optics 6 (1967) 1825

44   J. R. Platt, J. Chem. Phys. 34 (1961) 862

45   H. Labhart in I. Prigogine (ed.), 'Advances in Chemical Physics', Vol.. 13, Inter-
     science (1967) 179

46   L. A. Blyumenfel'd, F. P. Chernyakovskii, V. A. Gribanov and I. M. Kanevskii, J.
     Macromol. Sci. Chem. A6 (1972) 1201

47   W. Liptay, 'Modern Quantum Chemistry', Vol. 3, Academic Press, New York
     (1965) 45

48 J. E. Kuder and D. Wychick, Chem. Phys. Lett. 24 (1974) 69

49 O. W. Kolling and J. L. Goodnight, Anal. Chem. 46 (1974) 482

50 L. G. S. Brooker, J. Am. Chem. Soc. 63 (1941) 3214

51 W. Kühn, H. Dürkop and H. Martin, Z. Phys. Chem. B45 (1939) 121

52 L. G. S.. Brooker and W. T. Simpson, Am. Rev. Phys. Chem. 2 (1951) 121

53 L. G. S. Brooker, F. L. White, D. W. Heseltine, G. H. Keyes, S. G. Dent, Jr. and E. J. Van Lare, J. Photogr. Sci. 1 (1953) 173

54 L. G. S. Brooker, Kirk Othemer Encycl. Chem. Technology 5 (1964) 763

55 J. R. Platt. J. Chem. Phys. 25 (1956) 80

56 J. Kumamoto, J. C. Powers, Jr. and W. R. Heller, J. Chem. Phys. 36 (1962) 2893

57 J. C. Powers, Jr., W. R. Heller, J. Kumamoto and W. E. Donath, J. Am. Chem. Soc. 86 (1964) 1004

58 F. P. Chernyakovskii and V. A. Gribanov, J. Struct. Chem. 9 (1968) 387

59 V. A. Gribanov, I. M. Kanevskii, F. P. Chernyakovskii and L. A. Blyumenfel'd, J. Phys. Chem. 45 (1971) 3

60 I. M. Kanevskii, V. A. Gribanov, F. P. Chernyakovskii, A. V. Ryazanova and L. A. Blyumenfel'd, J. Phys. Chem. 45 (1971) 232

61 I. M. Kanveskii, V. A. Gribanov, F. P. Chernyakovskii and L. A. Blyumenfel'd, J. Phys. Chem. 45 (1971) 483

62 V. A. Gribanov, I. M. Kanevskii, F. P. Chernyakovskii and L. A. Blyumenfel'd, J. Phys. Chem. 45 (1971) 894

63 J. L. Stevenson, S. Ayers and M. M. Faktor, J. Phys. Chem. Solids 34 (1973) 235

64 S. Ayers, M. M. Faktor, D. Marr and J. L. Stevenson, J. Mater. Sci. 7 (1972) 31

65 E. A. D. White, Proc. 1st European Electrooptics Conf., Geneva, Switzerland, September 1972, IPC Press, London, 161

66 I. F. Chang, B. L. Gilbert and T. I. Sun, IBM Report RC 5179, December 1974 and J. Electrochem. Soc. 122 (1975) 955

67 L. Michaelis and E. S. Hill, J. Am. Chem. Soc. 55 (1933) 1481

68 L. Michaelis, Chem. Rev. 16 (1935) 243

69 R. M. Elofson and R. L. Edsberg, Can. J. Chem. 35 (1957) 646

70 C. S. Johnson, Jr. and H. S. Gutowsky, J. Chem. Phys. 39 (1963) 58

71 E. M. Kosower et al., J. Am. Chem. Soc. 86 (1964) 5515, 5524 and 5528

72 P. B. Sweetser, Anal. Chem. 39 (1967) 979

73 J. W. Strojak, G. A. Gruver and T. Kuwana, Anal. Chem. 41 (1969) 481

74 T. Osa and T. Kuwana, J. Electroanal. Chem. Interfacial Electrochem. 22 (1969) 389

75 N. Winograd and T. Kuwana, J. Am. Chem. Soc. 92 (1970) 224

76 M. Ito and T. Kuwana, J. Electroanal. Chem. Interfacial Electrochem. 32 (1971) 415

77 R. Visco, Ph.D. Thesis, University of Illinois (1963)

78 W. Schwarz, Ph.D. Thesis, University of Wisconsin (1961)

79 T. Kuwana, R. K. Darlington and D. W. Leedy, Anal. Chem. 36 (1964) 2023

80 W. H. Hansen, R. A. Osteryoung and T. Kuwana, J. Am. Chem. Soc. 88 (1966) 1062

81 J. W. Strojek and T. Kuwana, J. Electroanal. Chem. Interfacial Electrochem. 16 (1968) 471

82 T. Kuwana and J. W. Strojek, Discussions Faraday Soc. 45 (1968) 134

83 E. F. Speranskaya and D. B. Mambeeva, Sov. J. Phys. Chem. 39 (1965) 977

84 F. Stonehart, Anal. Chim. Acta 37 (1967) 127

85   F. Chauvean, M. Boyer and B. LeMeur, C. R. Acad. Sci. 268C (1968) 479

86   G. A. Tsigdinos and C. J. Hallada, Climax Molybdenum Co. Bulletin Cdb-14 (1969)

87   J.-P. Launey, C. R. Acad. Sci. 269C (1969) 971

88   J.-P. Launey, P. Souchay and M. Boyer, Collect. Czech. Chem. Comm. 36 (1970) 740

89   I. Hodara and I. Balouka, Electrochim. Acta 15 (1970) 283

90   M. Boyer, J. Electroanal. Chem. Interfacial Electrochem. 31 (1971) 441

91   J. F. Keggin, Proc. Roy. Soc. 144A (1934) 75

92   P. P. Souchay, Ann. Chemie 18 (1943) 61 and 169

93   J. H. Kennedy, J. Am. Chem. Soc. 82 (1960) 2701

94   V. E. Simmons, Ph.D. Thesis, Boston University (1963)

95   P. Souchay and G. Herve, C. R. Acad. Sci. 261 (1965) 2486

96   P. Stonehart. Anal. Chim. Acta 37 (1967) 350

97   M. T. Pope and G. M. Varga, Jr., Inorg. Chem. 5 (1966) 1249

98   M. T. Pope and E. Papaconstantinou, Inorg. Chem. 6 (1967) 1147

99   E. Papaconstantinou and M. T. Pope, Inorg. Chem. 6 (1967) 1152

100  G. M. Varga, Jr., E Papaconstantinou and M. T. Pope, Inorg. Chem. 9 (1970) 662

101  E. Papaconstantinou and M. T. Pope, Inorg. Chem. 9 (1970) 667

102  G. A. Tsigdinos, Climax Molybdenum Co. Bulletin Cdb-12a (1969)

103  M. T. Pope, Inorg. Chem. 11 (1972) 1973

104  P. Stonehart, J. G. Koren and J. S. Brinen, Anal. Chim. Acta 40 (1968) 65

105  J. D. H. Strickland, Inorg. Chem. 74 (1952) 862, 868 and 872

106  G. A. Tsigdinos and C. J. Hallada, Inorg. Chem. 7 (1968) 437

107  C. J. Hallada, G. A. Tsigdinos and B. S. Hudson, J. Phys. Chem. 72 (1968) 4304

108  G. A. Tsigdinos and C. J. Hallada, Inorg. Chem. 9 (1970) 2488

109  G. A. Tsigdinos, Climax Molybdenum Co. Bulletin Cdb-15 (1971)

110  V. V. Treier and B. I. Shapiro, Sov. J. Instrum. & Control 8 (1970) 66

111  A. G. Gray (ed.), 'Modern Electroplating', John Wiley (1953)

112  S. Zaromb, J. Electrochemical Soc. 109 (1962) 903

113  S. Zaromb, J. Electrochemical Soc. 109 (1962) 912

114  J. Mantell and S. Zaromb, J. Electrochem. Soc. 109 (1962) 992

115  I. M. Kolthoff, 'Acid-Base Indicator', Translated by C. Rosenblum, 2nd Ed., MacMillan (1953)

116  J. R. Alburger, Electronic Industries and Tele-Tech. February 1957, 50

117  C. J. Schoot, P. T. Bolwijn, H. T. Van Dam, R. A. Van Doorn, J. A. Ponjee and G. Van Houten, S.I.D. Int. Symp. Digest 4 (1973) 146

118  C. J. Schoot, J. J. Ponjee, H. T. Van Dam, R. A. Van Doorn and P. T. Bolwijn, Appl. Phys. Lett. 23 (1973) 64

119  H. T. Van Dam and J. J. Ponjee, J. Electrochem. Soc. (1974) 121 and 1555

120  T. Kawata, M. Yamamoto, M. Yamana, M. Tajima and T. Nakano, Jap. J. Appl. Phys. 14 (1975) 725

121  D. Meyerstein, F. M. Hawkridge and T. Kuwana, J. Electroanal. Chem. Interfacial Electrochem. 40 (1972) 377

122  A. Ronlan and V. D. Parker, J. Chem. Soc. Chem. Comm. 1 (1974) 33

123  R. D. Giglia, S.I.D. Int. Symp. Digest 6 (1975) 52

124  P. N. Moskalev and I. S. Kirin, Sov. J. Phys. Chem. 46 (1972) 1019

125 P. N. Moskalev and I. S. Kirin, Sov. J. Phys. Chem. 16 (1971) 57

126 P. N. Moskalev and I. S. Kirin, Opt. & Spectrosk. 29 (1970) 414

127 P. N. Moskalev and I. S. Kirin, Opt. & Spectrosk. 29 (1970) 1149

128 A. G. Mackay, J. F. Boas and G. J. Troup, Aust. J. Chem. 27 (1974) 955

129 R. Hurditch, Electron. Lett. 11 (1975) 142

130 J. H. McGee, W. E. Kramer and H. N. Hersh, S.I.D. Int. Symp. Digest 6 (1976) 50

131 H. Braekken, Z. Krist. 78 (1931) 484

132 S. Tanisaki, J. Phys. Soc. Japan 15 (1960) 566

133 S. Tanisaki, J. Phys. Soc. Japan 15 (1960) 573

134 E. Salje and K. Viswanathan, Acta Cryst. A31 (1975) 356

135 B. T. Matthias, Phys. Rev. 76 (1949) 430

136 B. T. Matthias and E. A. Wood, Phys. Rev. 84 (1951) 1255

137 E. Salje, J. Appl. Cryst. 7 (1974) 615

138 R. Ueda and T. Ichinokawa, Phys. Rev. 80 (1950) 1106

139 T. Iwai, J. Phys. Soc. Japan 15 (1960) 1596

140 R. J. D. Tilly, Mater. Res. Bull. 5 (1970) 813

141 J. G. Allpress and P. Gado, Crystal Lattice Defects 1 (1970) 331

142 A. Magneli, Acta Cryst. 6 (1953) 495

143 M. Sundberg and R. J. D. Tilly, J. Solid State Chem. 11 (1974) 150

144 J. M. Berak and M. J. Sienko, J. Solid State Chem. 2 (1970) 109

145 M. J. Sienko and J. Berak in L. Eyring and M. O'Keefe (ed.), 'The Chemistry of Extended Defects in Nonmetallic Solids', Elsevier, Amsterdam (1970) 541

146 H. Hirose, I. Kawano and M. Nino, J. Phys. Soc. Japan 33 (1972) 272

147 W. Meyer, Z. Physik 85 (1933) 278

148 B. L. Crowder and M. J. Sienko, J. Chem. Phys. 38 (1963) 1576

149 M. J. Sienko, Advances in Chemistry 39 (1963) 224

150 P. G. Dickens and R. J. Hurditch in L. Eyring and M. O'Keefe (ed.), 'The Chemistry of Extended Defects in Nonmetallic Solids', Elsevier, Amsterdam (1970) 555

151 B. S. Hobbs and A. C. C. Tseung, Nature 222 (1969) 556

152 B. S. Hobbs and A. C. C. Tseung, J. Electrochem. Soc. 119 (1972) 580

153 B. S. Hobbs and A. C. C. Tseung, J. Electrochem. Soc. 120 (1973) 766

154 B. S. Hobbs and A. C. C. Tseung, J. Electrochem. Soc. 122 (1975) 1174

155 O. Glemser, J. Weidelt and F. Fruend, Z. Anorg. & Allg. Chem. 332 (1964) 299

156 Handbook of Chemistry and Physics, The Chemical Rubber Co., 46th Ed. (1965)

157 J. J. Berzelius, Abh. Chem. Min. 4 (1815) 293 and Schweigger's Journal 16 (1816) 476

158 G. Torossian, Am. J. Sci. 38 (1914) 537

159 A. J. Hegedüs, T. Millner, J. Neugebauer and K. Sasvard, Z. Anorg. Chem. 281 (1955) 64

160 L. G. Austin, Ind. Eng. Chem. 53 (1961) 660

161 S. Khoobiar, J. Phys. Chem. 68 (1964) 411

162 H. W. Kohn and M. Boudart, Science 145 (1964) 149

163 J. E. Benson, H. W. Kohn and M. Boudart, J. Catalysis 5 (1966) 307

164 O. Glemser and Chr. Naumann, Z. Anorg. & Allg. Chem. 265 (1951) 2881

165 H. S. Taylor, Chem. Rev. 9 (1931) 1

166 P. G. Dickens and R. J. Hurditch, Nature 215 (1967) 1267

167  N. Kobosew and N. I. Nebrassow, Z. Electrochem. 36 (1930) 529

168  U. S. Bagotsky and S. A. Jofa, Dokl. Akad. Nauk. USSR 53 (1946) 439

169  M. Grenness and M. W. Thompson, J. Appl. Electrochem. 4 (1974) 211

170  R. Albrecht, H. Häusler and R. Möbius, Z. Anorg. & Allg. Chem. 377 (1970) 310

171  D. T. Hawkins and W. L. Worrell, Metall. Trans. 1 (1970) 271

172  G. C. Bond and J. B. P. Tripathi, J. Less-Common Metals 36 (1974) 31

173  B. W. Faughnan, R. S. Crandall and P. M. Heyman, RCA Review 36 (1975) 177

174  H. N. Hersh, W. E. Kramer and J. H. McGee, private communication, July 1975

175  A. M. Guzman, private communication. An examination of core orbital electron spectrum shows evidence of $W^{4+}$ and $W^{6+}$ but no $W^{5+}$.

176  J. W. Rabalais, R. J. Colton and A. M. Guzman, Chem. Phys. Lett. 29 (1974) 131

177  M. A. Vannice, M. Boudart and J. J. Fripiat, J. Catalysis 17 (1970) 359

178  P. G. Dickens, D. J. Murphy and T. K. Halstead, J. Solid State Chem. 6 (1973) 370

179  B. W. Faughnan, R. S. Crandall and M. A. Lampert, Appl. Phys. Lett. 27 (1975) 275

180  M. S. Whittingham and R. A. Huggins in W. Van Gool (ed.), 'Fast Ion Transport in Solids', North Holland (1973)

181  M. S. Whittingham, J. Electrochem. Soc. 122 (1975) 713

182  R. S. Crandall and B. W. Faughnan, Appl. Phys. Lett. 26 (1975) 120

183  H. A. Schwarz, J. Phys. Chem. 73 (1969) 1928

184  Perry's Chemical Engineers Handbook, 4th Ed. (1963)

185  R. Touillaux, P. Salvador, C. Van Meersche and J. J. Fripiat, Israel J. Chem. 6 (1968) 337

186  K. N. Schmidt and W. L. Buck, Science 151 (1966) 70

187  V. K. Sikka and C. J. Rosa, Determination of $O_2$ Diffusion Coefficient in Tungsten Oxide, AD 715006 (1970)

188  M. Boudart, M. A. Vannice and J. E. Benson, Z. Phys. Chem. 64 (1969) 171

189  M. Green and D. Richman, Thin Solid Films 24 (1974) 545

190  I. Ota, J. Ohnishi and M. Yoshiyama, Proc. IEEE 61 (1973) 832

191  A. R. Kmetz, IEEE Trans. Electron Devices ED-20 (1973) 954

192  J. Kirton and R. W. Sarginson, Opto-Electronics 6 (1974) 349

193  S. K. Deb, Proc. 24th Electronic Components Conf., Washington D.C., May 1974, 11

194  A. M. Marks, Appl. Optics 8 (1969) 1397

195  A. Davis and I. M. Thomas, S.I.D. Int. Symp. Digest 6 (1975) 88

196  T. P. Brody, J. A. Asars and G. D. Dixon, IEEE Trans. Electron Devices ED-20 (1973) 955

197  M. Rindl, S. African J. Sci. 11 (1916) 362

198  J. C. Ghosh and J. Mukherjee, J. Indian Chem. Soc. 6 (1929) 231

199  P. Fantl and M. H. Nance, Austrian J. Sci. 9 (1946) 116

200  A. B. Biswas, J. Indian Chem. Soc. 24 (1947) 103

201  A. Bolliger and N. T. Hinks, Austrian J. Exper. Biol. & Med. Sci. 27 (1949) 569

202  L. Chalkley, J. Opt. Soc. Am. 44 (1954) 699

203  L. Chalkley, J. Phys. Chem. 56 (1952) 1084

## APPENDIX A

### Patents Related to Electrochromic Displays

A1  S. K. Deb and R. F. Shaw, Am. Cyanamid, US 3 521 941, Electrooptical device having variable optical density (1970)

A2  S. K. Deb, Am. Cyanamid, US 3 829 196, Variable light transmission device (1974). Also DAS 1 589 429 (1970)

A3  G. A. Castellion, Am. Cyanamid, US 3 578 843, Control of light reflected from a mirror (1971)

A4  G. A. Castellion and D. P. Spitzer, Am. Cyanamid, US 3 712 710, Solid state electrochromic mirror (1973)

A5  J. F. Dreyer, Polacoat Inc., US 3 106 743, Light modulation device employing a scotophoric light valve (1965)

A6  J. J. Robillard, US 3 355 290, Electrocatalytic photography (1967)

A7  J. J. Robillard, US 3 457 069, Electrocatalytic photography utilizing photodeactivated catalysis (1969)

A8  J. A. Morton and G. E. Smith, BTL, US 3 497 286, Variable reflectance display device (1970)

A9  J. A. McIntyre and R. O. Hansen, Dow Chemical Co., US 3 560 078, Color reversible light filter utilizing solid state electrochromic substance (1971)

A10  D. E. Wilcox, USAF, US 3 589 896, Electrooptical article employing electrochromic and photoconductive materials (1971)

A11  E. C. Letter, Bausch and Lomb, US 3 620 866, Method for making an electrochemical cell (1971)

A12  E. C. Letter, Bausch and Lomb, US 3 745 044, Optical device (1973)

A13  Z. J. Kiss, Optel Corp., US 3 840 286, Electrochromic device (1974)

A14  P. Rosenberg, Research Frontiers Inc., US 3 743 382, Method, material and apparatus for increasing and decreasing the transmission of radiation (1973)

## APPENDIX B

### Patents Related to ECC Displays (Electro-Redox Systems)

B1  G. D. Jones and R. E. Friedrich, Dow Chemical Co., US 3 283 656, Color reversible electrochemical light filter utilizing electrolytic solution (1966)

B2  F. Findl, Electro-Optical Systems Inc., US 3 373 091, Data storage device and method (1968)

B3  P. Manos, Dupont Co., US 3 451 741, Electrochromic devices (1969)

B4  E. Kissa et al., Dupont Co., US 3 453 038, Compartmented electrochromic device (1969)

B5  J. A. Hall and J. J. McCann, Polaroid Corp., US 3 692 388, Electrically responsive light filter (1972)

B6  H. G. Rogers, Polaroid Corp., US 3 873 185, Variable light filtering device (1975)

B7  I. F. Chang and B. Gilbert, Electrochromic liquid display, IBM Technical Disclosure Bulletin 17 (1974) 1050

B8    I. F. Chang and T. I. Sun, Electrochromic materials, IBM Technical Disclosure Bulletin 17 (1975) 3144

B9    I. F. Chang, Electrochromic display device operable in both passive and active mode, IBM Technical Disclosure Bulletin 17 (1975) 3148

B10   W. J. Horkans and L. T. Romankiw, Electrochemical display through use of a redox complex in solution, IBM Technical Disclosure Bulletin 17 (1975) 2517

## APPENDIX C

## Patents Related to ECC Displays (Electroplating Systems)

C1    F. H. Smith, British 328 017, Apparatus for use as anti-dazzle and fog-penetrating device for use in conjunction with motor car head lamps (1929)

C2    W. C. Flanagan and C. L. Miles, Eastman Kodak, US 3 153 113, Electroplating light valve (1964)

C3    S. Zaromb, Philco Corp., US 3 245 313, Light modulating means employing a self-erasing plating solution (1966)

C4    E. H. Land and H. G. Rogers, Polaroid Corp., US 3 443 855, Variable polarized light-filtering apparatus (1969)

C5    H. G. Rogers, Polaroid Corp., US 3 443 859, Variable light-filtering device (1969)

C6    A. Ambrosia and C. J. Sambucetti, Electrochemical display, IBM Technical Disclosure Bulletin 13 (1970) 669

C7    H. de Koster, General Time Corp., US 3 626 410, Moving indicator electrochemical displays (1971)

C8    A. H. Zechman, Alphanumeric electrochemical display, IBM Technical Disclosure Bulletin 14 (1971) 201

C9    C. J. Sambucetti, IBM, US 3 736 043, Electrochemical molecular display and writing (1973)

C10   E. H. Land, US 2 237 567 and US 2 328 219

## APPENDIX D

## Patents Related to ECC Displays (pH Variation Induced Color Change)

D1    G. C. Sziklai, RCA, US 2 632 045, Electrochemical color filter (1953)

D2    B. Rosenberg, Westinghouse Electric Corp., US 3 015 747, Fluorescent Screen (1962)

D3    J. R. Alburger, US 3 123 806, Composition of matter, process and apparatus for visually indicating and storing electrical data (1964)

D4    J. F. Donnelly and R. C. Cooper, Donnelly Mirrors Inc., US 3 280 701, Optically variable one-way mirror (1966)

D5    W. J. Levine, S. A. Manning and C. J. Sambucetti, IBM Technical Disclosure Bulletin 15 (1972) 367

## APPENDIX E

### Patents Related to ECC Displays (Electrochemical Reaction Coupled with Chemical Reaction)

E1   N. V. Philips Gloeilampenfabrieken, Dutch Pat. Appl. 7 009 521, Image reproducing device (1971)

E2   M. S. Simon, Polaroid Corp., US 3 641 034, Polymers of dipyridyl (1972)

E3   H. G. Rogers, Polaroid Corp., US 3 652 149, Variable light-filtering device with a redox compound which functions as its own electrolyte (1972)

E4   H. G. Rogers, Polaroid Corp., US 3 774 988, Variable light filtering device (1973)

E5   J. G. Kenworthy, Imperial Chemical Industries Ltd., US 3 712 709, Variable light transmission device

E6   I. F. Chang, Scannable refresh-type electrochromic display, IBM Technical Disclosure Bulletin 17 (1974) 3146

E7   I. F. Chang, Controllable persistence in electrochromic effect and its use in beam adressable displays, IBM Technical Disclosure Bulletin 17 (1975) 3151

E8   M. D. Shattuck, G. T. Sincerbox and H. W. Werlich, Photovoltaic switching of electrochromic materials, IBM Technical Disclosure Bulletin 17 (1974) 2778

E9   D. S. Bailey and M. D. Shea, Eastman Kodak, US 3 880 659, Triazolium salt photoreductive imaging (1975)

E10  C. J. Schoot, J. J. Ponjee, R. A. Van Doorn and P. T. Bolwijn, N. V. Philips Gloeilampenfabrieken, US 3 864 589, Matrix picture display device using liquid that is reversibly reducible and oxidizable by electric current (1975)

## APPENDIX F

### Patents Related to ECC Displays (WO$_3$ ECC Devices)

F1   P. Talmey, Radio Inventions Inc., US 2 319 765, Electrolytic recording (1943). Also US 2 281 013, Electrolytic recording paper (1942)

F2   L. C. Beegle, Am. Cyanamid, US 3 704 057, Electrochromic device having identical display and counter electrode materials (1972)

F3   M. D. Meyers and T. A. Augurt, Am. Cyanamid, US 3 708 220, High conductivity electrolyte gel materials (1973). See also German patent application based on US serial No. 41 155 (1970)

F4   M. D. Meyers and H. P. Landi, Process for preparation of an electrode structure containing WO$_3$, use of the structure in an electrochromic cell, Am. Cyanamid, German patent application based on US serial No. 105 882 (1971)

F5   M. D. Meyer, Am. Cyanamid, Electrochromism, German patent application based on US serial No. 41 154 (1970)

F6   D. J. Berets, Am. Cyanamid, Electrochromic devices, German patent application based on US serial No. 41 153 (1970)

F7   G. A. Castellion, Am. Cyanamid, US 3 807 832, Electrochromic mirror which rapidly changes reflectivity (1974)

F8    R. D. Giglia, Am. Cyanamid, US 3 810 252, Additives that increase the stability of electrochromic films in electrochromic devices (1974)

F9    R. D. Giglia and R. H. Clasen, Am. Cyanamid, US 3 827 784, Simple bonded graphite counter electrode for electrochromic devices (1974)

F10   E. Saurer, Ebauches S.A., US 3 836 229, Electrooptical display device (1974)

F11   D. J. Berets, G. A. Castellion and G. Haacke, Am. Cyanamid, US 3 839 857, Electrochromic information display (1974). Also in German application

F12   H. Witzke, S. E. Schualterly, Optel Corp., US 3 840 287, Symmetrical electrochromic cell (1974)

F13   S. E. Schualterly, Optel Corp., US 3 840 288, Electrochromic display having electrocatalyst (1974)

F14   D. J. Berets, Am. Cyanamid, US 3 843 232, Electrochromic light modulating devices having a palladium counter electrode (1974)

F15   D. J. Berets, Am. Cyanamid, US 3 879 108, Electrochromic devices (1975)

SHORT COMMUNICATION

## MECHANISM OF ELECTROCHROMISM IN $WO_3$

G. E. WEIBEL

Zenith Radio Corporation Research Laboratories, Chicago, Illinois, USA

We have recently offered evidence that the electrochromic effect in $WO_3$ is due to the formation of a new compound produced by the addition of a positive ion and an electron, rather than the extraction of a negative ion.[1] The compound formed electrochemically has the same empirical formula as an oxide bronze rather than that of an oxygen-deficient oxide. In the original work by Deb[2] on "dry" films of $WO_3$, and even in a recent publication by Chang and others,[3] the electrochromic effect is assumed to be due to extraction of oxygen; Faughnan, Crandall and Lampert,[4] however, have come out with an interpretation supporting the one presented here.

Evidence supporting the thesis of an addition compound is developed as follows.

1) The first three figures summarize combined electrochemical-optical experiments. Figure 1 shows a battery composed of two half-cells in which the positive ion $A^+$ is $H^+$, the metal M is Zn and n = 2. During discharge the $WO_3$ electrode acts as reversible cathode and turns blue, the optical absorption being proportional to the number of coulombs discharged. Figure 2 is a plot of the normalized absorption spectra for low and high coloration of the film. At high coloration there is a shift of the peak absorption to shorter wavelength and considerable line broadening.

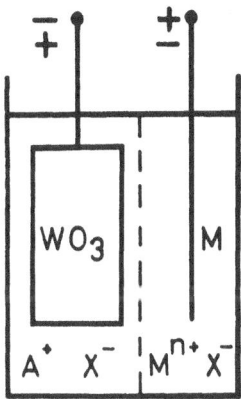

Fig. 1. Schematic cell structure.

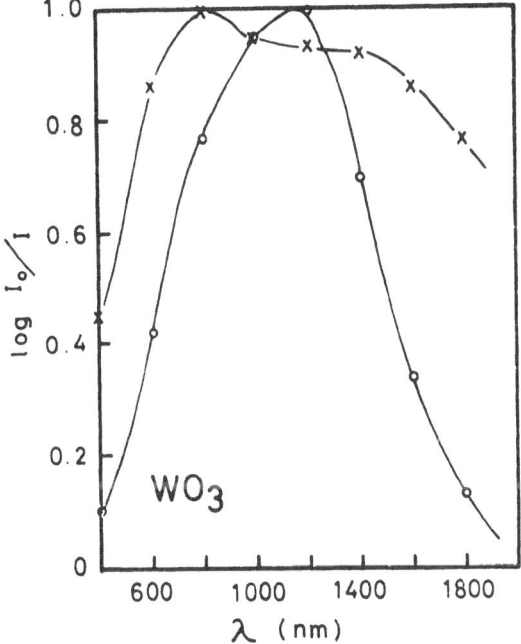

Fig. 2. Normalized absorption spectra of low and high coloration WO$_3$ films.

The open-circuit reversible potential decreases with the increasing depth of discharge. The shape of the curve shown in Fig. 3 is indicative of a single-phase reaction at the $WO_3$ electrode. This conclusion is also supported by x-ray diffraction studies that indicate that there is no change in crystallography. In general the emf developed by such a battery varies with the concentration of $H^+$ and $Zn^{2+}$ ions, but in this experiment the concentration of the latter is kept constant. The shape and slope of the emf vs. [$H^+$] curve are diagnostic of an electron reaction involving an equivalent number of $H^+$ ions. Specifically it can be shown thermodynamically that the data are consistent only with the one-electron reaction

$$WO_3 + xH^+ + xe^- \xrightleftharpoons[\text{bleach}]{\text{color}} H_x^+(WO_3)e_x^-$$

but not with a reaction involving oxygen extraction. In our work $H^+$ is frequently replaced by $D^+$, $Li^+$ or $Na^+$, and the reactions taking place are described by simply substituting in the above expression for $H^+$ the ion actually involved.

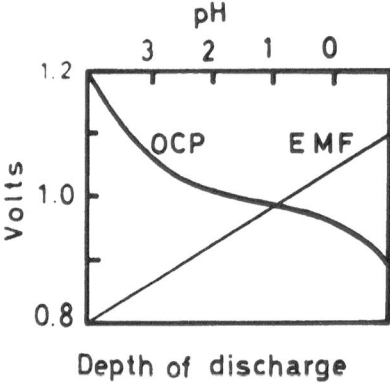

Fig. 3.  Open-circuit reversible potential and battery emf vs. depth of discharge indicated by pH.

2) Direct monitoring of the reversible changes that occur in the WO$_3$ film during coloration and bleaching has been possible through the study of valence band and core-orbital photoelectron spectra. ESCA and Auger spectroscopy data have established the following results.

In the valence band region a new electron distribution is detected which can be turned on and off electrochemically. The shape of the distribution is characteristic of the species of positive ions injected. This establishes that both electrons and positive ions participate in the process. The density of states near the Fermi level is correlated with the presence of electron centers created with the participation of the injected ions. The incorporation and extraction of lithium and sodium ions into the film has also been directly observed through the appearance and disappearance of lines at 55 eV for Li$^+$ and 497 and 1070 eV for Na$^+$, giving proof of the existence of these elements in their fully ionized state. The observed coexistence of high densities of stored electrons and positive ions (up to the order of $10^{22}$ cm$^{-3}$) may be explained by exchange forces between the injected positive ion and O$^{2-}$ that give rise to the formation of chemical bonds.

In the absence of ions, the film does not color. Using electrolytes containing H$^+$, D$^+$, Li$^+$ and Na$^+$, the same optical absorption band is produced with the same coulometric efficiency.

Figure 4 shows the valence band region of ESCA spectra of an uncolored amorphous WO$_3$ film and three films colored by the injection of H$^+$, Li$^+$ and Na$^+$, respectively. All scans show the slope of the 2p oxygen line on the left side of the graph. The "fine structure" is an artifact due to integration noise of the instrument; however, the double peak evident in the Li$^+$ spectrum is believed to be real and tentatively explained by the spurious presence of H$^+$ in the high-purity, non-protic (organic) solvent used. In contrast to this the core orbital ESCA spectra, in particular the 4f line of the W$^{6+}$ and the 1s line of the O$^{2-}$, do not show any significant change in the ratio of the W to O. If the electrochromism of WO$_3$ were based on the formation of a substoichiometric oxide a drastic change of composition would have to occur to account for the high density of charge injected.

3) The empirical formula of A$_x$WO$_3$ for the well-known tungsten bronzes and the electrochemical reaction product A$_x^+$(WO$_3$)e$_x^-$ postulated for the electrochromic mechanism (A=H, D, Li or Na) are analogous. For this reason a direct comparison was made between the ESCA spectra of a chemically prepared bulk sample of a hydrogen bronze, H$_{0.5}$WO$_3$ (dotted curve in Fig. 4) and that of a WO$_3$ film colored by electrochromic action in an aqueous electrolyte (solid curve marked H$^+$ in Fig. 4). It can be seen that the spectrum of the bulk bronze sample follows closely the noise-smoothed curve of the electrochromic film.

In conclusion the above results are considered evidence for the existence of a reversible two-carrier electrochromic mechanism involving the injection and extraction of positive ions and electrons. The colored product is closely related to a tungstic oxide bronze rather than a substoichiometric oxide. Even in the presence of a high density of injected electrons the positive ions

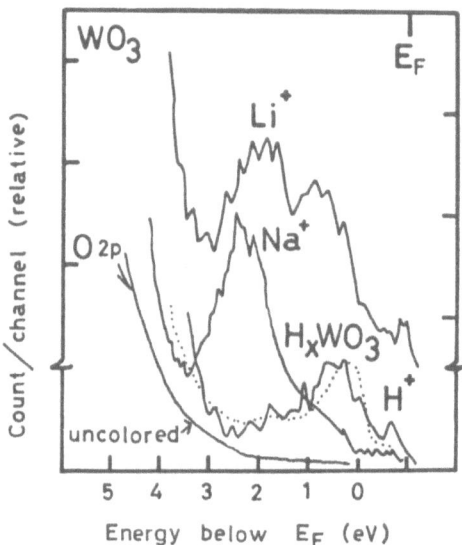

Fig. 4. ESCA spectra of amorphous $WO_3$ films: uncolored, colored by injection of $H^+$, $Li^+$ and $Na^+$.

remain fully ionized, and no drastic changes appear to happen to the tungstic oxide matrix.

The findings reported are based on work at Zenith Radio Corporation by Drs. H. N. Hersh, W. E. Kramer and J. H. McGee.

## REFERENCES

1   W. E. Kramer, J. H. McGee and H. N Hersh, S.I.D. Int. Symp. Digest 6 (1975) 45
2   S. K. Deb, Appl. Optics Suppl. 3 (1959) 192 and Phil. Mag. 27 (1973) 801
3   I. F. Chang, B. L. Gilbert and T. I. Sun, J. Electrochem. Soc. 122 (1975) 955
4   B. W. Faughnan, R. S. Crandall and M. A. Lampert, Appl. Phys. Lett. 27 (1975) 275

# ELECTROCHROMIC DISPLAY DEVICES

J. BRUININK

Philips Research Laboratories, Eindhoven, The Netherlands

## SUMMARY

Several electrochromic displays have been proposed in the last ten years. In general they differ from other non-emissive display devices in that they are based on electrochemical reactions. Most of their principles are briefly mentioned. More attention is devoted to electrochromic display devices dealing with tungsten oxides and viologens, with special emphasis on the latter. In order to obtain good display performance it is necessary to look for an optimal combination of electrochemical conditions and normal display requirements. As a result, special demands for display construction and practical examples are discussed in detail. Correct display performance can be obtained by the use of proper electronic addressing systems. A review of the possibilities is given. Theoretical as well as practical performance characteristics are discussed critically and present results are evaluated.

## 1. INTRODUCTION

Electrochromic (EC) displays are based on an electrochemical reaction which occurs within a specific voltage range. The visual changes due to this electrochemical reaction form the basis for such EC displays. This idea is not very new. In 1929 a patent was granted to Smith for his invention concerning an Anti Dazzle and Fog-penetrating Device for use in Motor Car Headlamps.[2] Later several other systems were applied with little success because of high energy consumption, poor performance and short lifetime. Further developments since 1970[3,7] have resulted in systems based on the reduction of transition metal oxides and viologens. In the last five years a considerable effort was made to produce an acceptable electrochromic device based on these

201

reactions. However, for technological and fundamental reasons, only a few prototypes have been developed.

In this paper we try, on the basis of presently available information, to evaluate the crucial factors involved in making a working display device. In the first section, general electrochemical requirements for good display performance are discussed. This is followed by a more specific discussion of the electrochemical and optical behavior of both convenient display systems. In following sections display construction and driving methods are discussed. Finally performance of electrochromic devices is reviewed.

## 2. CHEMICAL ASPECTS OF ELECTROCHROMIC DISPLAY PRINCIPLES

### 2.1. General Electrochemical Requirements

Only those electrochemical processes which produce a clear and noticeable optical effect can be applied to information display purposes. A large number of electrochemical reactions cause such optical effects, so this requirement gives no serious limitations. Since such an electrochemical reaction is supposed to operate in a device more than a million times at the same electrode without refreshing of the small amount of electrolyte, it is clear that we are faced with very special problems.

Two main requirements have to be fulfilled for an electrochemical reaction to be useful in display applications. First the current efficiency in the potential region of interest has to be 100 % to prevent complications. Apart from trivial leakage paths in electrical circuitry, this requirement means that no side reactions can be permitted.[1] In EC displays side reactions can be expected from solvents, electrode materials and other substances like oxygen, supporting electrolyte, materials used for cell construction, reaction products from the counter-electrode, additives like thickeners, opacifiers etc. One has to verify in each case whether the chemical component is involved in electrochemical reactions in the potential region of interest. Special attention is needed for possible reactions between reaction products from the counter-electrode and the working electrode. Therefore great care should be taken with the choice of reaction at the counter-electrode.

The second requirement is that the electrochromic process be reversible without severe limitations in both directions. This condition is very essential. The reversibility of reactions can be studied using several electrochemical methods. An example of an electrochromic process in which diffusion limitation occurs, making the process less suitable, is a pH-indicator as the electrochromic material.[22,23] The main reasons for this limitation are convection and the onset of diffusion of the colored electrochromic species into the solution directly after formation. This causes a smearing of the image. This phenomenon is observed in all electrochromic reactions in which the colored species is soluble in the solvent, e.g. $Fe^{2+}/Fe^{3+}$ in the presence of thiocyanate anions.[24]

To overcome these problems several artificial barriers involving thickeners and semi-permeable membranes have been proposed in order to reduce both diffusion and convection. However in principle it is very difficult to combine a desired easy transport of uncolored species and fast display action together with blocking conditions for the soluble colored species. Another solution of the diffusion problem was found in the addition of an oxidizing chemical such as $H_2O_2$. For example:

$$PTA + ne \longrightarrow PTA' \text{ (blue)}$$
$$PTA' + H_2O_2 + 2H^+ \longrightarrow PTA + 2H_2O$$
$$2H_2O \longrightarrow H_2O_2 + 2H^+ + 2e$$

The reduced colored polytungstate anions (PTA') are oxidized to the uncolored PTA by the hydrogen peroxide shortly after formation. In this case the current efficiency is low, so that it is only convenient to use these oxidizing chemicals for special applications.[5]

### 2.2. Practical Chemical Processes

Systems based on redox reactions of a solid film yielding a color change, eg. $mWO_3 + 2nH^+ + 2ne \rightarrow W_mO_{3m-n}$ (blue) $+ nH_2O$, are more practical. Another quite different system which also operates without diffusion problems after writing is based on a redox reaction coupled with a fast chemical reaction which produces an intensely colored insoluble film at the electrode. This film can be oxidized electrochemically even after some time. Several members of the large group of 4,4'-dipyridinium compounds (viologens) exhibit an electrochemical behavior of this type. In general viologens can be represented as

$$R -^+N\bigcirc\!\!-\!\!\bigcirc N^+ - R \quad , \quad 2X^-$$
$$\underbrace{\hphantom{R -^+N\bigcirc\!\!-\!\!\bigcirc N^+ - R}}_{A^{2+}}$$

and they exhibit the following two reduction steps:

$$A^{2+} + e \rightleftarrows \overset{\bullet}{A}{}^+ ,$$
$$\overset{\bullet}{A}{}^+ + e \rightleftarrows A\downarrow .$$

The first step is a reversible, one-electron reduction giving a blue-colored stable radical ion. Furthermore it is very fast, being diffusion limited, and therefore is very well suited for display applications.

Schoot et al.[7] have succeeded in preparing 4,4'-dipyridinium compounds which form an insoluble product directly after reduction. The electrochemical reaction proceeds in the following way:[8]

$$A^{2+} + e \rightleftarrows \overset{\bullet}{A}{}^+ \text{ (blue) },$$
$$\overset{\bullet}{A}{}^+ + X^- \rightleftarrows AX\downarrow \text{ (solid) .}$$

The solid deposit has a purple color. Its solubility product depends on the nature of R as well as on the anion $X^-$. The solubility product for the heptylviologen bromide radical film in water is $3.9 \times 10^{-7}$. Heptylviologen bromide ($R = -C_7H_{15}$; $X^- = Br^-$) fulfills the two important requirements of a small solubility product of the radical layer as well as sufficient solubility of the uncolored form needed for a fast display action. The reduction potentials of the first and second steps depend strongly on experimental conditions being about -0.45 Volts and -0.9 Volts (vs. Ag/AgCl) respectively for $10^{-3}$ M heptylviologen bromide in an aqueous solution of 0.3 M potassium bromide.

The separation of the two reduction steps is desirable since it has been observed that the occurrence of the second reduction step can cause poor display performance, in spite of the back reaction $A + A^{2+} \rightarrow 2\overset{\bullet}{A}^+$. An example of a small separation is given by p-cyanophenyl viologen chloride that has a difference of 0.2 Volt between the first step (green product) and the second reduction step.[9]

## 3. ELECTROCHEMICAL BEHAVIOR

Three stages can be distinguished in display operation; writing, memory and erasing. The first is production of color by electrochemical reduction. Memory is the conservation of the produced colored layer without energy consumption. The last step, erasing, means the removal of color by electrochemical oxidation.

Writing as well as erasing can be obtained in two ways. The first way is to apply a voltage to the working electrode (with respect to a reference electrode) such that the desired redox reaction proceeds with sufficient speed. The voltage will be constant in this case, while the current will vary with time. The second way is to maintain constant current between working electrode and counter-electrode during writing or erasing. In this case the voltage between working and counter-electrode will be time dependent and at such a level that the redox reaction can occur. If, for instance, diffusion limitation is reached after some time, the voltage of the working electrode with respect to the reference electrode increases until other electrochemical processes can take over in order to maintain the constant current. Electrochemical aspects of processes mentioned above will be discussed in the next sections.

### 3.1. Constant Potential Writing

Current-time characteristics of redox systems due to a voltage step have been given by Delahay.[1] Their applicability to the writing process of several electrochromic systems is treated by Chang et al.[10] If we confine ourselves to the formation of insoluble substances such as the viologen radical film, certain conditions are assumed; namely diffusion is the only mass transport process, and migration and convection may be neglected. These conditions are fulfilled under normal circumstances. Furthermore it is assumed that the diffusion process is semi-infinite and occurs at a planar electrode and finally that the reduction reaction is reversible.

The current-time characteristics after a monolayer has been deposited are given by Delahay:[1]

$$I = nFA\left(\frac{D_O}{\pi t}\right)^{1/2} [C^O - f(E)] . \tag{1}$$

$D_O$   diffusion coefficient of the reducible species O
$C^O$   bulk concentration of the reducible species O
A   surface area
F   Faraday's constant
n   number of electrons involved
t   time
E   potential.

f(E) is given by Nernst's law for the concentration $C_O(0,t)$ of the reducible species at the electrode (x = 0) at time t:

$$f(E) \equiv C_O(0,t) = \exp\left[\frac{nF}{RT}(E - E^O)\right], \tag{2}$$

where $E^O$ is the reduction potential. It is assumed that $C_R(0,t) = 1$ for the concentration of the reduced species at the electrode after time t, and further that the activity coefficients of O and R are equal. If the current is diffusion-limited (e.g. by applying a sufficiently large voltage), optimal speed conditions for the electrochromic system have been reached. The current-time relationship becomes in that case

$$I = n F A C^O \sqrt{D_O/\pi t} \tag{3}$$

The electrochemically induced optical density OD is proportional to the electric charge:

$$OD(t)|_\lambda = 0.43 \frac{\epsilon_\lambda Q}{AnF} = \frac{0.43\epsilon_\lambda}{AnF} \int_o^t I \, dt \tag{4}$$

Q   amount of charge used to produce color
$\epsilon$   extinction coefficient.

The maximum attainable OD for the colored layer under diffusional limitation is

$$OD(t)|_\lambda = 0.86 \, \epsilon_\lambda \, C^O \sqrt{D_O t/\pi} \, . \qquad (5)$$

This equation is also valid for the $WO_3$-$H_2SO_4$ system.[15] The factor $0.43 \, \epsilon_\lambda/nF$ is sometimes referred to as electrochromic efficiency parameter $\alpha_\lambda$.

Spectra of a heptylviologen bromide radical film and a reduced $WO_3$ layer are presented in Fig. 1. The absorption spectrum of the viologen radical

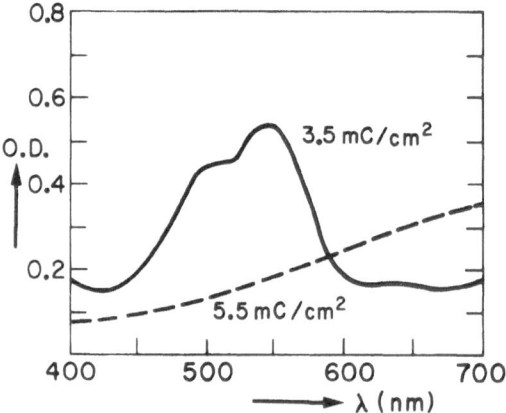

Fig. 1. Absorption spectra of heptylviologen bromide radical film on $SnO_2^{16}$ (—) and a reduced $WO_3$ layer[10] (— — —). OD is optical density.

layer shows two relatively broad peaks around 500 nm.[8,16] The absorption of a colored $WO_3$ film occurs at about 1000 nm with a very broad band.[15] $\alpha_\lambda$ amounts to, at most, 0.15 at 550 nm and 0.082 at 1000 nm measured in absorption for the viologen and $WO_3$-$H_2SO_4$ systems respectively.

We found that in practical devices the influence of the large series resistance due to the resistivity of electrodes and solution should be taken into account. Therefore the current-time (and OD-t) behavior is often found to be different from (3) and (5). The I-t response calculated from (1) for an applied potential $V = E - E^O$ with introduction of an ohmic drop in f(E) is shown in Fig. 2. It can be seen that the ideal behavior of (3) is approached with a low series resistance, e.g. 1 $\Omega$. Introduction of larger values of R yields a deviation from this behavior. For practical values of R (10 - 50 $\Omega$) the current is almost constant and determined by the series resitance $I = V/R$. Voltage-time characteristics as presented in Fig. 8 can also be explained with the aid of this theory.[18]

Chang and Howard[15] reported a current-time response of $I = At^{-0.5} + B$ with $A = 0.48$ and $B = 27.3$ for the heptylviologen-KBr system with 5 V

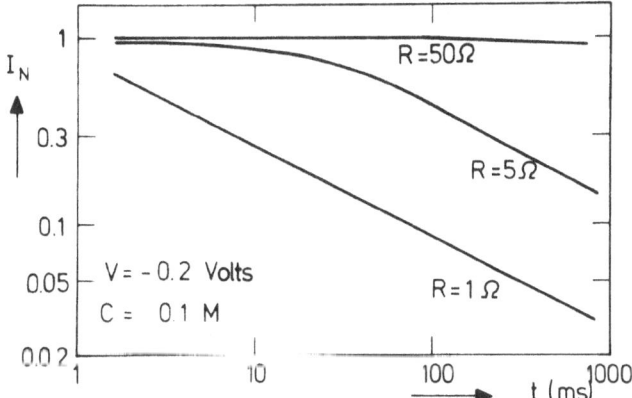

Fig. 2. Calculated normalized $I_N$ vs. t response to voltage step $V = E-E^O$ with series resistance R. $I_N = I/(V/R)$.

applied voltage. This behavior was explained by a disturbing convection process which gave rise to the extra contribution B.

Finally it can be concluded that constant potential writing is a method to prevent the occurence of redox reactions at higher voltages. The predicted time dependence for diffusion limiting is not always obeyed due to an ohmic drop. With the latter, the current is resistance limited and constant with time.

### 3.2. Constant Current Writing

On applying a constant current, diffusion limitation for writing can be reached, after which other reactions will take part in the reduction process at the electrode in order to maintain constant current. For a fixed concentration $C^O$ of a reducible species and an imposed current density i, the maximum writing time $\tau$ when other reduction steps at higher potentials can be neglected is given by Sand's equation[17,21]

$$\tau^{1/2} = \frac{\pi^{1/2} \, n \, F \, D_O^{1/2} \, C^O}{2 \, i} \, . \tag{6}$$

This means that with $\tau = 100$ ms, n = 1, $D_O = 10^{-5}$ cm$^2$/s and i = 0.02 A/cm$^2$, $C^O$ has to be more than 0.02 mol/l, as was verified experimentally. In this case the time dependence of the optical density OD is given by

$$OD(t)|_\lambda = 0.43 \frac{\epsilon_\lambda Q}{AnF} = 0.43 \frac{\epsilon_\lambda \, it}{nF} \, . \tag{7}$$

The maximum attainable OD with constant current writing is given by setting $t = \tau$ in (7).

### 3.3. Memory

The colored product formed during writing is preserved without energy consumption during memory. From the redox potentials it can be seen however that the viologen radical layer can in principle be oxidized by oxygen. It has been observed that this reaction is fast enough to cause a large decrease in current efficiency together with a bleaching of the layer during memory. Although not reported as such, it can be concluded from cyclic voltammetric measurements[10] that $WO_3$-$H_2SO_4$ systems are also sensitive to oxidation by oxygen.[11] It is therefore a stringent requirement to exclude oxygen from the display envelope.

### 3.4. Erasing

During erasing the colored layer is oxidized electrochemically. Erasing is the most critical process in electrochromic display performance, although generally in the literature extra attention is not paid to this step. This can be illustrated by the case of the $WO_3$-$H_2SO_4$ electrochromic device driven with a constant current pulse.[11] During erasing the constant current conditions cannot be fulfilled due to some limiting process; therefore voltage is increased up to source limitation. Often the erasing time is longer than the writing time. This also holds for the viologen-KBr system as can be seen from Fig. 3. On erasing there is a small but noticeable tailing off of the reflectance.

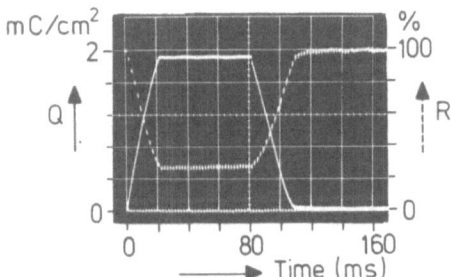

Fig. 3. Charge Q (solid line) and reflectance R at 540 nm (dashed line) as a function of time during one write-memory-erase cycle. Applied voltage U = 0.4 V; Au electrodes, potentiostatic drive.[7]

The underlying problem can be better illustrated with the voltammetric measurements in Fig. 4. Curve A shows the ideal I-V behavior under optimal conditions. In this case the dissolution behavior of very thin metal layers is approached.[20] Only systems with this type of voltammetric behavior will

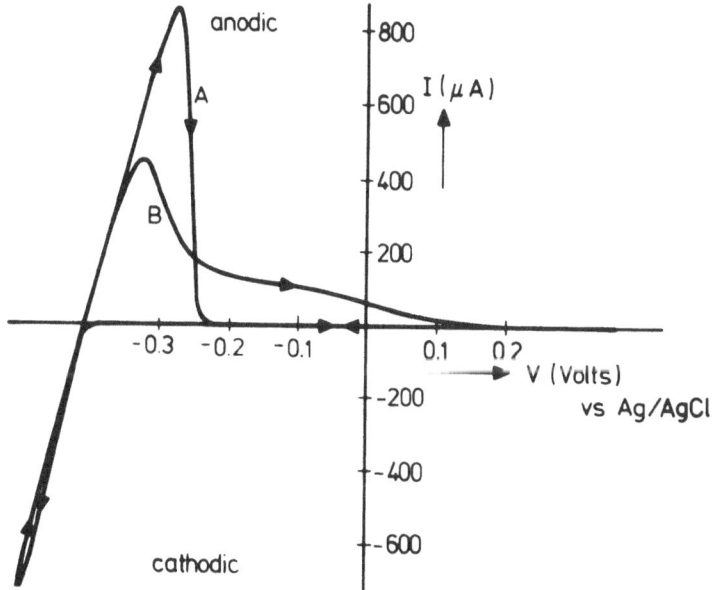

Fig. 4. Single sweep voltammetric measurements with 0.1 M heptylviologenbromide, 0.3 M KBr at $25^{\circ}$ C. $SnO_2$ electrodes. Scan rate 0.1 Volt/s. Curve A: ideal performance. Curve B: poor performance.

exhibit sufficient speed of erasing without auxiliary redox couples. Curve B is a voltammogram obtained with a viologen radical layer in which alterations have occured, as can be seen from the appearance of an extra erasing peak at a more cathodic potential. This is thought to be due to aging of the radical layer. One of the reasons for the occurence of these alterations can be the use of inferior electrode materials. Also the use of a longer memory time is less favorable. Displays with type B voltammograms erase very slowly and remains of the colored layer are formed after some time which are difficult to erase. In the case of viologens it can be advantageous to use an auxiliary redox couple that causes an indirect chemical erasure. An example of such a redox couple is $Fe^{2+}/Fe^{3+}$. The oxidized form $Fe^{3+}$, which is formed by oxidation of $Fe^{2+}$ during erasing, will oxidize the radical layer by means of the reaction $Fe^{3+} + ABr \rightarrow Fe^{2+} + A^{2+} + Br^{-}$

### 3.5. Counter-Electrode Processes

So far, the electrochemical process at the counter-electrode has not been taken into consideration. Although not much attention is paid to the counter-electrode reaction in the literature (except for Ref. 12), a good understanding of it is essential for optimal display performance. If we confine ourselves to the case of viologens, there are several possibilities for the counter-electrode reaction, depending on circumstances. It has been observed that obvious candidates such as $Ag+Cl^- \rightleftharpoons AgCl+e$ can give problems due to the deposition of small silver spots on the working electrode after some time.

In present display devices with heptylviologen bromide and potassium bromide, the combination of the viologen redox reaction and the bromine redox reaction is used. In the first writing cycle viologen radical is deposited at the working electrode and bromine (as an insoluble bromine-viologen complex) is formed at the counter-electrode. The total cell voltage $\Delta E$ needed for the first writing cycle is equal to the difference between working electrode and counter-electrode potential, $E_w - E_c$. In order to obtain the reduction of viologen $E_w < -0.5$ V (Ag/AgCl); in the same way, $E_c > 1.5$ V (Ag/AgCl) to oxidize bromide. Therefore the total cell voltage $|\Delta E|$ has to be larger than 2 V for the first writing cycle.

It has been observed that the bromide/bromine redox couple is irreversible at tin dioxide electrodes which are normally used in displays. Therefore reduction of bromine plays a minor role at the counter-electrode during erasing and the viologen reduction becomes the dominating process. The oxidation of the viologen radical layer is the electrode process at the working electrode in that case. In this situation, when the same redox processes occur at working and counter-electrode, $|\Delta E| = |E_w - E_c| > 0$. Therefore, after a few cycles when equilibrium is attained, only small voltages (a few tenths of a volt) are needed for erasing and writing under ideal conditions.

In the case of the $WO_3$-type display, no details about the anodic reaction for the first operating cycle are available. It has to be assumed that for this first cycle either the electrolyte or the counter-electrode has to be decomposed. Under equilibrium conditions the $WO_3$-$H^+$ redox reaction is used for the cathodic as well as for the anodic reaction. It appears however from voltage-time characteristics during constant current operation that other electrochemical processes (at higher voltages) are also involved in the erasing process.[11]

### 3.6. Cell Solution

There are several possibilities for the composition of the cell filling. For the $H_2SO_4$-$WO_3$ system the main chemicals are sulfuric acid with thickeners such as glycerine, sometimes mixed with additional $WO_3$ to increase life. Corresponding fillings for the viologen type display are aqueous solutions of viologen salts with supporting electrolyte, e.g. 0.1 M heptylviologen bromide with 0.3 M potassium bromide. The first system ($WO_3$-$H_2SO_4$) is reported to be filled in air, although good sealing is recommended.[11] For the second system it seems more appropriate to start immediately with oxygen-free conditions.

## 4. DISPLAY CONSTRUCTION

### 4.1. Envelope

In display technology it is necessary to construct an envelope as a container for the cell solution that is compatible with the electrochemical display processes and with the solution. Many display constructions have been described in the literature. Most of them are based on a sandwich cell as represented in Fig. 5. The envelope consists of a front plate and a back plate separated by

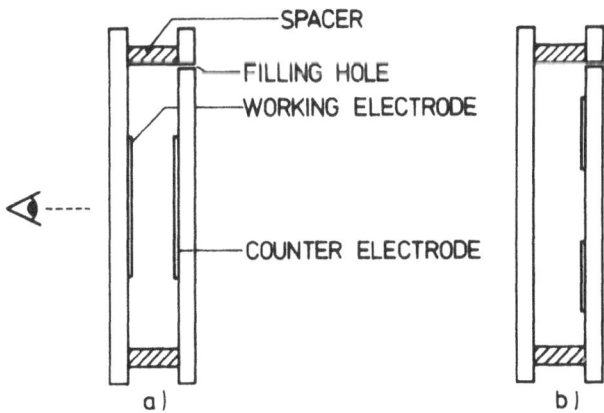

Fig. 5. Cross-section of an electrochromic display with a) working electrode opposite to counter-electrode, b) co-planar positioning of working electrode and counter-electrode.

means of a spacer with a thickness of 0.1-10 mm. At the moment, the dimensions of the electrodes vary between 5-10 mm. A list of some common materials for ECD construction is given in Table 1.

TABLE 1

Survey of Some Materials for Envelope Construction

| Part | Material |
| --- | --- |
| Front plate | Glass |
| Back plate | Glass, stainless steel |
| Spacer | Glass, teflon |
| Working electrode | Tin dioxide, platinum, gold |
| Counter electrode | Tin dioxide, platinum, gold, carbon |
| Filling hole seal | Epoxy resin |
| Spacer seal | Glass frit, epoxy, polymers |

The envelope has to be of such a construction that it can withstand considerable temperature changes, e.g. -20° C to +70° C. Expansion and contraction have to be absorbed by means of an appropriate cell technology. To insure long lifetime, leakage of oxygen has to be prevented as much as possible since this can induce undesired side reactions and their reaction products. The display is filled through one or two small holes. Hermetic sealing of these holes is also essential.

## 4.2. Electrodes

The choice of electrodes is rather limited because of the requirement of compatibility with the electrochemical systems used. Materials such as indium oxide cannot be used due to the corrosive nature of both $WO_3$-$H_2SO_4$ and viologens. More suitable is tin dioxide, a very stable electrode material. It is transparent in the visible region and can be deposited by pyrolitic decomposition of tin tetrachloride solutions. Antimony trichloride is added as a dopant to increase conductivity.[13] It is commercially available as tin dioxide coated glassplates. The thickness of tin dioxide coatings varies generally between 0.6 and 1 $\mu$m. Since it is necessary to combine optical transparency with a uniform low resistivity, homemade tin dioxide ($R_\square \cong 8\,\Omega$) is sometimes preferred. The latter is important since large resistance deviations cause IR-drops over the electrode, resulting in a non-uniform coloration.

The electrochemical stability of tin dioxide in aqueous solutions is limited on the cathodic side by the evolution of hydrogen, accompanied by the reduction of tin dioxide to metallic tin. This can be avoided either by use of a suitable voltage or pH stabilization. A crucial point with the $WO_3$-$H_2SO_4$ system is the fact that the reduction of protons, which is thought to play a major role in the electrode process, can also cause a chemical attack on the $SnO_2$ supporting layer, thus decreasing lifetime.

In the set-up shown in Fig. 5a, the use of tin dioxide as transparent working electrode is necessary. With a coplanar construction (Fig. 5b) the working electrode can in principle also be made of gold or platinum. In both cases, the counter-electrode can be either $SnO_2$ or other materials, e.g. platinum, gold or carbon. The combination viologen-platinum is less favorable because of the low overvoltage for hydrogen revolution. Furthermore a Pt working electrode tends to be less suitable when good memory capability is required. It is therefore not widely used as working electrode.

A practical example of electrodes for the $WO_3$-$H_2SO_4$ system is tin dioxide coated with a $WO_3$-film as working electrode and a carbon-$WO_3$ mixture on stainless steel support as counter-electrode.[11] With the viologen system tin dioxide may be used for both working electrode and counter-electrode.

Normally a reflecting medium is placed between working electrode and counter-electrode. This is done for several reasons:

a) to enhance the contrast ratio
b) to hide disturbing coloration due to the electrochemical reactions at the counter-electrode
c) to prevent unwanted reflections from the interior of the display.

Actually this can be achieved by use of an opacifier such as $TiO_2$ together with a thickener like polyvinylalcohol.

To prepare seven-segment numeric displays, tin dioxide films on glass are etched yielding a pattern as represented in Fig. 6. It may be noted that one

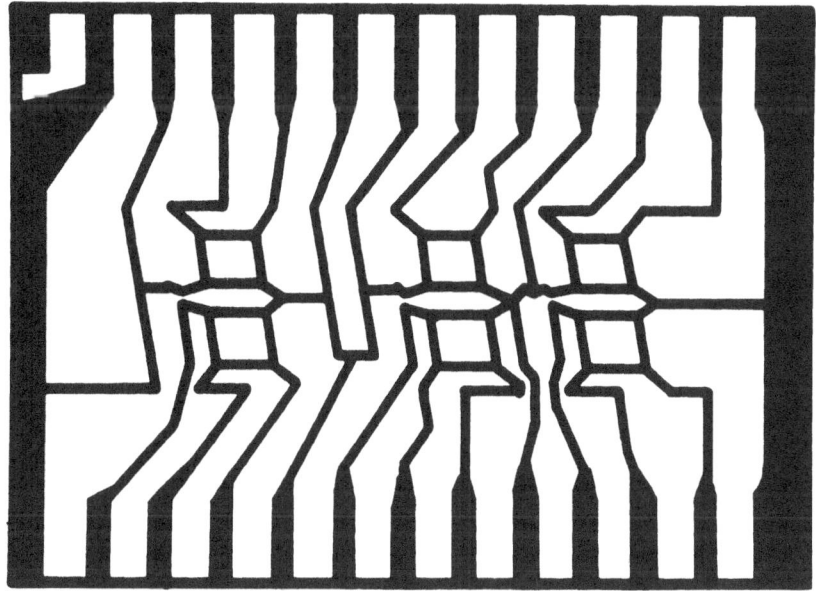

Fig. 6  Example of a 3 1/2 digit working electrode.

of the leads has been reserved for connection to the counter-electrode. The positioning of the electrodes is preferentially opposite to each other, again in order to avoid local high current densities. For the $WO_3$-$H_2SO_4$ system, formation of the active layer is obtained by vacuum deposition of a thin layer of tungsten oxide through a mask on selected parts of the etched tin dioxide. Preparation is completed by depositing an insulating mask of e.g. $SiO_2$, $CaF_2$ or $MgF_2$ on the leads. Since the edges of the segments are subjected to high currents compared with the more non-profiled parts, they are subjected to more electrochemical attack. To decrease this attack some overlap of the isolation mask can be used.[14]

## 5. DRIVING PRINCIPLES

The major task of the driving system is to transform information signals into electrical signals which can cause defined electrochemical reactions. From electrochemistry it is known that there are two main stimuli to cause electrochemical reactions at an electrode, namely a potential step or a current step.

The electrochemical responses of redox systems at the working electrode due to these stimuli and their complications have been discussed already in Section 3.1. In the next sections driving principles for electrochromic devices will be discussed.

## 5.1. Constant Potential Driving[4]

With this method, the formation of color is obtained by applying a certain potential $V_{WR}$ to the working electrode with respect to the reference electrode. This can best be done by means of a three-terminal potentiostatic driving circuit. The method is therefore often called potentiostatic driving. A very simple example of a potentiostatic driving circuit is shown in Fig. 7a.

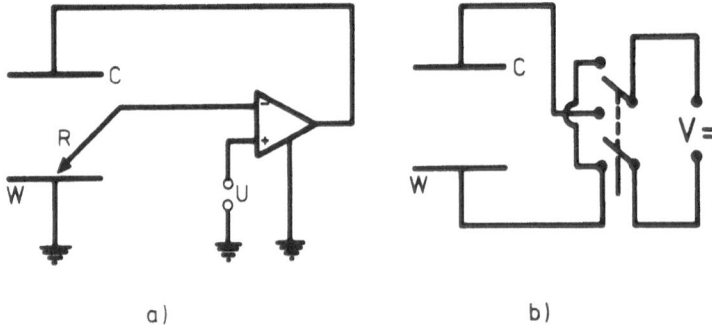

a)                  b)

Fig. 7. a) Potentiostatic driving circuit, b) Two terminal voltage driving. C: counter-electrode, W: Working electrode, R: reference electrode.

The operational amplifier of high gain supplies a current between the counter-electrode C and working electrode W such that the latter is maintained at a constant potential $V_{WR}$ with respect to the reference electrode R during the writing time. Since the operational amplifier has a very high input resistance, polarization of the reference electrode is kept to a minimum. In an ideal situation the distance between the reference electrode and working electrode is very small in order to minimize the IR drop. In a practical set-up however this IR drop is no longer negligible due to a large contribution from the resistance of the $SnO_2$ coating. This can cause a non-equipotential surface resulting in a non-uniform colored layer and/or corrosion phenomena at places with higher current densities. Therefore the resistivity of the $SnO_2$ must be kept as low as possible. The most important advantage of potentiostatic driving is that unwanted side reactions are avoided by proper selection of the working electrode potential. For experimental purposes normal reference electrodes such as Ag/AgCl with a salt bridge can be used. For display applications the reference electrode consists of a small Pt or $SnO_2$ electrode preferably close to the working electrode. The reference electrode couple depends on the application, e.g. for the viologen system the viologen reaction itself can be used.

Display operation proceeds as follows. During the writing period a pulse U is applied to the potentiostatic circuit (Fig. 7a; for a detailed description see Ref. 17, p. 420). This causes a current to flow through the display. The above mentioned reduction takes place at the working electrode producing a colored layer. At the counter-electrode another process will have to take place, e.g. bromide oxidation. The total cell voltage ($V_{CW}$) will be more than U, e.g. 1-2 V. During the memory period the working electrode is disconnected mechanically or electronically from the driving circuit. If no disturbing chemical or physical processes take place the colored layer can exist indefinitely without power consumption. Care should be taken that no spurious leakage paths exist between working electrode and counter-electrode since even a resistance as high as 1 MΩ can cause a current of 1 $\mu$A due to the presence of electrochemical potentials of both cathodic and anodic reaction products. For erasing, U is set to zero and the existing potential $V_{RW}$ due to the presence of a deposit, acts as driving force during erasing. If the working electrode process is reversible, it runs in the opposite direction until erasing is obtained. As mentioned before the availability of a counter-electrode process that is compatible with the working electrode process is essential. Application of potentiostatic driving in a device is not practical due to design and electronic driving considerations.[4]

Two terminal driving (Fig. 7b) is a simplified version of the three terminal driving. The reference electrode as sensor is no longer present and the external driving voltage is applied between working electrode and counter-electrode. The influence of electrode processes at the counter-electrode is important in this case, but not yet completely understood. Nevertheless, $2 \times 10^6$ write-erase cycles have been obtained with viologen systems.

One of the problems that one meets with constant potential driving is the difference in surface area of the various segments in a device. This causes a variation of thickness of the colored layer. This problem can be solved by means of constant current writing.

## 5.2. Constant Current Driving

The second addressing method is writing and erasing by means of a constant current. Especially since the various segments of alphanumeric displays can have different surface areas, the major advantage of the current pulse method is that the charge density of every segment can be simply controlled. In a practical design one can provide the electronic circuitry with current sources for each particular electrode segment. An important requirement with constant current driving can be the prevention of undesired side reactions or further reduction steps, as already pointed out in Section 3.2. Erasing can be obtained in the same way but with reversed current. Also in this case prevention of unwanted reactions may be necessary. This can be obtained (with loss of current constancy) by limitation of the total cell voltage.[11]

## 5.3. Matrix Addressing

The only literature available about electrochromic matrix addressed displays is a theoretical evaluation by Chang and Howard.[15] Several fundamental problems concerning matrix addressing with electrochromic systems have been discussed. The most serious one concerns the IR drop due to the resistance of transparent electrodes along the addressing lines, which causes a severe loss of discrimination with the larger display panels. On the other hand, matrix addressing is facilitated by the inherent memory capability of several electrochromic systems, which enables addressing with long frame times and/or large display panels.

## 6. PERFORMANCE OF ELECTROCHROMIC DISPLAYS

Potentiostatic measurements on a device with viologen-KBr are presented in Fig. 8. This figure shows voltage-time characteristics for the deposition of a viologen radical layer with a given reflectance at 540 nm. They can be explained by assuming a limiting IR-drop as already discussed in Section 3.

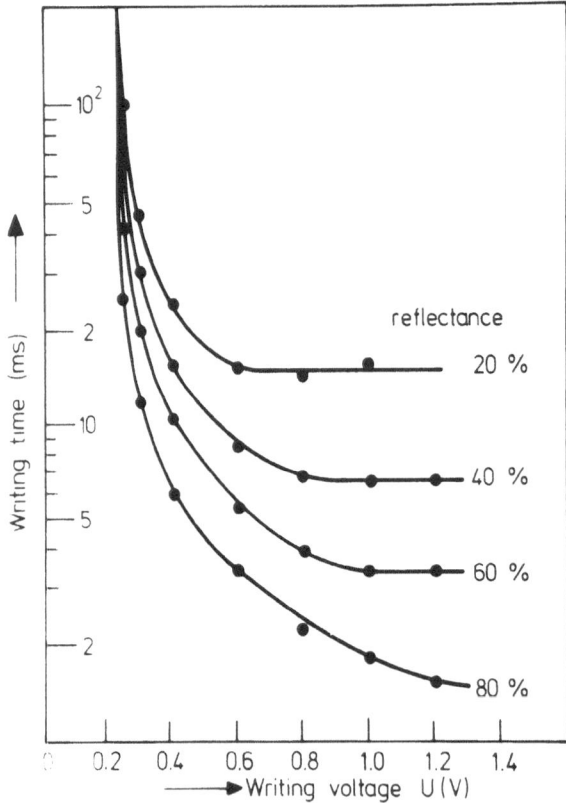

Fig. 8. Writing time as function of applied voltage U for various values of reflectance R at 540 nm.[7]

Display operation can be limited by diffusion as well as by series resistances. Figure 8 demonstrates clearly the existence of a threshold voltage at about 0.2 V. This threshold voltage depends on nature and composition of the electrolyte. U is the voltage between reference and working electrode; the total cell voltage is usually 1-2 V. It also appears that a reflectance of about 20 % (at 540 nm) can be obtained within 20 ms. This response can be better seen from Fig. 3. The charge Q varies linearly with time due to the influence of the series resistance. During the (short) memory time Q and R are constant, demonstrating the inherent memory.

Details concerning the writing and erasing speed of devices other than those based on viologen are very scarce. McGhee et al.[19] reported that 70 % contrast can be reached in more than one second with the $WO_3$-$LiClO_4$ system. This required a charge of about 15 mC/cm$^2$. Ciglia[11] reported that 30 - 40 % contrast can be reached in 0.5 s with 2 - 5 mC/cm$^2$.

The perceived quality of electrochromic displays[11] is very good because their contrast is independent of the viewing angle. If one assumes that 2 mC/cm$^2$ yields sufficient contrast, which is true for the viologen system, the energy required for writing is about $2 \times 10^{-3}$ J/cm$^2$. Combined with a writing time of 10 ms, this yields 200 mW/cm$^2$ as the required power for writing and about the same for erasing. With a memory time of say one minute we find about 70 $\mu$W/cm$^2$ as an average power.

The aging of the radical layer during memory is the crucial factor for devices with viologens. For that reason it is much more difficult to obtain correct device operation during a limited number of display cycles combined with a long memory time than during a large number of switching cycles with very short memory time. At the moment, lifetime experiments with modified viologens are running, which up to now have withstood $10^5$ cycles with a memory time of 60 s and 1.5 mC/cm$^2$ charge density. The writing and erasing times are about 200 ms. With shorter memory time $2 \times 10^6$ cycles have been obtained. The stability of the electrolyte and electrodes does not give rise to any problems. The $WO_3$-$H_2SO_4$ type display can operate for $2 \times 10^6$ cycles without memory.[11] About 0.5 s is needed for writing and erasing. The contrast obtained is 30-40 %. After this time each of the segments has decreased to 80 % of its original size. This chemical attack of the $WO_3$ segment and/or the $SnO_2$ substrate is the main problem of the $WO_3$-$H_2SO_4$ type display. These problems become more severe with increasing contrast ratio.

## 7. CONCLUSIONS

After more than 45 years of interest in different sorts of electrochromic display devices, it can be concluded that only in the last five years has reasonable progress in the development of a practical device been made. Presently existing technological and electrochemical problems can be recognized which seem difficult but not insurmountable.

## ACKNOWLEDGEMENTS

I wish to thank my colleagues at the Philips Research Laboratories for valuable discussions and criticism.

## REFERENCES

1  P. Delahay, "New Instrumental Methods in Electrochemistry", Interscience, New York (1954)
2  F. H. Smith, British Patent 328017 (1929)
3  S. K. Deb, Appl. Optics Suppl. 3 (1969) 192
4  J. H. L. Lorteije, "Neuartige Bauelemente der Anzeigetechnik", NTZ Report 19 (1974) 34, VDE-Verlag, Berlin
5  I. F. Chang, IBM Technical Disclosure Bull. 17 (1975) 3146
6  W. M. Schwarz, Thesis, University of Wisconsin (1961)
7  C. J. Schoot, J. J. Ponjee, H. T. van Dam, R. A. van Doorn and P. T. Bolwijn, Appl. Phys. Lett. 23 (1973) 64 and, with S. van Houten, S.I.D. Int. Symp. Digest 4 (1973) 146
8  H. T. van Dam and J. J. Ponjee, J. Electrochem. Soc. 121 (1974) 1555
9  British Patent 1314049 (1970)
10  I. F. Chang, B. L. Gilbert and T. I. Sun, J. Electrochem. Soc. 122 (1975) 955
11  R. D. Giglia, S.I.D. Int. Symp. Digest 6 (1975) 52
12  M. D. Meyers and T. A. Augurt, U.S. Patent 3708220 (1970)
13  H. Kim and H. A. Laitinen, J. Am. Chem. Soc. 58 (1975) 23
14  Ebauches, e.g. Dutch Patent Appl. No. 7300213
15  I. F. Chang and W. E. Howard, S.I.D. Int. Symp. Digest 5 (1974) 55
16  S. van Houten, Proc. 3rd. Europ. Solid State Devices Res. Conf., Munich (1973) and Int. Phys. Conf. Series 19 (1974) 131
17  K. J. Vetter, "Electrochemical Kinetics", Academic Press, New York (1967)
18  H. T. van Dam, private communication
19  J. H. McGhee, W. E. Kramer and H. N. Hersh, S.I.D. Int. Symp. Digest 6 (1975) 50
20  M. M. Nicholson, J. Am. Chem. Soc. 79 (1957) 7
21  H. R. Thirsk and J. A. Harrison, "A Guide to the Study of Electrode Kinetics", Academic Press, New York (1972) 52
22  J. R. Alburger, Electronic Industries, Febr. 1957, 50
23  J. R. Alburger, U.S. Patent 3123806 (1957)
24  E. Kissa, P. Manos and C. F. Wahlig, U.S. Patent 3453038 (1966)

Electrochromic Displays

## JOINT DISCUSSION

M. L. Hitchman (RCA Zurich)
In discussing the mechanism of coloration of $WO_3$, Dr. Chang raised the question of whether hydrogen atoms are involved, i.e. whether

$$xH^+ + xe^- \rightarrow xH^0, \qquad xH^0 + WO_3 \rightarrow H_x WO_3$$

or

$$xH^+ + xe^- \rightarrow H_x WO_3$$

occurs. Two pieces of experimental evidence that I have obtained strongly support the $H^0$ atom model. First, an indium wire held $\sim 1$ mm above a $WO_3$ film submerged in an acid electrolyte causes the film to color slowly. This is interpreted as $H^0$ atoms, formed as the In dissolves in the acid, diffusing to the $WO_3$ surface; $H^0$ atoms are known to have half lives $\sim 5$ s in aqueous acid media. Second, the degree of coloration of the system $SnO_2 |WO_3| M, H^+$ (where M is a metal in contact with the $WO_3$) for a constant amount of charge depends on the nature of the metal. For three metals I have used, the amount of coloration is Hg > Ag > Pt. The adsorption energy of $H^0$ atoms on these metals is in the order Pt > Ag > Hg, suggesting that the more strongly they bind an $H^0$ atom the less the degree of coloring. Dr. Weibel pointed out that in Li- and Na-tungsten bronzes the metals are present as fully ionized species. I believe this is also true for hydrogen in a hydrogen-tungsten bronze. Analysis of the equilibrium potential of bronzes of various compositions in terms of activities and activity coefficients shows good agreement between theory and experiment when a model assuming complete dissociation is used, but not when no dissociation is assumed.

I. F. Chang

The ESCA spectra reported by Rabelaise et al. and by Weibel must be interpreted with caution because this technique measures only a surface layer about 50 Å thick.

G. Weibel

Our ESCA measurements were made through the thickness of the film by sputtering off successive layers.

J. Kirton (Royal Radar Establishment)

With non-conducting deposits it is said to be necessary to add species such as iron ions in order to produce a satisfactory erase process. By how much does this restrict memory?

J. Bruinink

Resistive materials like the viologens are indeed more difficult to erase than metal deposits. We are still studying this problem, but it appears that viologen layers which are colored adequately for display requirements can be erased directly without ionic additives. I think such additives are "medicine for a sick system".

G. Weibel

Refresh cycles of $WO_3$ electrochromic displays are in the order of hours. Experimentally we get memory times between 30 minutes and days.

I. F. Chang

Memory behavior is very much dependent on the device configuration. In an isolated cell under open-circuit conditions the retention time can be fairly long, but a matrix-addressed device is an entirely different matter. Howard and I discussed this case in IEEE Trans. Electron. Devices ED 22 (1975) 749.

D. Ross (RCA Princeton)

Would you please comment on the relationship between the retention of the colored film and the life of the device?

I. F. Chang

In some devices the operating life is found to comprise many more write/erase cycles than if an appreciable memory time were included for write/store/erase cycling. While a cell is being driven, the concentration of the electrochromic species in the electrolyte near the surface tends toward zero, whereas it is restored to a higher level by diffusion during the memory time. I believe this may influence the solubility and stability of the deposited layer.

M. L. Hitchman

The comparison of electrochemical electrochromic devices to batteries perhaps high-lights one of the serious problems that has to be overcome in an electrochemical display, namely that of obtaining a system which is capable of being "charged" and then completely "discharged" for something in excess of $10^6$ cycles. Although there are a number of obvious differences between a true battery and an electrochemical display, nevertheless the fact that no battery exists which is capable of undergoing so many charge/discharge cycles illustrates the inherent difficulty in obtaining a long-life device.

J. Kirton

Have any difficulties been experienced as a result of the working substance attacking the transparent electrode?

G. Weibel

In lifetime studies up to $10^6$ cycles, we have found the failure mechanism to be related to the $SnO_2$ electrode rather than to the $WO_3$ film.

J. Bruinink

We find no difficulties with tin oxide in the viologen system. The $SnO_2$ layers we make ourselves have low resistance ($10\ \Omega/\square$) and are very stable.

M. L. Hitchman

The degree to which one achieves only the desired reaction of the working electrode depends on the degree to which one controls the potential at that electrode, and this in turn depends on the reference electrode in the system. Using a combined auxiliary/reference electrode at which there is no stable reversible couple could lead to serious degradation in any electrochromic cell if if there is a potential drift of the reference potential level. Have you encountered such problems in your display device and have you endeavored to use a combined reference/auxiliary electrode in the form; say, of Ag/AgBr, since you have $Br^-$ ions present already in the system?

J. Bruninik

The choice of the auxiliary/reference electrode system is a very important consideration. In practical displays we use the viologen couple on $SnO_2$ which behaves satisfactorily. Ag/AgBr is not used since silver deposition occurs on the working electrode after some time.

C. Hilsum (Royal Radar Establishment)

Users complain that liquid crystal devices have a limited temperature range of $-10^\circ$ C and $+60^\circ$ C. What is the temperature range of electrochromic devices? Have any measurements been made of their stability to ultraviolet exposure?

J. Bruinink

The upper temperature limit depends on cell technology. We now work up to $+30^\circ$ C, but there should be no difficulty in extending this to $+60^\circ$ C. Freezing, however, is a serious problem. We have made no UV measurements, but I don't anticipate a problem.

# ELECTROPHORETIC DISPLAYS

J. C. LEWIS

Allen Clark Research Centre, The Plessey Company Limited
Caswell, Towcester, Northants., England

## SUMMARY

The electrophoretic display is a light modulating display in which changes in reflectivity and color occur as a result of the migration of pigment particles in a dye solution under the influence of an electric field. It combines a large number of desirable features, including simplicity of construction, high contrast ratios at wide viewing angles and in a wide range of ambient light conditions, low power consumption, wide operating temperature range, availability of a variety of color combinations, memory, multiplexing capability and low or zero toxicity of working materials. The construction and mode of operation of the electrophoretic display are described, and the physicochemical principles governing important characteristics of the display (operating voltage, response speed, contrast ratio, stability and life) are briefly discussed, with special reference to the criteria applied in the selection and modification of materials. Materials selection has a particularly important bearing on the lifetime, which also depends on the display design and duty cycle. The lifetimes of laboratory samples of simple displays are now close to satisfying the requirements for some of the applications envisaged. The feasibility of producing a range of display formats, including bistable ("on-off"), seven-segment digital, analog and matrix types, has been demonstrated. Possible applications of electrophoretic displays of each of these types, and the advantages over other light modulating displays, are discussed.

# 1. INTRODUCTION

Although the phenomenon of electrophoresis has been applied for many years in such processes as photocopying, electrophoretic painting and the deposition of rubbers, plastics and even some metals, the first reported work on its use in a reversible manner to form the basis of a passive electrooptic display appears to be that of Ota[1] in 1972. At least three or four laboratories around the world are now working on electrophoretic displays (EPID), but publications have so far been few and brief: for this reason, this paper is based almost entirely on work carried out at the Allen Clark Research Centre, although reference will be made to the results reported by other groups where appropriate. The main characteristics of our electrophoretic displays are:

- High visual contrast at wide viewing angles and over a wide range of ambient light levels
- Response speeds varying from hundreds of milliseconds at 2.5 V to as little as 10 ms at 50 V; low currents, typically about 1 $\mu$A/cm$^2$
- Lifetimes up to $10^7$ operations; long shelf life expected
- Memory (i.e. the ability to retain an image after the removal of the applied voltage) variable in a controlled manner up to at least several months
- A threshold voltage, with adequate discrimination for use in matrix displays based on the "half-select" principle.

# 2. MODE OF OPERATION

As shown in Fig. 1, the electrophoretic display consists essentially of a sandwich cell formed between two electrodes, at least one of which is transparent, with a working fluid composed of a suspension of small (sub-micron) pigment particles in a densely colored liquid. If the two phases are suitably

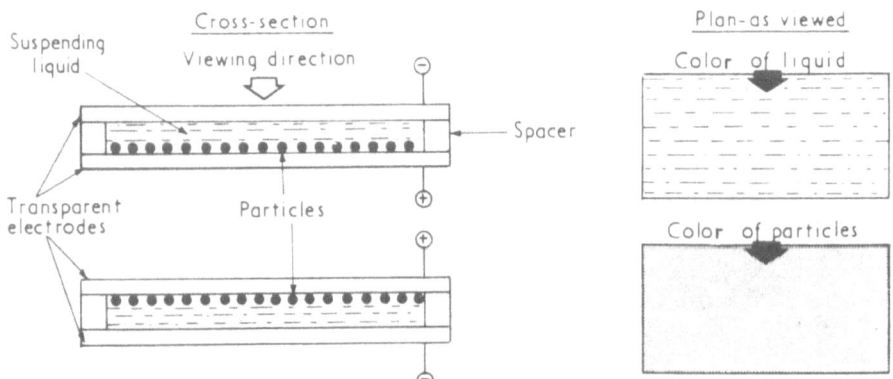

Fig. 1. Simple electrophoretic display device.

selected, the pigment forms a lyophobic (i.e. solvent-repellent) sol or suspension in which the particles are electrically charged, probably as a result of the adsorption of impurity ions from the liquid phase. Application of an electric field causes migration of the particles to one electrode, which therefore assumes the color of the pigment; reversal of the field causes the particles to move in the opposite direction, and the electrode color changes to that of the liquid phase. The use of shaped or segmented electrodes, together with pigment particles and dye solution of contrasting colors, enables on-off, digital and alphanumeric displays to be made.

An important feature of the display is its memory, the particles remaining on the electrode for periods varying from a few seconds to many months after removal of the field. The length of the memory, like many other properties of the display, depends critically upon materials selection, and much of our work has been aimed at a greater understanding of the relationship between working fluid composition and display characteristics. As a result of this work, it is now possible to select materials (or even, in some cases, to develop new materials) in order to optimize particular properties of the display, although some of these properties are interrelated and compromises may therefore be necessary. For example, a reduction in the operating voltage will inevitably be accompanied by a drop in response speed unless the thickness or composition of the display is also changed.

## 3. SELECTION OF MATERIALS

The liquid, the dye and the pigment all contribute to the overall properties of the system and cannot really be considered in isolation; however, in the interests of clarity, let us look briefly at the importance of each of these materials separately before going on to consider their combination in practical systems and the performance of the resultant displays.

Electrophoresis, the motion of the suspended particles under the influence of an electric field, derives from the tendency of the particles to adsorb or generate ions and thus to become positively or negatively charged. This effect is illustrated in Fig. 2, in which we see a negatively charged particle surrounded by a tightly held sheath of positive counter-ions (the Stern layer). A sharp drop in viscosity occurs between the Stern layer and the shear plane, outside which is a more diffuse layer of counter-ions in gradually decreasing concentration. On the application of an electric field the particle, together with its sheath of counter-ions and the liquid within the shear plane, moves in one direction, while the cloud of counter-ions outside the shear plane moves in the opposite direction. The distance at which the potential falls to $1/e$ (i.e. 37%) of its value at the shear plane is called the double layer thickness $1/\kappa$ and the potential drop across this double layer (i.e. 63% of the potential difference between the shear plane and the bulk of the liquid) is the electrokinetic or zeta potential, which is one of the factors controlling the mobility of the particles and hence the response speed of the display.

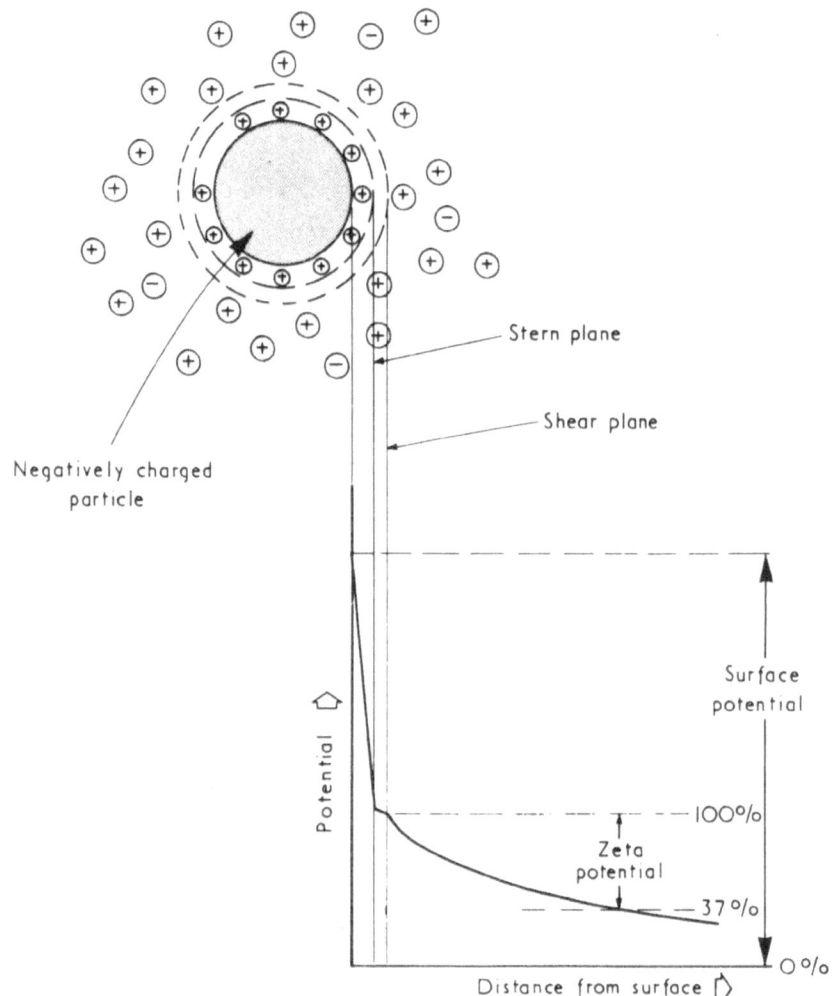

Fig. 2. Schematic diagram of the Stern double layer.

The electrophoretic mobility of a particle is in fact given by the equation

$$u = \frac{\epsilon \zeta}{k \pi \eta}$$

where u: mobility, $\epsilon$: dielectric constant of the liquid, $\zeta$: zeta potential, $\eta$: viscosity of the liquid, and k: a numerical factor which can vary from 4 to 6 depending upon the size and shape of the particle. As a guide to the quantities involved, the ratio $\epsilon/\eta$ is about 9000 poise$^{-1}$ for water and 200 - 1000 poise$^{-1}$ for most organic liquids. Zeta potentials range up to about 150 mV and particle mobilities are in the range $10^{-4}$ to $10^{-3}$ cm$^2$/Vs in water and about $10^{-5}$ cm$^2$/Vs in organic fluids of the type used in electrophoretic displays.

The size of the particle does not appear in the mobility equation; the only variables are the zeta potential and the ratio $\epsilon/\eta$ for the liquid. For high mobilities, and thus fast response at low voltages, it therefore appears desirable to choose a liquid with a high $\epsilon/\eta$ ratio. In practice, however, many low-viscosity liquids are also rather volatile, and liquids with high dielectric constants are not usually very stable chemically or electrochemically, are very sensitive to contaminants and are difficult to purify because of their strong solvent action on ionic materials. For these reasons, liquids such as certain esters, ethers and ketones with moderate dielectric constant values in the range 4 - 10 have proved more suitable. They appear to offer the best overall combination of speed, chemical stability, ease of purification and freedom from severe problems of contamination. Excessive concentrations of ionic contaminants are particularly undesirable, since any current carried by charge carriers not associated with the electrical double-layer is purely parasitic and merely increases the power consumption; very impure materials have also been found to produce gas bubbles and visible attack at the electrodes. Resistivity may be taken as some guide to purity, and the minimum satisfactory value appears to be about $10^9$ $\Omega$cm. It is usually possible to achieve this figure by standing the liquids over molecular sieves in a closed vessel, although more sophisticated methods of purification have been used for some liquids.

Conductivity is also an important factor in the choice of a dye. Many thousands of dyes are listed in the "Colour Index", but most of them are chemically reactive, ionic and water-soluble and are therefore unsuitable for the production of stable, high resistivity, non-aqueous systems. However, dyes described as "Oil (Color)" or "Solvent (Color)" are of low ionicity and are soluble in non-aqueous solvents. A selection of these has been evaluated, with emphasis on the blacks, blues and reds, which are capable of producing solutions of high optical density.

Finally, we have to choose a material for the disperse phase, which can be either an organic or an inorganic pigment. The latter have the advantages of being very stable, insoluble and of high refractive index (i.e. efficient light scatterers), but they are usually much denser than any suitable liquid phase. Organic pigments are less dense but are often soluble to some extent in the liquid phase and of lower stability and refractive index. Both types have been evaluated in the present work.

## 4. STABILIZATION OF LYOPHOBIC SOLS

As a result of the high density of preferred pigments such as titania (anatase 3.84 g/cm$^3$, rutile 4.26 g/cm$^3$), probably the most important factor limiting the life of an electrophoretic display is the inherent thermodynamic instability of the lyophobic sol to gravitational settling and flocculation. These two effects are related, since flocculation often leads to sedimentation. (According to Stokes' Law the settling rate is proportional to the square of the effective radius.) Flocculation of lyophobic sols results from the very high surface area of the disperse phase and the repulsive interaction between it and the surrounding liquid phase: these two factors combine to produce a high interfacial energy, which can be reduced by aggregation of the particles with a consequent reduction in the surface area per unit mass of the disperse phase. However, although lyophobic sols are thermodynamically unstable, dispersions which are stable over periods of many months or even years can be obtained by taking steps to reduce the frequency of particle-particle collisions and thus the rate of aggregation: i.e. kinetic stability can be achieved.

Almost all the basic work on the stabilization of colloids reported in the literature has used water as the dispersion medium, not only because of large scale industrial applications (e.g. soaps, detergents, adhesives and emulsion paints) but also because theories relating to the properties of water are more fully developed than for other solvents. Theoretical correlation of colloid behavior in non-aqueous media with behavior in water is difficult because of differences in properties. Practical work on non-aqueous systems has generally been characterized by lack of reproducibility, probably due (at least in part) to the presence of traces of contaminants such as water; as a result comparatively little work, generally limited to entirely non-polar liquids such as xylene and heptane, has been reported in this field.

The main repository of knowledge on non-aqueous colloids is the paint industry which, rather than produce fluid paints which are absolutely stable, has found it easier to produce either fluid paints which settle out relatively quickly (i.e. in weeks) to give a "soft" sediment easily redispersible on agitation, or the so-called "gel" paints which rely for their stability on thixotropy which develops after agitation has ceased and which strongly retards the sedimentation of pigment particles. Obviously, these types of stability will not suffice for the electrophoretic display, which requires almost absolute stability over periods of years, combined with low viscosity to ensure adequate speed of response. Before discussing possible ways of achieving this result, it may be of value to consider the nature of the forces already acting on the particles in a lyophobic sol. As we shall see, some of these forces will help us, while others will hinder us, in our objective of keeping the particles apart and stabilizing the sol.

Firstly, we have attractive (van der Waals') forces, namely:

a) London dispersion forces between molecules having no net dipole

$$V_{London} = -\frac{3\alpha^2 h\nu}{4H_o^6}$$

where $\alpha$: polarisability, h: Planck's constant, $\nu$: characteristic frequency of the molecule and $H_o$: separation distance.

b) The Debye attraction between a permanent dipole and a dipole induced by it in another molecule

$$V_{Debye} = -\frac{\alpha\mu^2}{H_o^6}$$

where $\mu$: dipole moment.

c) The Keesom attraction between permanent dipoles in different molecules

$$V_{Keesom} = -\frac{\mu^4}{k_B T H_o^6}$$

where T: absolute temperature, $k_B$: Boltzmann constant. As the equations show, the interaction energies for a pair of atoms or simple molecules are in all cases proportional to the inverse sixth power of the separation distance and the van der Waals' forces effectively operate over only a few Angstrom units. However for larger particles it may be shown that, when the separation is much less than the particle size, the London energy varies only as the inverse first power of the separation. The attractive force therefore becomes a comparatively long-range one given by

$$V_A = -\frac{A^*a}{12H_o},$$

where $A^*$ is a constant known as the Hamaker constant (with a value of about $5 \times 10^{-13}$ erg for titania in water) and $a$ is the particle radius.

In order to achieve stable dispersions, it is necessary to produce repulsive forces at least equal in magnitude to the attractive ones. There is, of course, an electrostatic repulsion between the particles by virtue of the electrical double-layer and the zeta potential associated with it. According to theories developed by Derjaguin and Landau,[2] and by Verwey and Overbeck,[3] the form of the interaction energy equation depends on the ratio between the particle radius a and the double-layer thickness $(1/\kappa)$. Moreover, the situation is complicated by the effect of the diffuse atmospheres of counter-ions which partially screen the surface charges from each other while themselves being mutually repulsive. For the simplest case, when $\kappa a < 3$ and $H_o \ll a$,

$$V_R = \frac{\epsilon a \zeta^2}{2} e^{-\kappa H_o}.$$

The overall energy of interaction is obtained by summing the attractive and repulsive energies:

$$V = \frac{\epsilon a \zeta^2}{2} e^{-\kappa H_o} - \frac{A^*a}{12H_o}.$$

Two possible results of this summation are shown in Fig. 3. The form of the
potential energy curve depends on the shapes of the two individual (attrac-

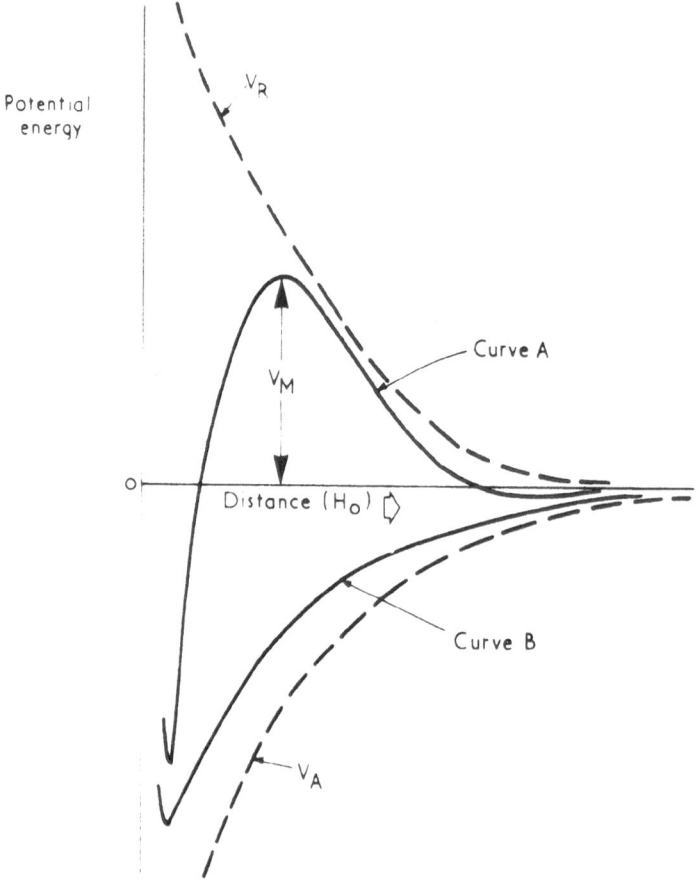

Fig. 3. Schematic plot of repulsive and attractive forces between two particles.

tion and repulsion) curves, which in turn depend on the numerical values of
the various parameters in the equation. Curve A is an example of a system
showing some degree of stability because of the potential energy barrier $V_m$
which must be surmounted before two particles can approach closely enough
to coalesce. For long-term stability, a value of $V_m$ in excess of 10 - 20 kT is
essential to prevent the Maxwell-Boltzmann distribution of thermal energies
from leading to a significant rate of coalescence. When $V_m$ is comparable
with, or less than, thermal energies, coalescence and sedimentation will
occur. In curve B there is no energy barrier, so coalescence is rapid and the
sol is very unstable.

The theory thus suggests that electrostatic repulsion, resulting from the
repulsive interaction between electrical double-layers, can in certain circum-

stances prevent (or, more accurately, markedly reduce the rate of) coalescence of two particles. However, the situation is greatly complicated by the fact that in the practical system the interaction forces between our two particles will be modified by the close proximity of other particles ($H_o \cong 2a \ll 1/\kappa$). Extension of the theory to cover this practical case quantitatively has not been attempted. Nevertheless the simple theory is of value in indicating qualitatively how electrostatic repulsion may be optimized, e.g. by increasing $\epsilon$ and/or $\zeta$ as far as is compatible with the low ionicity required for satisfactory operation of the display. This latter limitation is such that electrostatic repulsion alone is not sufficient in practice to stabilize the sol. Other means of stabilization therefore have to be employed, and the following techniques have been used, with varying degrees of success.

### 4.1. Reduction of Particle Size and Density

Stokes' Law tells us that the rate of settling of a particle of radius a and density $\rho$ in a liquid of density $\rho'$ and viscosity $\eta$ is given by

$$\frac{dx}{dt} = \frac{2ga^2(\rho-\rho')}{9\eta}$$

where g is the acceleration due to gravity. Reduction of the settling rate can be achieved, to some extent, by reducing the particle size; however, as has been explained, this reduces the thermodynamic stability of the sol and must therefore be accompanied by separate action to retard flocculation. A further limitation is that the light-scattering efficiency of the particles falls rapidly when the radius is reduced below the wavelength of light: a practical limit for pigments in the electrophoretic display is about 0.1 $\mu$m.

Major improvements in stability may be obtained by reducing the density mismatch $(\rho-\rho')$ between the pigment and the liquid medium, but only at the expense of some loss of contrast. This is because high light-scattering efficiency is dependent on a high refractive index, which tends to be associated with a high density. Promising results have been obtained with titania particles incorporated into low-density materials to produce a composite with a density of about 2 g/cm$^3$ and also with certain organic pigments. In the case of composite pigments, it is important to ensure that the materials and preparative techniques used are not such as to lead to a drastic reduction in the zeta potential. This may happen, for example, if a titania pigment is effectively encapsulated in an inert plastic such as polyethylene: the rather low response speeds observed in earlier work by Ota[4] using composite pigments may be due to this effect.

### 4.2. Rheological Control

It is common practice in the cosmetic and pharmaceutical industries to delay settling by the addition of thickening agents. The disadvantage of this method for electrophoretic displays is that it merely delays the settling process and does not provide absolute stability. Moreover, the increase in viscosity leads to a reduction in mobility and thus in response speed. It should in principle be possible to use network-formers which are strong enough to prevent the particles from settling under gravity, but weak enough to yield and allow the passage of particles under the much stronger forces resulting from the applied electrical field.

### 4.3. Steric Stabilization

Steric (or entropic) stabilization provides an alternative method of providing a repulsive energy. It occurs when long chain lyophilic groups are firmly attached to the particle and extend an appreciable distance, i.e. 20 - 100 Å, into the medium. The interaction between the ends of chains terminated on different particles prevents the particles from approaching one another closely enough for coagulation to occur. It is thought[5] that when two particles approach (Fig. 4) the chains interpenetrate causing a loss of configurational

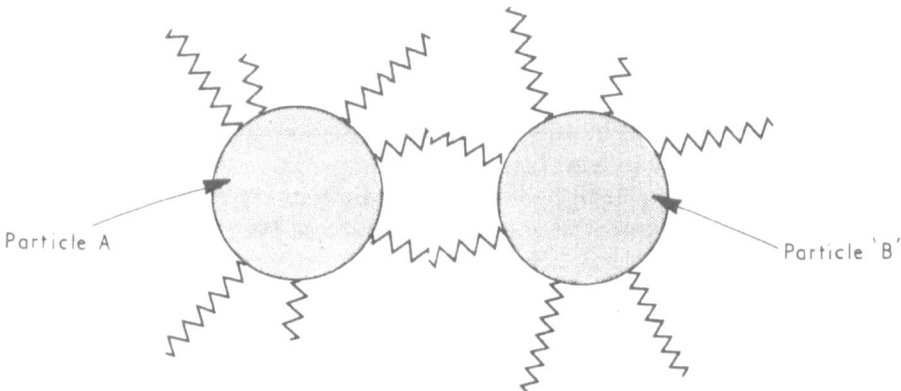

Fig. 4. Schematic representation of steric repulsion between two particles.

entropy (i.e. a loss in freedom of movement). The resultant increase in the free energy of the system gives rise to a repulsion. Attempts have been made to formulate a mathematical treatment of this type of stabilization[5,7,8] and it is likely that the overall shape of the energy curve will be similar to curve A of Fig. 3, but it must be admitted that, compared to electrostatic stabilization, it is very poorly understood. A major advantage over electrostatic stabilization is that the steric repulsive force remains strong and operates over a nearly constant distance (the maximum distance at which it can act is approximately twice the maximum extension length of the stabilizing chains) instead of becoming very diffuse and extended as does the electrical double-layer in solvents of low polarity. Although the first sterically stabilized non-aqueous dispersion was described in 1950,[6] it is only recently that commercial interest has been aroused. It seems likely that this will become the major stabilization mechanism for non-aqueous systems (it is also applicable to aqueous systems).

## 4.4. Applicability of Stabilizing Mechanisms to the EPID

Since the particles in the EPID suspension must be charged to give electrophoretic movement, then electrostatic stabilization will be present in the display. However, as explained above, electrostatic stabilization in non-aqueous liquids suitable for use as display media is weak owing to the very diffuse and extended nature of the electrical double-layer in systems of low polarity and ionic strength. Such systems are also limited to concentrations of particles below about 1% by volume, and it seems likely that the repulsive forces exerted by the very extended double-layer, even if they were strong enough to give stabilization in the suspension, would not be strong enough to prevent coagulation when the particles are brought into close proximity on the surface of the cell after switching.

Steric stabilization is fairly well documented in non-aqueous systems. It provides stabilization even of highly concentrated (< 60%) dispersions, so it should persist even when particles are brought into close contact at the electrodes. However some reduction of zeta potential has been noted in sterically stabilized dispersions. This has been ascribed to the effect of the long adsorbed chains effectively pushing the 'shear plane' further away from the particles, thus reducing the thickness of the section of the double-layer which gives rise to the zeta potential.

Little can be said of rheological control owing to the difficulty of finding 'plastic flow'-inducing materials compatible with suitable non-aqueous media. Less effective but similar 'thixotropic' media have been found to exhibit good stabilization but often with undesirable side effects such as reduction of contrast, promotion of adhesion, etc. These systems, however, do have the advantage of lengthening the memory of the device considerably.

## 5. DISPLAY PROPERTIES

As already stated, the properties of electrophoretic displays vary to some extent with the working fluid used and are often interdependent. For example, low voltage operation can only be achieved at present at the expense of some reduction in response speed and probably also in life. Continuous improvements are being made, so the performance data quoted below should certainly not be regarded as the optimum obtainable. However some of the properties of electrophoretic displays are clear from the results already obtained.

The contrast is in excess of 30 : 1 for titania-based displays, although rather less for the more stable low-density pigments. Viewing angles approach ± 90° in normal conditions, i.e. in the absence of abnormally bright specular reflections from the front face. Dalisa and Delano[9] found that the effect of viewing angle on contrast was comparable with the performance of ink on paper, and much less than for liquid crystal displays. The most stable (lower contrast) displays are easily legible at illumination levels of 10 lux and improve in brighter conditions (up to bright sunlight, $10^5$ lux).

Operating voltages in the range ± 2.5 V to ± 50 V have been used in experimental work. The indications so far are that lifetimes are longer near the upper end of this range. Response times are generally of the order of tens of milliseconds: the fastest response so far observed is 10 ms at 15 V. Switching currents are in the range 0.1 to 10 $\mu$A/cm$^2$, giving low instantaneous power consumptions (usually tens of microwatts per cm$^2$). The memory permits a further reduction in the mean power consumption for displays which are switched comparatively infrequently. For example, a display switched once per minute by means of a 100 ms pulse might have a mean power consumption of 0.1 $\mu$W/cm$^2$ or less. However it has not yet been established that this pulsed mode of operation is compatible with long life.

EPID devices remain in their last driven state (i.e. 'particles up' or 'particles down') for relatively long periods of time when the field is removed. The memory depends mainly on the precise nature of the working fluid: some begin to lose contrast as soon as the applied field is removed and are completely redispersed in 1 - 2 minutes, while others retain their memory for many months, the pigment particles remaining on the electrode with no detectable change in contrast until a field is applied in the reverse direction. As a result, electronic drive circuitry requirements are slightly different from those needed for displays which revert rapidly to the 'off' state on removal of the exciting voltage. The EPID requires to be driven 'off' as well as 'on'. In practice this requirement reduces to holding the front electrode of the display at nominal 0 V and driving the patterned areas to ±P where P is the nominal operating voltage. This can be easily done by the use of the circuit shown in Fig. 5. The display device is connected across AB. A is connected to the front electrode and becomes the reference potential point (at +P). With $S_1$ open, point B assumes a potential close to +2P, i.e. +P with respect to A, because of the low current consumption of the device. With $S_1$ closed, B is

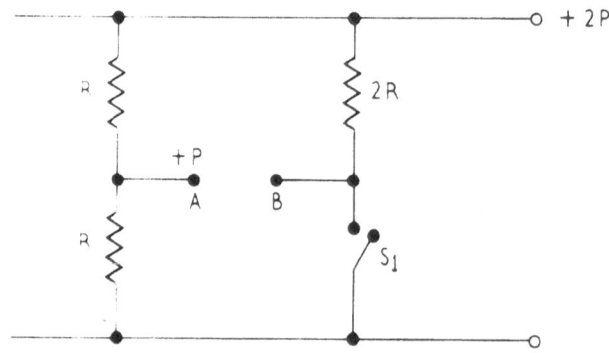

Fig. 5. Basic drive circuit for electrophoretic display.

connected to the 0 V line and thus assumes a potential −P with respect to A. Thus simply opening and closing S$_1$ produces the reversal of polarity at the patterned electrodes required to drive the device 'on' and 'off'. This circuit can of course be extended to drive 7-segment displays using conventional circuitry. It has been found possible, for example, to drive a single 7-segment digit at ± 15 V using a standard SN 7446 integrated circuit decoder/driver as shown in Fig. 6. In these cases an additional connection is made to the back-

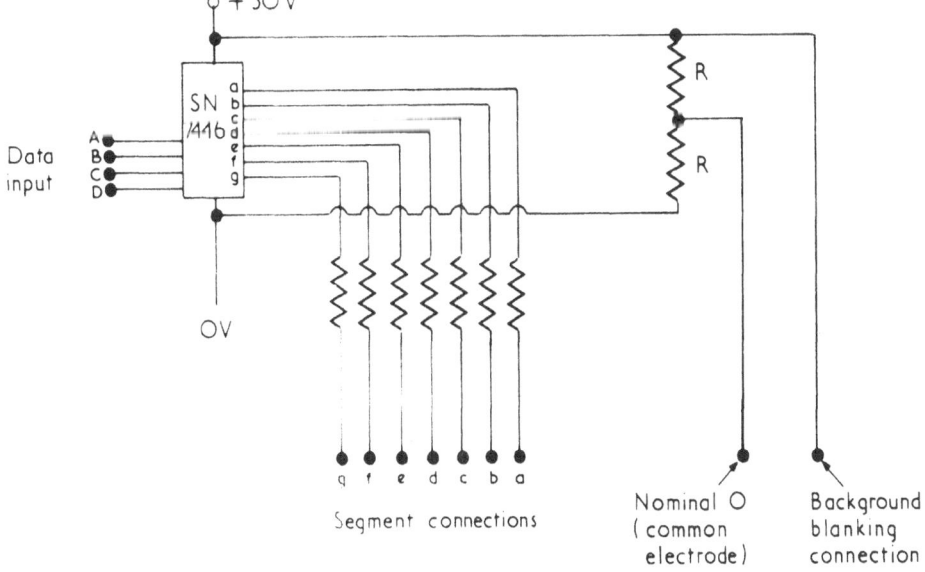

Fig. 6. Drive circuit for single 7-segment digit (± 15 V).

ground areas of the seven-segment bar pattern to ensure that this area is always in the 'off' state. If higher drive voltages are required the decoder/driver may be used with individual high-voltage switching transistors in accordance with normal practice; however, this increases the cost and complexity of the circuitry (Fig. 7).

The versatility of electrophoretic display format results in a wide range of potential applications. Display formats so far investigated (Fig. 8) include:

— Plain panels, in which a legend inscribed on the front face is made visible by changing the background color electrophoretically from that of the legend (e.g. black) to a contrasting color (e.g. white or yellow), used as fixed-message warning devices for aircraft cockpit, car dashboard and roadside use.

— Seven-segment digits (on which most of our stabilization and life-test work has been concentrated), for similar applications and also for clocks and watches.

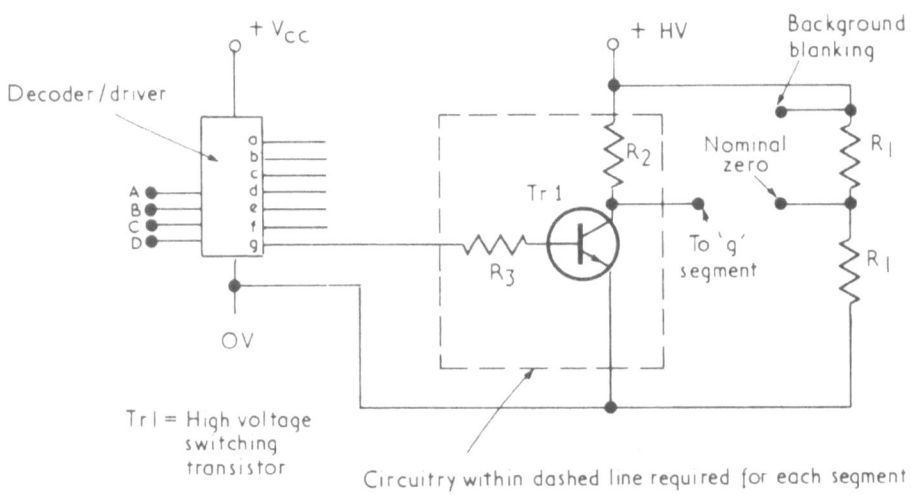

Fig. 7. Drive circuit for single 7-segment digit (high voltage).

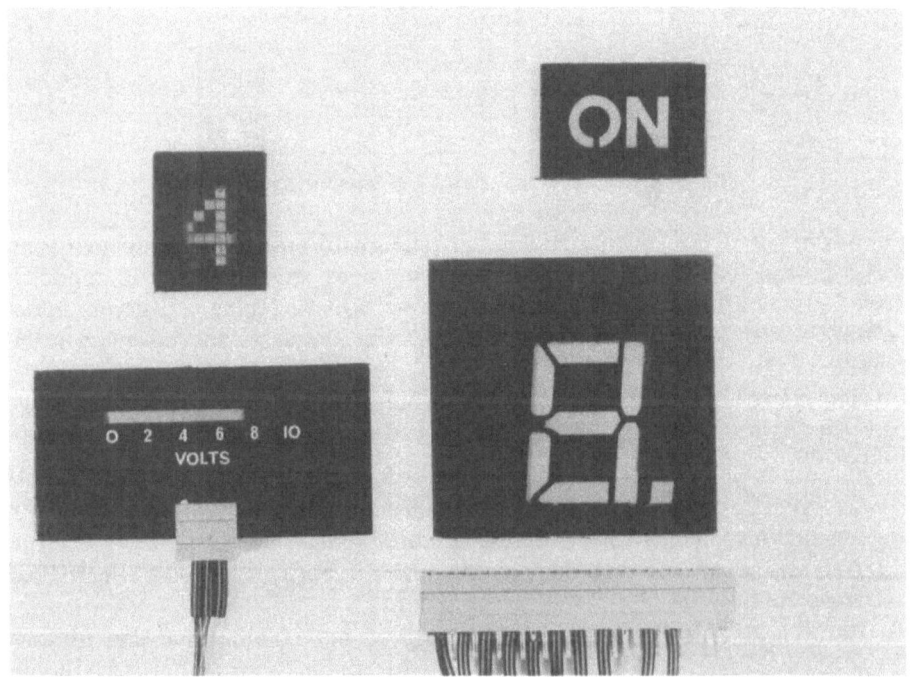

Fig. 8. Working prototypes of various electrophoretic display formats.

— Simple, low-cost analog bar-type displays for applications such as fuel and pressure gauges, thermometers and voltmeters.

— Matrix displays addressed, as a result of the development of a voltage threshold, on the 'half-select' principle and applicable to flat-panel displays for more complex alphanumeric and perhaps graphical information. (Since the response speed is limited by particle rather than electron mobilities, the possibility of use in rapidly changing 'real-time' dynamic displays — the flat TV screen — is remote.)

Display dimensions have varied so far from 5 mm digits to an experimental plain panel $300 \times 150$ mm. Ota et al.[10] have developed an impressive $184 \times 32$ dot matrix panel capable of displaying 78 alphanumerics, with (at present) the disadvantage of a long scan time of 40 seconds. Most of our work has been based on blue/white and black/yellow color combinations, but others, including a visually pleasing red/yellow combination, are possible.

Life depends on various factors, including working fluid, drive voltage and duty cycle. Dalisa and Delano[9] reported a lifetime of $10^6$ cycles in 1974 and Ota et al.[10] have recently reported $3 \times 10^7$ cycles ($> 10\,000$ hours), on unspecified display formats (probably plain panels). The largest number of operations we have achieved is $10^7$ ($\sim 3000$ h) on a 7-bar digit switched continuously at 1 Hz, $\pm 45$ V. Other displays have been switched irregularly a few times a week for many months with little deterioration, while yet others have remained switchable after several months in the memory state. As yet the effect of duty cycle is inadequately known, so further life-testing in appropriate conditions may be necessary before an electrophoretic display can be recommended for any particular application. Ota et al.[10] report considerable deterioration of their displays, including dye decoloration, after six weeks in sunlight. This effect has not been observed in our work, probably because of the different suspension fluids and dyes used; some of our experimental displays have been operating, in a laboratory with glazed walls facing south and west, for many months without decoloration of the dye.

Electrophoretic displays are compact and safe (working fluids are of low or zero toxicity). They are generally agreed to be esthetically pleasing — an important property, though difficult to express quantitatively. Tests so far have been limited to the temperature range $-10$ to $+70^\circ$ C and have not been of long duration, but the use of working fluids having a wide liquid temperature range without excessive volatility should permit extension of this operating range to meet all normal display requirements.

## 6. CONCLUSION

Comparison with other passive displays may at this stage be somewhat premature, in view of the more advanced state of development of liquid crystal displays and particularly of the older forms of passive display such as electromechanical devices. The latter are bulky and costly, requiring precision engineering in order to ensure reliability, and are comparatively high power consumers. Electrochromics, still at an early stage of development, offer high contrast at wide viewing angles but again power consumption is rather high (tens of milliwatts per $cm^2$) due to the ionic mechanism involved. Liquid crystal displays are still meeting considerable resistance from users, perhaps unjustly in view of some recent developments in the technology. It is interesting to note that of the four most widespread criticisms of liquid crystals, namely inadequate contrast and field of view and doubts about their reliability and life, only the last two are normally levelled at electrophoretics. Clearly, neither technology entirely satisfies users' needs at present; whether either can do so remains to be seen. Until both electrophoretics and the new higher contrast liquid crystal displays have been fully evaluated, it is difficult to say whether either will prove to be the 'ideal' passive display. Indeed, just as in the field of active displays, we shall no doubt find that the 'ideal' does not exist, that there is no clear-cut single answer to the demands of all users, and that in ten years from now a wide range of passive displays will be in use. In the author's view, electrophoretics will be among them.

## ACKNOWLEDGEMENTS

I should like to express my thanks to: colleagues at the Allen Clark Research Centre, especially Mr. G. M. Garner, Mr. R. T. Blunt and Dr. C. F. Carter, whose work has formed the basis of this paper; the Procurement Executive, Ministry of Defence, for financial support of part of the work; Dr. J. C. Bass and Dr. J. Kirton for their interest and encouragement; and the Plessey Company Limited for permission to publish.

## REFERENCES

1  I. Ota, J. Ohnishi and M. Yoshiyama, IEEE Conference on Display Devices, 72 CH 0707-0-ED (1972) 46
2  B. V. Derjaguin and L. Landau, Acta Physicochem. URSS 14 (1941) 633
3  E. J. W. Verwey and J. Th. G. Overbeek, "Theory of the Stability of Lyophobic Colloids", Elsevier, Amsterdam (1948)
4  I. Ota, private communication via P. C. Newman, May 1973
5  E. L. Mackor, J. Colloid Sci. 6 (1952) 492
6  M. van der Waarden, J. Colloid Sci. 5 (1950) 317
7  E. W. Fischer, Kolloid Z. 160 (1958) 120
8  E. J. Clayfield and E. C. Lumb, J. Coll. Interface Sci. 22 (1966) 269
9  A. L. Dalisa and R. A. Delano, S.I.D. Int. Symp. Digest 5 (1974) 88
10  I. Ota, T. Sato, S. Tanaka, T. Yamagami and H. Takeda, Laser 75 Seminar, Munich, June 1975

## DISCUSSION

C. von Planta (Hoffmann-La Roche)

Is the ratio of particle radius to Debye length such that particle mobility is independent of size?

J. C. Lewis

The Debye length in many of our systems is larger than the electrode separation. For particles between 0.2 $\mu$m and 2 $\mu$m, we have experimentally confirmed the theoretical expectation that mobility is independent of size.

M. L. Hitchman (RCA Zurich)

As I understand it, the Coulombic repulsive potential $V_R$ between two interacting particles is related to the surface potential $\psi_o$ by $V_R \propto \psi_o^2$ only for small values of $\psi_o$ ($< \sim 100$ mV). At higher values the relationship is $V_R \propto \tanh^2 k\psi_o$, where k is a constant. Thus it would seem that one cannot hope to achieve greater colloidal stability simply by increasing the potential in the particles, but rather that one must rely more on stabilization by steric effects. I believe the pharmaceutical industry makes use of this technique to obtain colloids which are stable for four or five years.

J. C. Lewis

I certainly agree that electrostatic stabilization is suspect in the concentrations we use. Two particles which are mutually repelled by Coulombic forces will be pushed together by the same forces from a surrounding cloud of additional particles. With respect to the pharmaceutical technique mentioned, it must also be remembered that a colloid which is stable in a test tube may not be so stable when switched in a display. Particles are taken out of suspension by the electric field and pressed several layers deep against the electrode where aggregation can occur.

F. J. Kahn (Hewlett-Packard)

Does your matrix display have a true quasi-static threshold?

J. C. Lewis

We have produced threshold voltages which vary from a few volts to a few tens of volts. The optical response/voltage curve is sigmoid and somewhat time-dependent. By appropriately adjusting both pulse amplitude and duration, we were able to address a 7×5 matrix for risetimes as short as 0.1 - 0.2 s, and several seconds were required for half-selected points to lose contrast. I should emphasize that we have concentrated on simpler display formats; these results were obtained in a few man-weeks of work almost two years ago. The 32-line panel recently reported by Ota shows that complex matrix displays are possible, albeit with slow response.

P. Wild (Brown Boveri)
Would you please elaborate on the electrooptic performance and lifetime of electrophoretic displays operated at very low voltages?

J. C. Lewis
There is a slight voltage dependence of contrast, but legibility is satisfactory even at 2.5 V. Switching times then are hundreds of milliseconds. Low-voltage operation is a recent development for which life tests have not yet been done. There is an indication that reducing the voltage from 45 V to 10 V shortens life, but that is based on fluids particularly developed for use at 45 V. It may be that further development of low-voltage systems will give long life.

F. J. Kahn
What is the nature of the degradation you observe?

J. C. Lewis
Once the initial problem of settling has been overcome, other modes of degradation become apparent after $10^5$ or $10^6$ operations. One of these, on segmented displays, involves migration of pigment particles from one segment to another. However when a certain amount of this migration is observed, it does not necessarily mean that the end of display life is near because the migration can spontaneously reverse with a consequent improvement of contrast during continued operation.

C. von Planta
What do you suggest for new materials?

J. C. Lewis
We would like to extend the color range available, but long life presently requires low-density pigments so we will concentrate initially on organic pigments. Ultimately the best solution will be some method of stabilizing a high-density pigment like titania. We have made progress by using small particles and stabilizing coatings, but it may take two years to achieve lifetimes equal to organic pigments.

# FERROELECTRIC DISPLAYS

K. H. HÄRDTL

Philips GmbH. Forschungslaboratorium, Aachen, Fed. Rep. Germany

## SUMMARY

In general ferroelectric materials show large electrooptic effects. The decisive breakthrough, however, for the large scale application of these materials in electrooptic display devices did not come until 1970, when Haertling succeeded in preparing ferroelectric ceramics of high transparency. These ceramics belong to the lead-zirconate-titanate family with the perovskite structure, in which lanthanum ions are partly substituted for the lead ions (PLZT materials). Electrooptic effects in PLZT ceramics may originate from field-controlled birefringence, light scattering and surface deformation. Which of these three contributions predominates depends primarily on the composition (amount of La substitution and Zr/Ti ratio) and on the microstructure (e.g. grain size) of the ceramics. The strong electrooptic effects are mainly caused by electrically induced changes in the orientation of ferroelectric domains. Compositions can be prepared that show linear or quadratic electrooptic behavior, depending on the crystal structure which governs domain structure and thus domain switching. Other compositions exhibit memory properties due to the remanent polarization of domains. At present a large number of electrooptic devices using PLZT compositions are being developed in many different laboratories. These devices include shutter-type devices for flash and thermal eye protection, ferroelectric picture devices (FERPIC) operating in the birefringent or scattering mode, and alphanumeric displays.

## 1. INTRODUCTION

Electrooptic effects are observed in many solids. They manifest themselves in a field-induced change of the refractive index, of the light-scattering behavior, of the absorption, or of some other optical parameter. Investigations of ferroelectric materials have shown that the electrooptic activity of this class of substances is extraordinarily high. This must be due to an additional electrooptic effect in ferroelectrics. Besides the common, linear Pockels and quadratic Kerr effects, which change the refractive index by $10^{-3}$ to $10^{-4}$, switching of the ferroelectric domains occurs, which may change the refractive index by more than $10^{-2}$. Whereas in the former processes the optical indicatrix is only deformed by the electric fields, domain switching leads to a reorientation of the indicatrix with interchanging of its principal axes, which makes the great changes in the refractive index understandable. Moreover transparent ferroelectric ceramics show a further, novel electrooptic effect which is based on a field-induced phase transition from a non-birefringent to a birefringent phase. This effect, too, may lead to significant changes in the refractive index of more than $10^{-2}$.

Since practically all electrooptic applications require high optical transparency, it has been the practice in the past to use exclusively ferroelectric single crystals mainly from two groups of substances: KDP (potassium dihydrogen phosphate, $KH_2PO_4$) and its isomorphs; and oxides with perovskite or related structures like $KTaO_3$, $LiNbO_3$, or $Bi_4Ti_3O_{12}$. Although many applications of these substances are described in the literature, there has not yet been any large-scale application of such single-crystal materials.

The situation seems to have changed decisively since Haertling in 1970, succeeded in preparing ferroelectric ceramics of high transparancy.[1-3] These ceramics belong to the lead zirconate titanate family with perovskite structure. The lead ions are substituted partly by lanthanum ions[4,5] and the chemical composition is $(Pb,La)(Zr,Ti)O_3$, abbreviated to PLZT.

Polycrystalline PLZT materials are superior to single-crystal substances mainly for two reasons. The first is purely technical and commercial: large pieces can be produced cheaply with the usual ceramic technologies. The second reason is related to the formation of mixed crystals: since an arbitrary choice of the composition is possible, both with respect to the zirconium/titanium ratio and with respect to the lanthanum content, the different electrooptic effects can be optimized for any particular application. Because of these favorable properties, transparent ferroelectric PLZT ceramics have nowadays acquired great importance in research and development, and they may well be on the threshold of industrial mass production. Apart from a brief consideration of competing ferroelectric single crystals, this review will therefore be concerned with the technological, physical and application-related problems of transparent PLZT ceramics.

## 2. PREPARATION OF TRANSPARENT PLZT CERAMICS

Even today the manufacture of polycrystalline substances of high transparency requires considerable technological effort. The transmittance of pure PLZT ceramics is mainly reduced by scattering and not by absorption, since the fundamental lattice absorption starts only at about 3 eV. Scattering of light is due to a number of different effects.

— The first is the presence of a second phase. If the second phase is PbO, for example, the result is an appreciable lowering of the transmittance even at the lowest concentrations. In PLZT, second phases are easily avoided for two reasons: first the existence region of the perovskite phase is relatively broad, and second, an excess of lead oxide evaporates if the sintering times are long enough.

— If second phases are avoided, then residual pores which occur in any normal ceramic are the predominant scattering centers.[6] This means that sintering techniques must be used that eliminate the residual porosity completely. Hot pressing procedures, i.e. sintering with simultaneous application of high pressures, are currently being successfully used.[7,8]

— Grain boundaries occur in all ceramics. Because the crystallographic orientation of the individual grains of a ceramic is random, regions with different orientation will meet at a grain boundary. In a birefringent material a change in orientation generally results in a discontinuity in the refractive index, and this gives rise to light scattering.

— If the material is ferroelectric, there will additionally be domain walls separating regions of different crystallographic orientation, and these also give rise to light scattering by causing discontinuities in the refractive index.[9] Since each grain is divided into a multitude of domains, there are considerably more domain walls than grain boundaries. In a ferroelectric material in which the grain size is not too small, scattering effects due to grain boundaries may therefore be neglected.

Avoiding these transmittance-lowering effects as far as possible, and using carefully controlled processing steps, Haertling, in the Sandia Laboratories, was the first to succeed in making PLZT ceramics of high transparency, and this was followed by successes in many other laboratories. Some typical PLZT ceramics made at Sandia are shown in Fig. 1; the high transparency even in thicknesses of several millimetres can clearly be seen. These ceramics are generally hard, insoluble in water, easy to cut and to polish, and capable of taking an optical finish. In spite of the glassy appearance, the bodies are crystalline, as revealed by the micrograph in Fig. 2. In Fig. 3 the transmittance of a commercially available PLZT ceramic is shown as a function of wavelength. Since plate thicknesses of about 50–300 $\mu$m (depending on the particular application) are sufficient, the optical losses within the specimens are negligible.

## 3. PHASE RELATIONS IN THE PLZT SYSTEM

The various electrooptic effects in PLZT materials are closely correlated with the crystallographic properties of the PLZT system. Knowledge of the phase relations is therefore necessary in order to be able to interpret the physical behavior of PLZT ferroelectrics.

Fig. 1. Transparent PLZT ceramics made by Haertling.[2]

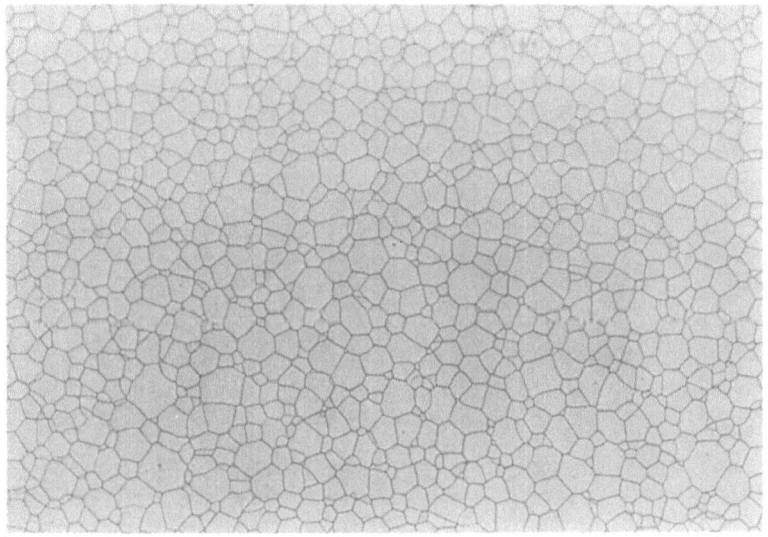

Fig. 2. Micrograph of a transparent PLZT ceramic (molecular ratio La/Zr/Ti = 7.5/72/28). Thermal etching renders the grain boundaries visible.

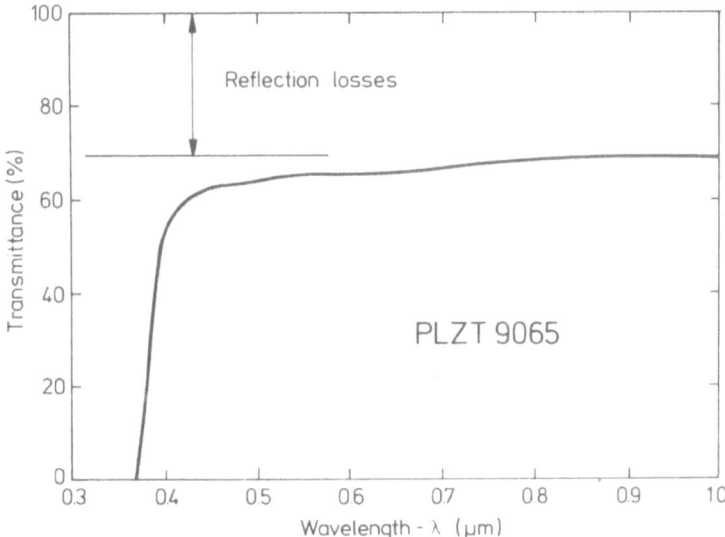

Fig. 3. Transmittance as a function of wavelength for a PLZT plate of 0.14 mm thickness. Data sheet, Motorola Inc., Opto-Ceramic Products.

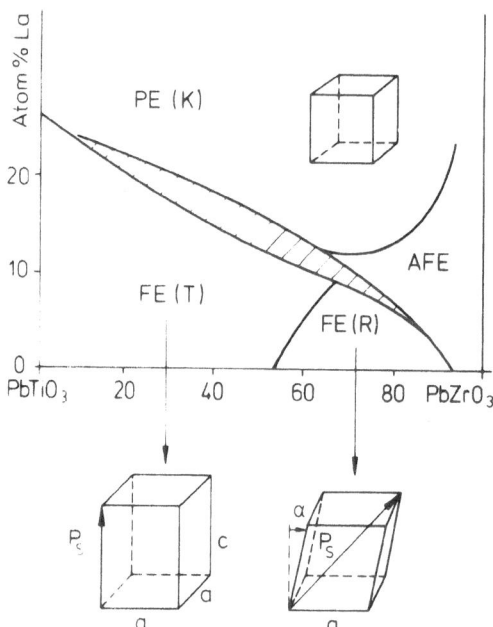

Fig. 4.   Phase diagram of the (Pb,La)(Zr,Ti)O$_3$ or PLZT system, in simplified form.

Figure 4 shows the room temperature phase diagram of the PLZT system in a simplified form.[1] At high lanthanum contents one obtains the ideal cubic, optically isotropic, paraelectric perovskite structure with cubic unit cell. At low lanthanum contents, two ferroelectric phases are observed, one with a tetragonally distorted perovskite cell for titanium-rich compositions, the other rhombohedrally distorted in materials rich in zirconium. The two phase regions are separated by a morphotropic phase boundary. The direction of spontaneous polarization, which coincides with the direction of the optic axis of these two optically anisotropic materials, lies along the c axis in the tetragonal case and along the body diagonal in the rhombohedral case. The c/a ratio of tetragonal specimens decreases from 1.05 in titanium-rich compositions containing no lanthanum to about 1.01 for compositions rich both in zirconium and in lanthanum. A similar behavior is exhibited by the rhombohedral distortion angle $\alpha$, which also decreases with increasing proportions of lanthanum and reaches values of about 10 minutes of arc at the morphotropic phase boundary. The absolute value of the spontaneous strain of rhombohedral specimens is therefore only about one third of that of adjacent tetragonal specimens. This fact will be important in later considerations.

The application of high electric dc fields exceeding the coercive field E$_c$ of the ferroelectric materials has the effect of switching the direction of the spontaneous polarization, and of the optic axis spatially coupled to it. This switching behavior is shown in Fig. 5. Starting from a thermally depoled

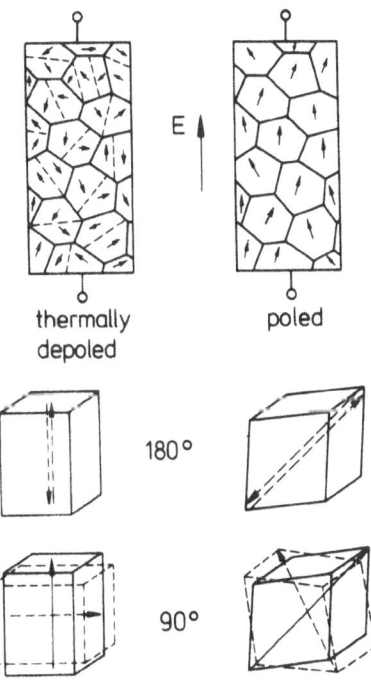

Fig. 5. Switching behavior of PLZT ceramics.

state with the directions of polarization randomly distributed, one obtains, after application of a field $E > E_c$, the poled remanent state with markedly anisotropic properties which are conserved when the field is turned off. Switching can take place via 180° or 90° processes. 180° rotations of the direction of spontaneous polarization leave a uniaxial crystal optically unchanged. Only 90° processes, which in the tetragonal case, for example, interchange the c axis and the a axis, are optically effective. All 90° processes are accompanied by dimensional changes in each grain. The domains are therefore clamped, and switching processes take place only to a reduced degree. Accordingly their number in tetragonal materials with large strain is very small, but in rhombohedral materials with comparatively low strain the number increases rapidly and, particularly at high lanthanum contents, may almost attain the theoretically possible value.[10,11]

Between the cubic phase at high lanthanum contents and the two ferroelectric phase regions there is a lens-shaped transition region, hatched in Fig. 4. The most interesting compositions occur near the intersection of the morphotropic phase boundary with this transition region. These compounds have therefore been investigated in detail at several laboratories.[12-16] Figure 6 represents a part of the phase diagram. The left-hand picture illustrates the phase relations in the thermally depoled state; the right-hand diagram represents the phase relations in the poled, remanent state. The crystallographic structure of the investigated compounds is symbolized by T for tetragonal, R for rhombohedral, C for cubic, and O for orthorhombic. It will be seen that in the poled remanent state there are specimens which exhibit tetrago-

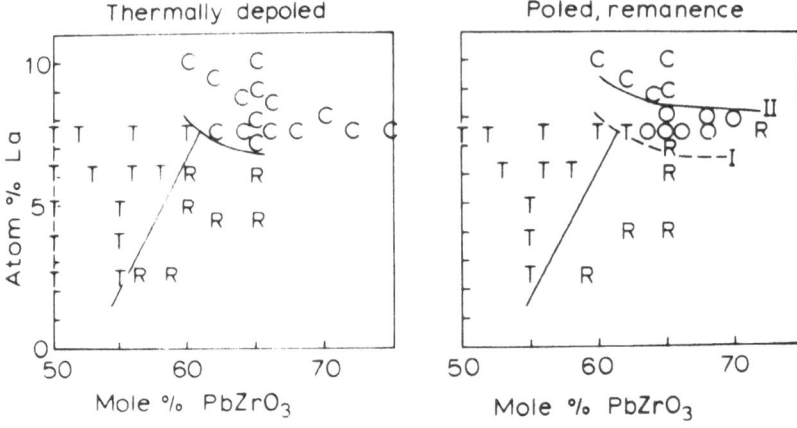

Fig. 6. Partial PLZT phase diagram.[15]

nal, rhombohedral, or even orthorhombic symmetry, whereas in the previous
thermally depoled state they were cubic and therefore optically inactive.

The crystallographic relationships become clearer when the compositions
are observed along a section through the phase diagram, i.e. on examination
of materials with constant Zr/Ti ratio and variable lanthanum content.[15]
The results for the series of compositions with a fixed Zr/Ti ratio of 65/35
and increasing La content x represented by x/65/35 are shown in Fig. 7. At

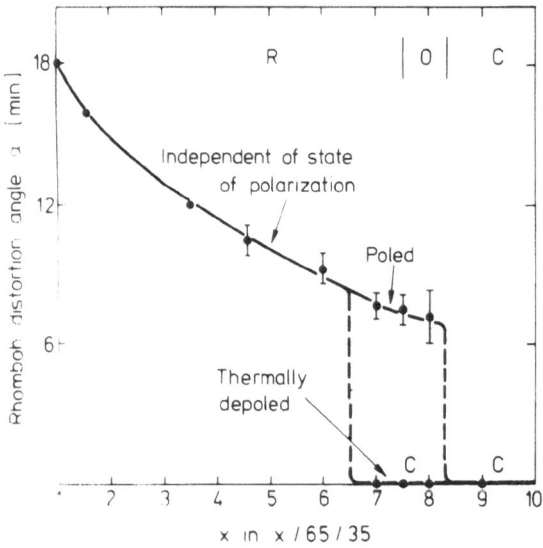

Fig. 7. Rhombohedral distortion angle α as a function of lanthanum content x in
x/65/35. For the orthorhombic phase, the angle plotted corresponds to an equivalent
distortion.[15]

low lanthanum contents the materials are rhombohedral, and their deviation from the cubic high-temperature phase is given by the rhombohedral distortion angle $\alpha$. As the lanthanum content rises, this angle decreases. In the thermally depoled state, all specimens with $x \geqslant 7$ are cubic. In the poled remanent state, however, these specimens are rhombohedral for $x = 7$, and orthorhombic for $x = 7.5$ and $x = 8$. Keve[16] describes this property as a field-induced phase transition from an isotropic cubic $\alpha$-phase to an optically anisotropic $\beta$-phase which remains even after the field has been turned off. Carl and Geisen[13] have suggested that this uncommon behavior is caused by an irreversible transition from an "undercooled" short-range order of microdomains ($\alpha$-phase) to a long-range order with usual ferroelectric properties ($\beta$-phase) brought about by the application of an electric field. Materials of higher lanthanum contents, e.g. $x = 9$, are cubic both in the thermally depoled and in the poled remanent state. Only the application of high fields of about 10 kV/cm induces a transition to a polar state, but for this composition the polar state decays immediately after the field has been turned off. This is characteristic of all compositions within the hatched region of the phase diagram.

## 4. ELECTROOPTIC EFFECTS IN THE PLZT SYSTEM, AND APPLICATIONS

The electrooptic phenomena in PLZT ceramics may be due to three different effects:

- — Electrically controlled birefringence,
- — Electrically controlled light scattering,
- — Electrically controlled surface deformation.

Which of these three contributions predominates will depend primarily on the composition and phase relationships, and also to some extent on the microstructure (e.g. grain size) of the ceramics.

### 4.1. Electrically Controlled Birefringence

The electrically controlled birefringence can be observed in PLZT ceramics only in a transverse-mode operation of electric field and optical beam. In longitudinal-mode operation, birefringence effects are possible only under special experimental conditions, e.g. using fringing fields. Figure 8 illustrates the birefringence behavior of a series of characteristic compositions. Figure 8a shows the conventional test set-up with crossed polarizer and analyzer, and Fig. 8b gives the characteristic compositions A, B, C and D within the phase diagram. The lower part of Fig. 8 shows the hysteresis behavior and the $\Delta n(E)$ dependence of the four characteristic compositions.

Material A, e.g. 14/65/35, is parelectric and thus behaves as a linear dielectric. Its electrooptic behavior corresponds to the quadratic $\Delta n(E)$ dependence of the classic Kerr effect.

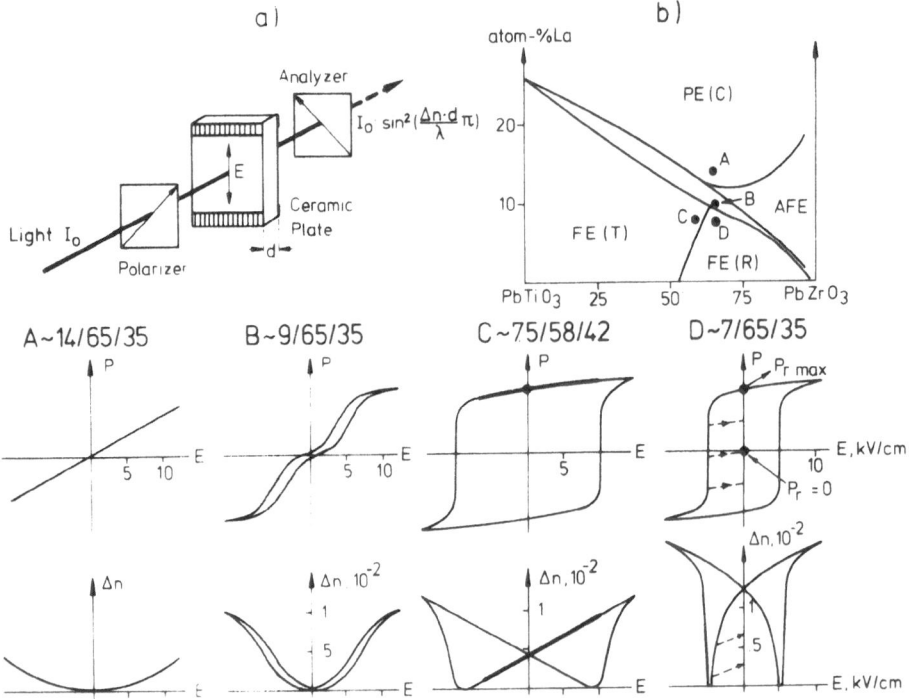

Fig. 8. Electrically controlled birefringence of characteristic PLZT materials.

Material C, e.g. 7.5/58/42, lies in the region of tetragonal compositions with high spontaneous strain, so ferroelectric switching processes take place only at relatively high fields. At low fields such a composition exhibits the linear $\Delta n(E)$ dependence of the Pockels effect.

Material B in the transition region, e.g. 9/65/35, shows the typical slim-loop hysteresis. This may be considered as being derived from the linear behavior of the cubic material at small fields which at higher fields changes abruptly to the nonlinear behavior of the field-induced ferroelectric phase. Accordingly the $\Delta n(E)$ behavior too can be derived from the quadratic Kerr effect at low fields and the linear Pockels effect at high fields.

Finally material D, e.g. 7/65/35, is rhombohedral with slight spontaneous strain. Switching processes here are achieved at relatively low fields of about 5 kV/cm. Remanent states between $+(P_r)_{max}$ and $-(P_r)_{max}$, including the electrically depoled state $P_r = 0$, can be obtained and stored. Accordingly it is also possible to adjust remanent values of $\Delta n$ between 0 and the maximum value of $\Delta n$. These materials thus exhibit unique memory properties.

The clear advantage of transparent PLZT ceramics compared to other electrooptic materials is that, with appropriate selection of the compositions, all intermediate compositions of the four representative materials chosen here can be manufactured with optimized properties for the application in question. The attainable induced birefringence in all specimens is about $10^{-2}$ except for specimens of type A, which means that a halfwave retardation can be attained in plates only 25 $\mu$m thick.

Applications. No applications of type-A materials have yet been reported. Applications have, however, been published for materials of types B, C and D, which will now be discussed in more detail.

The transverse linear electrooptic effect of type-C materials can be utilized to construct an optical modulator used for sensing low voltages. This was done by Thacher with transparent 12/40/60 PLZT.[17] The arrangement and the test results are shown in Fig. 9. The expected linear change of the birefringence is found over a wide voltage range.

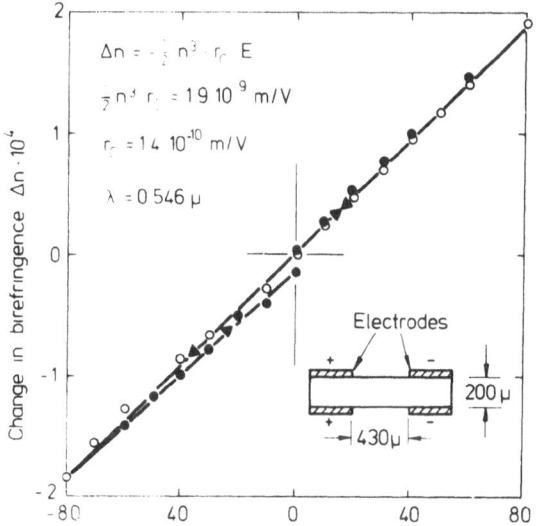

Fig. 9. Change in birefringence $\Delta n$ as a function of voltage for 12/40/60 PLZT. Open and closed circles indicate data applicable to increasing and decreasing voltages, respectively.[17]

Materials of the transition region, e.g. type B (9/65/35) have the great advantage, from a practical point of view, of being optically isotropic in the absence of an electric field.[12,13] This zero-field isotropy gives an excellent off-state for any device placed between crossed polarizer and analyzer and allows large contrast ratios. This material has no intrinsic memory. Nevertheless the electrical resistivity of the materials is greater than $10^{14}$ $\Omega$cm, so that capacitive storage of the displayed information is possible. The duration of storage depends on the insulating ability of the entire set-up and may be several hours with optimum insulation. Short-circuiting the electrodes erases the information immediately. Examples of application are the so-called goggles[18] and alphanumeric displays.[19] Goggles are protective spectacles with a transmittance controlled by the incident light via the transverse electrooptic effect. They are intended primarily for aviation, but welding and three-dimensional television[20] are also of interest. Here large-scale applications seem to be possible.

As shown schematically in Fig. 10, transverse addressing takes place with unilateral interdigital structures by means of transparent electrodes. The retardation thus achieved is proportional to $E^2$ multiplied by the effective thickness. In the arrangement shown here the effective thickness will be roughly equal to the interdigital distance s, so that the retardation will be proportional to $E^2$s or $U^2$/s. With a small interelectrode distance (20–50 $\mu$m) this permits in principle the achievement of high contrast ratios even at low voltages (< 50 V).[21] The plate thicknesses used may be a few hundred $\mu$m. The optical effect is restricted to a surface layer of thickness s, the rest serving as an optically inactive substrate. The switching times decrease rapidly with rising field strengths and are of the order of microseconds. At higher fields the switching times are generally limited by the addressing electronics.[22] Since the mechanical strains during the switching process are relatively small, these materials withstand many switching cycles (> $10^9$) without evidence of degradation.[23]

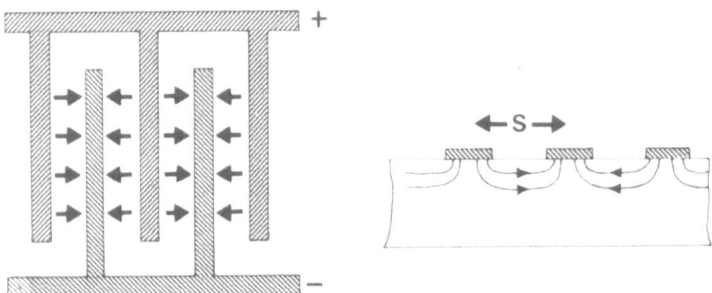

Birefringence is proportional to $E^2 \cdot$(effective thickness)

i.e.     $\Gamma \propto E^2 \cdot s$

$\propto \dfrac{V^2}{s}$

Fig. 10. Interdigital electrode structure.[19]

Materials of the transition region appear to be suitable for alphanumeric displays as well. Figure 11 shows schematically the construction of a crossbar-addressed display of this type.[19] It consists of two plates of transparent ceramic between three crossed polarizers. The first plate serves to address the horizontal rows, the second to address the vertical columns. The transparent addressing electrodes again consist of unilaterally applied interdigital struc-

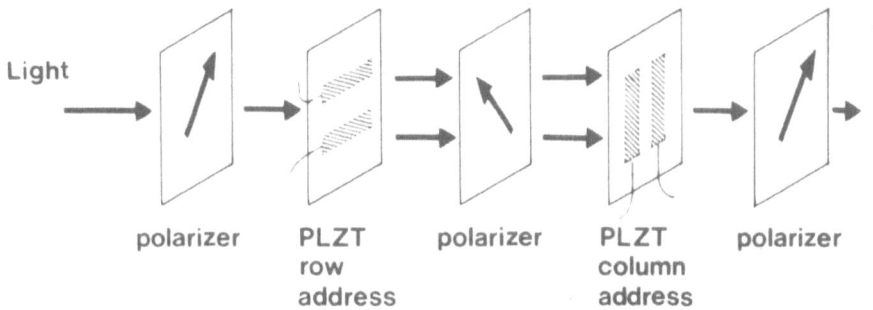

Fig. 11. Crossbar-addressed display element.[19]

tures. With an appropriate voltage, only that part will transmit light which is simultaneously addressed by a row and a column. If the usual 7 rows and 5 columns are used, any numeral or letter can be represented. With only one ceramic plate used between crossed polarizers, the simple structures of the standard 7-bar numeric arrays can be used. The same conditions as those just mentioned apply to switching voltage, switching time and number of switching cycles. The energy required for one switching cycle is about 0.5 $\mu J/mm^2$. The important factor for large-scale application will be whether it will be possible to lower the switching voltage to a magnitude compatible with integrated circuits by optimizing the PLZT composition, the microstructure, and the electrode geometry.

For applications requiring true storage properties, materials of type D with memory properties are necessary. Here the difficulty will rather be the erasure of the information, which can be done by thermal or electrical depolarization.

The FERPIC shown in Fig. 12, which was developed by Meitzler and Maldonado,[24] uses electrical depolarization for erasing. It consists of a ceramic plate of transparent 7/65/35 PLZT, 75 $\mu m$ thick. The plate is provided with a photoconducting film and transparent electrodes and is bonded to a plexiglass substrate. Whereas the devices discussed so far are electrically adressable, those equipped with photoconductors are addressed optically. Flexing of the plexiglass substrate leads to a tensile stress in the ceramic plate. This causes the well-known domain switching processes, and the plate becomes birefringent with the principal axis of the optical indicatrix along the strain axis. In operation, the image to be stored is projected onto the photoconductive film. A voltage applied to the transparent electrodes causes a field in the ceramic with an intensity modulated by the photoconductive film. When the field is removed, the desired image is stored as a spatial modulation of the birefringence of the ceramic plate. To erase the image, the entire structure is flooded by light in the presence of an electric field in the reverse direction, and the plate returns to its initial state. A half-wavelength

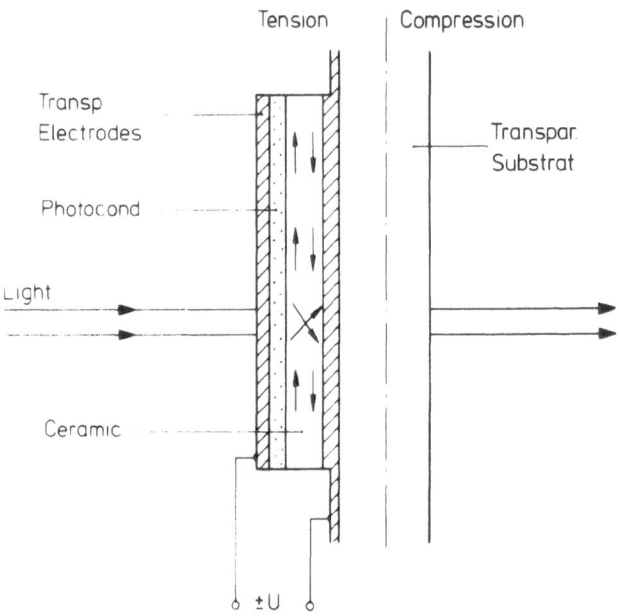

Fig. 12.  Strain-biased image storage and display device called FERPIC.

change of retardation is achieved in a plate of 75 $\mu$m thickness by variation
of the remanent polarization from its maximum value to zero. The FERPIC
is capable of an image resolution of 40 line pairs/mm, it has a contrast ratio
of 15 dB, and it requires about 100 V for operation. A FERPIC device
should in principle be capable of large-screen TV projection in real time. One
problem seems to be the lifetime of the device, which is required to be very
long and must exceed $10^{10}$ operating cycles. Besides this, a number of other
technical problems still remain to be solved.

### 4.2. Electrically Controlled Light Scattering

In contrast to electrically controlled birefringence, electrically controlled
light scattering can be operated in the much more favorable longitudinal
mode. According to Dalisa and Seymour,[9] scattering ability depends first of
all on the number and magnitude of refractive index discontinuities. If the
production processes of the transparent ceramic are controlled so well that
scattering by second phases and by pores is absent, then refractive index discontinuities occur only at 90° domain walls. Electrically controlled light
scattering can therefore be attributed mainly to a change in the number
and/or the magnitude of these discontinuities at the domain walls. If this
basic explanation is taken together with the structural information given in
Fig. 7, the electrically controlled light scattering behavior shown in Fig. 13
for compositions x/65/35 can be deduced. Figure 13 shows schematically
the expected change in the intensity of scattered light $I_s$ as a function of the
lanthanum content x of a series of x/65/35 specimens in the poled, electri-

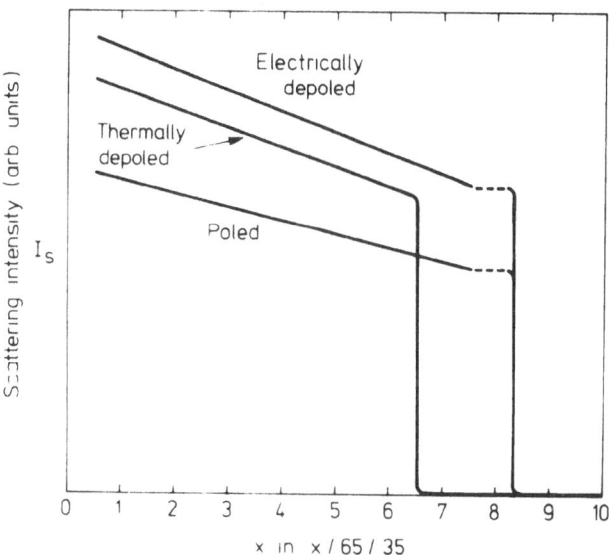

Fig. 13. Schematic representation of expected variation of light scattering $I_s$ of PLZT ceramics of composition x/65/35 with the material in the poled, in the electrically depoled, and in the thermally depoled states.[13]

cally depoled, and thermally depoled states, as postulated by Keve and Bye.[15] In all three states the scattered intensity falls continuously with rising lanthanum content since the magnitude of the refractive index discontinuity is related directly to the rhombohedral angle $\alpha$, which falls with x in accordance with Fig. 7. This is the basic reason for the increased transparency of lanthanum-doped ferroelectric ceramics: only with lanthanum contents in the region of $x \geqslant 6$–7 have the rhombohedral angle $\alpha$, the birefringence and the refractive index discontinuities become so small that the specimens exhibit scattered intensities low enough that one can speak of transparent materials.

Thermally depoled specimens with $x < 7$ show high scattering. With $x \geqslant 7$, scattering is negligible because the material has the structure of the cubic $\alpha$-phase with no discontinuity in the refractive index. If specimens with $x = 7$–8 are poled or electrically depoled, the polar $\beta$-phase is induced and scattering sets in again. In this case an electrically depoled specimen exhibits a greater number of refractive index discontinuities than the poled specimen on account of its higher concentration of 90° domain walls, and its scattered intensity is correspondingly higher.

Applications. The two modes of operation of electrically controlled light scattering can easily be derived from Fig. 13. A change of the scattered intensity can be expected when the specimen switches from the poled to the electrically depoled state. A second change will occur on switching from the thermally depoled to the poled state in the case of specimens with $x \geqslant 7$. In the latter case, high contrast ratios are to be expected because the isotropic state of the thermally depoled material represents an excellent off-state, but thermal erasing involves many technical difficulties.

The change of the scattered intensity on switching from the poled to the electrically depoled state was used by Smith and Land in the construction of their "scattering-mode ferroelectric-photoconductor image storage and display devices".[25] As in the case of the FERPIC, the structure consists of a sandwich arrangement of ceramic, photoconductor and transparent electrodes. Figure 14 shows the test results obtained with a helium-neon laser and a detector with an aperture angle of about $2^{\circ}$. As expected, the intensity of the transmitted light is greatest in the two poled states A or B, and least in the electrically depoled state C. The two test curves were obtained with 7/65/35 PLZT plates 0.375 mm thick, made by different production processes. The contrast ratio of more than 100 is very high but depends sensitively on the aperture angle, the preparation method and on the microstructure of the ceramic used.

With increasing aperture the contrast falls appreciably, so these scattering devices do not seem to offer prospects of alphanumeric displays. However, one should keep in mind that scattering-mode devices have some advantages in comparison with birefringent-mode devices:

—   They exhibit a longitudinal effect
—   Polarizer and analyzer are not necessary
—   Plate thickness is not critical
—   Sources of white light can be used for viewing and projection.

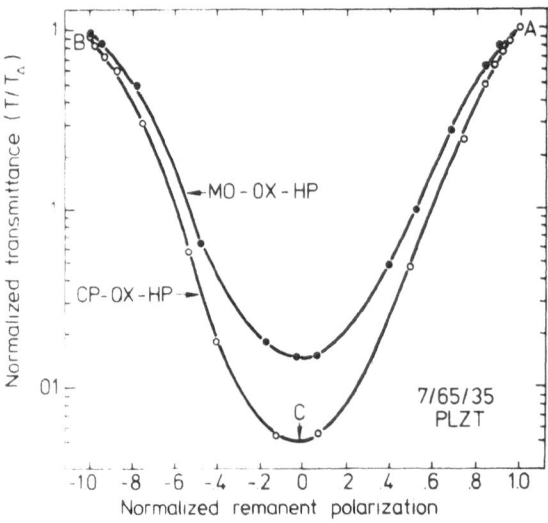

Fig. 14. Normalized transmittance as a function of normalized remanent polarization of two PLZT plates 0.375 mm thick, produced by different methods.[15]

## 4.3. Electrically Controlled Surface Deformation

In 1973 Land and Smith reported another electrooptic effect observable in PLZT ceramics, namely electrically controlled surface deformation.[26] The effect is based on the phenomenon that certain PLZT ceramics exhibit a pronounced change in volume on cycling of the hysteresis loop. The volume is greatest at maximum remanent polarization and smallest at negligible remanent polarization. The maximum change in volume is more than $10^{-3}$.

Applications. As shown in Fig. 15a, the surface deformation device called FERICON is a five-layer structure with a reflective back electrode and a transparent front electrode.[27] It further comprises two photoconductive layers to produce deformation on both surfaces of the ceramic plate so as to

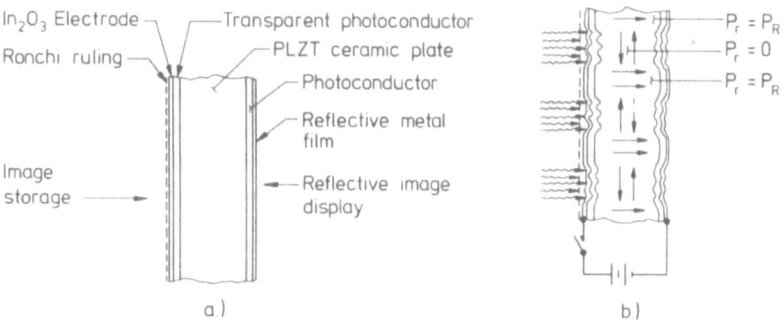

Fig. 15. a) The FERICON structure for image storage and display.[25] b) Surface deformation resulting from a bar pattern stored through the Ronchi ruling.[25]

avoid distortion of the planar configuration. Initially the entire ceramic plate is at maximum remanent polarization. In Fig. 15b, a light intensity pattern is focussed upon the transparent electrode side until the areas illuminated with the highest intensity are switched to zero remanent polarization. The displayed image is made visible with a Schlieren optical system in reflective-mode operation. The incident collimated light is modulated spatially by a Ronchi ruling; without this modulation, only the edges of the bars would be observable. The maximum resolution is about half the number of line pairs/mm of the ruling. If the reflecting back electrode of the FERICON is replaced by a transparent electrode, the surface deformation effect is also useful in transmission-mode operation. If a transparent ceramic with a good longitudinal scattering effect is used, e.g. a coarse-grained 7/65/35 PLZT, both effects combine to create the CERAMPIC device, which has a resolution of 35–40 line pairs/mm and a contrast ratio of 17–20 dB. Tests have shown that at least $2 \times 10^4$ storage/erasure cycles are possible with no sign of degradation.[27]

## 5. SINGLE CRYSTAL MATERIALS

A decisive improvement of all ferroelectric materials in electrooptic applications would be a lowering of their coercive field and hence a reduction of the switching voltage. Transparent PLZT ceramics, which have so many advantages commercially, technologically and physically, unfortunately have rather high $E_c$ values. Given an appropriately chosen composition the coercive fields can be reduced to 4 to 5 kV/cm, yet this value seems to be a limit in the PLZT system which is not easy to overcome. Some ferroelectric substances of other systems have lower $E_c$ values and should be more useful from the point of view of lower switching voltages. Since optical transparency in the polycrystalline state of ferroelectrics has so far been achieved only with PLZT systems, expensive single crystals are needed in the other systems, if indeed it is at all possible to grow single crystals of sufficient size and homogeneity.

One of the single crystal materials under consideration is KTN, which is a solid solution with perovskite structure represented by the formula $KTa_xNb_{1-x}O_3$.[28,29] KTN has a transition temperature which is linearly dependent on the composition ratio x. In particular with $x \sim 0.6$, the Curie temperature is just below room temperature. At room temperature this composition is cubic and exhibits a very large quadratic electrooptic effect. Unfortunately, however, the material suffers from inhomogeneities in composition and from the strong temperature dependence of the necessary driving voltage. These disadvantages have prevented its use in practical devices.

The so-called "Titus" tube for large-screen TV projection uses large single crystals of ferroelectric KDP.[30] Though this crystal has a number of drawbacks — in particular, having to be cooled down to $-50°$ C, the severe risk of degradation if heated above $100°$ C, and sensitivity of humidity — results obtained with experimental devices have shown that these difficulties can be overcome.

Single crystals of $Gd_2(MoO_4)_3$ can be produced in good quality and with large dimensions.[31,32] The switchable birefringence of $Gd_2(MoO_4)_3$ is fairly low ($\Delta n = -4 \times 10^4$), so that in order to obtain half-wave retardation a thickness in the switching direction of about 400 $\mu$m is required. Since $E_c$ of $Gd_2(MoO_4)_3$ is about 2 kV/cm, a voltage of about 100 V is needed for switching, which cannot be provided by integrated circuits.

Of all single crystal materials, $Bi_4Ti_3O_{12}$ seems to be most promising.[33-35] Its structure at room temperature is monoclinic, and it is spontaneously polarized in the monoclinic a-c plane with a big component of $50\ \mu C/cm^2$ along the a axis and a small component of $4\ \mu C/cm^2$ along the c axis. Switching along the c axis takes place by means of rocking the polarization vector through an angle of $9°$. This rocking is accompanied by a $58°$ rotation of the axis of the optical indicatrix. Experiments with epitaxial $Bi_4Ti_3O_{12}$ films of 16 $\mu$m thickness, grown by sputtering on MgO substrates, shows polarization values equal to those of bulk crystals. However the coercive fields are increased. Along the c axis, $E_c$ was 12 kV/cm, which is approximately 3.5 times larger than the bulk value, and along the a axis it was approximately 90 kV/cm, which is about twice that of the bulk value. Since birefringence is rather high ($\sim 10^{-2}$), thin slices can be used, thus allowing switching voltages which are compatible with integrated circuits.

During the last few years in particular, the technology of displays has become very competitive and cost conscious. In this respect, it seems clear that ferroelectric ceramics are of considerable advantage compared to single-crystal ferroelectrics, since single crystals of the relatively large sizes required for a direct-view display are simply too expensive.

## REFERENCES

1   G. H. Haertling and C. E. Land, J. Am. Cer. Soc. 54 (1971) 1

2   G. H. Haertling, J. Am. Cer. Soc. 54 (1971) 303

3   G. S. Snow, J. Am. Cer. Soc. 56 (1973) 91

4   D. Hennings and K. H. Härdtl, Phys. Status Solidi a 3 (1970) 465

5   K. H. Härdtl and D. Hennings, J. Am. Cer. Soc. 55 (1972) 230

6   J. G. J. Peelen and R. Metzelaar, J. Appl. Phys. 45 (1974) 216

7   G. H. Haertling, Am. Cer. Soc. Bull. 43 (1964) 875

8   K. H. Härdtl, Philips tech. Rev. 35 (1975) 65; Amer. Cer. Soc. Bull. 54 (1975) 201

9   A. L. Dalisa and R. J. Seymour, Proc. IEEE 61 (1973) 981

10  K. Carl, Dissertation, Univ. Karlsruhe (1972)

11  G. Krüger, Dissertation, Univ. Karlsruhe (1975)

12  A. H. Meitzler and H. M. O'Bryan, Jr., Proc. IEEE 61 (1973) 959

13  K. Carl and K. Geisen, Proc. IEEE 61 (1973) 967

14  E. T. Keve and A. D. Annis, Ferroelectrics 5 (1973) 77

15  E. T. Keve and K. L. Bye, J. Appl. Phys. 46 (1975) 810

16  E. T. Keve, Appl. Phys. Lett. 26 (1975) 659

17  P. D. Thacher, Ferroelectrics 3 (1972) 147

18  G. H. Haertling, IEEE Symposium "Applications of Ferroelectrics", Albuquerque, N. M., USA (June 1975)

19  R. W. Cooper, Mullard Res. Lab., Redhill, England, private communication

20  A. S. Khallafalla and J. A. Roese, IEEE Symposium "Applications of Ferroelectrics", Albuquerque, N. M., USA (June 1975)

21  K. L. Bye and R. W. Cooper, IEEE Symposium "Applications of Ferroelectrics", Albuquerque, N.M., USA (June 1975)

22  G. Wolfram, Elektronik Praxis 10 (February 1975)

23  K. Carl, NTG-Diskussionstagung, Frankfurt, Germany (March 1973)

24  J. R. Maldonado and A. H. Meitzler, Proc. IEEE 59 (1971) 368

25  W. D. Smith and C. E. Land, Appl. Phys. Lett. 20 (1972) 169

26  C. E. Land and W. D. Smith, Appl. Phys. Lett. 23 (1973) 57

27  W. D. Smith, J. Solid State Chem. 12 (1975) 186

28  J. E. Geusic, S. K. Kurtz, L. G. van Uitert and S. H. Wemple, Appl. Phys. Lett. 4 (1964) 141

29  A. J. Fox and P. W. Whipps, Electron. Lett. 7 (1971) 139

# MATRIX ADDRESSING OF NON-EMISSIVE DISPLAYS

A. R. KMETZ

Brown Boveri Research Center, Baden, Switzerland

## SUMMARY

Matrix addressing affords reductions in the number of interconnections and drivers, and sometimes simplification of fabrication, which are needed for all but the simplest displays. Static addressing can be advantageously employed for displays such as bargraphs and reticles whose pattern set is restricted a priori, but dynamic (multiplexed) matrix addressing is needed for general-information displays. Techniques for analyzing the trade-off between multiplexing capacity and optical performance are reviewed for specific examples of display devices with four categories of transient behavior: rms response, persistence, memory and instantaneous response. Finally the status, problems and potential for multiplexed operation of all of the principal non-emissive electrooptic display technologies are surveyed. The most promising candidates for further development are judged to be twisted nematic displays, cholesteric-nematic phase transition guest-host displays and electrophoretic displays.

## 1. INTRODUCTION

### 1.1. Matrix Displays

A multi-element display generally performs best when each element is individually driven, but this is feasible only when the number M of display elements is not too large. The first commercial calculators with liquid crystal displays[1] were successfully mass-produced despite 67 individually driven display leads. However the high system cost of so many drivers and intercon-

nections proved uncompetitive with calculators which use only a single integrated circuit. When M exceeds about 100, it becomes difficult even to fabricate a display with direct electrical access from the outside to every individual element. The motivation for matrix addressing is thus clear and com-
pelling: the number of external lines that must be driven is reduced to as
little as $2(M)^{1/2}$ when each line is shared by many elements, and fabrication is simplified if the M elements are defined by the intersections of two
orthogonal sets of conductors. The bulk of this paper examines the compromises in performance that must be made to achieve these gains.

Matrix addressing is not restricted to displays which are simple dot ras-
ters. Segmented numeric and alphanumeric characters exemplify displays
which convey information through the position and shape of their elements.
Other examples are pictorial displays for advertising and for process control.
In these cases too, the driver economies of matrix addressing can be obtained
by appropriately interconnecting the display elements. Where these interconnections are made within the display itself, fabrication as a matrix simplifies
packaging and system assembly.

A segmented numeric display font suitable for matrix addressing is
shown in Fig. 1a. Replicating the front and rear electrode patterns laterally
N times yields an N digit display with like segments interconnected horizontally and with each digit brought out vertically. Although the elements have
particular shapes and positions, they are conceptually equivalent to the intersections of the $8 \times N$ simple matrix shown. The manner in which display ele-

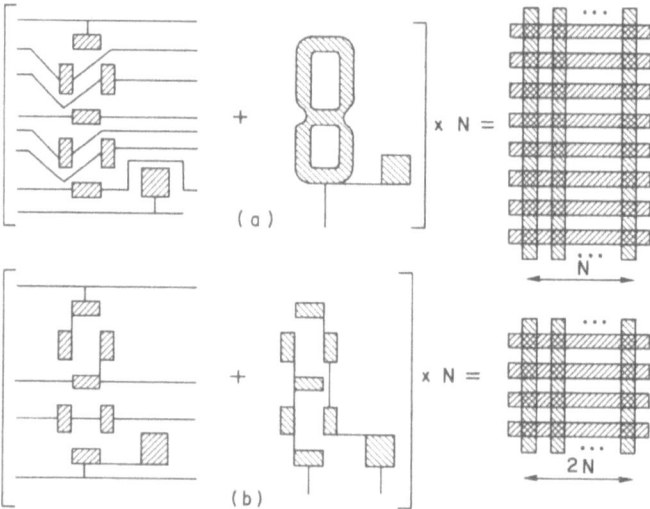

Fig. 1. Two examples (a and b) of interconnection patterns for front and rear electrodes of matrix-addressed numeric display.

ments are interconnected is in principle arbitrarily chosen. The minimum lead count is obtained if the elements are connected as a square matrix. Thus the pattern in Fig. 1a is optimal for an eight-digit display, 64 elements then being addressed by only 16 lines. Furthermore this arrangement is a natural choice if data are presented in digit-serial format. However the interconnections are photolithographically complex, and cosmetic requirements give rise to constrictions in the leads which raise their resistance. These problems are avoided by the alternative interconnection pattern[2] shown in Fig. 1b. This layout is likewise equivalent to a matrix, but one with a different aspect ratio. The lead count is higher if $N > 4$, and a non-standard decoder is required.

## 1.2. Static Matrix Addressing

The simplest way to address a matrix is just to apply various steady voltages to its row and column electrodes. It is apparent that each display element is connected through series/parallel combinations of other elements to every matrix electrode. Sobel[3] has analyzed the effect of these "sneak paths" on the voltage and current distributions in a matrix under various conditions of static addressing. Unless the elements have diode impedance characteristics, he finds that it is generally desirable to connect every address line to a low impedance voltage source, with none left unconnected and thus floating. In that case the potential difference across each matrix element is simply defined by the voltage difference between its own row and column electrodes.

The principal limitation of static addressing is its inability to display all possible patterns. For example no static scheme can display a closed figure with a hollow center. This would involve the contradiction that the same row and column voltages should cause adjacent elements to be simultaneously excited and unexcited. Nevertheless static addressing can sometimes be used advantageously in applications where the information to be displayed is restricted a priori. This is illustrated by the following examples.

Consider a matrix whose address electrodes are connected to a voltage either V or 0. Elements whose row and column voltages are the same see no potential difference and remain unexcited. The other elements are excited by a voltage ± V. If V is large enough and the device is, like liquid crystals, insensitive to drive polarity, then the excited elements uniformly display full contrast. Matrix drive voltages need be no higher than for direct driving of the individual elements. No electrooptic threshold is required, and matrix size and resolution are not limited by addressing considerations. Rather complex patterns can be displayed, subject to the restriction that each row be the same as any other row or its complement. This scheme has been employed for electrically positioned cross-hair graticules in optical instruments[4] and to generate the restricted pattern set which corresponds to an electrically selectable Walsh transform filter for an optical processing application.[5]

A bargraph is another example of a display whose lead count can be reduced by interconnecting its elements as a matrix and whose pattern set is restricted a priori. In particular, if groups of adjacent elements are considered as columns in an equivalent matrix, any bargraph pattern will comprise fully selected columns, fully unselected columns and at most one column with both selected and unselected elements. An elegant scheme[6] for static addressing of such a display uses unipolar squarewaves for all addressing waveforms. Signals in fully selected, mixed and unselected columns have phase shifts of $0°$, $90°$ and $180°$. The information content in the mixed column is specified as $\pm 45°$ phase shifts of the signals on the row electrodes. Consequently the effective voltage difference across any selected element is 1.73 times larger than that across any unselected element. Thus a bargraph with any desired resolution M can be addressed with only $2(M)^{1/2}$ leads if the display device has sufficient contrast at an rms voltage 73% above threshold.

The repertoire of patterns available by static addressing can be enlarged by stacking two or more matrix displays so they are optically in series.[7] The composite display then transmits light only where transmitting elements coincide on all levels. By thus extending the matrix into the third spatial dimension, the number of lines required to select one of M elements can be reduced.[8] With L matrix levels, the total number of address lines can be as little as $2L(M)^{1/2L}$. For example 4096 points require 128 address lines if a single level of selection is used, but the same selection can be accomplished by stacking three $64 \times 64$ displays, each interconnected to form a different $4 \times 4$ matrix, for a total of only 24 independent address lines. The driver count is thereby greatly reduced, but the number of external connections may actually increase. Moreover multilevel displays suffer from reduced transmission, non-uniform perceived depth and parallax. When these deficiencies are considered with the additional cost and complexity, the use of this technique is seldom warranted.

## 1.3. Dynamic Matrix Addressing

Static addressing is unable to display all possible matrix patterns because, by definition, it must satisfy selection requirements of all parts of the display simultaneously. It is natural then to consider satisfying the conflicting drive requirements of different locations sequentially. Figure 2 is a schematic illustration of such time-multiplexed matrix addressing. The columns are scanned in sequence, only one being selected at any instant by the strobe voltage $V_s$. The information to be displayed is multiplexed onto the row electrodes, which are shared by all the columns, in synchronism with the column scan. During the selection interval or dwell time for a given column, voltages corresponding to the data to be displayed in that column are applied to the row electrodes: the instantaneous potential difference across an excited element is $V_s - V_1$ and across an unexcited element is $V_s - V_0$. The choice of these voltage levels will be discussed later.

The figure shows column-at-a-time addressing wherein data are presented simultaneously to all rows. With a non-square matrix or with predetermined

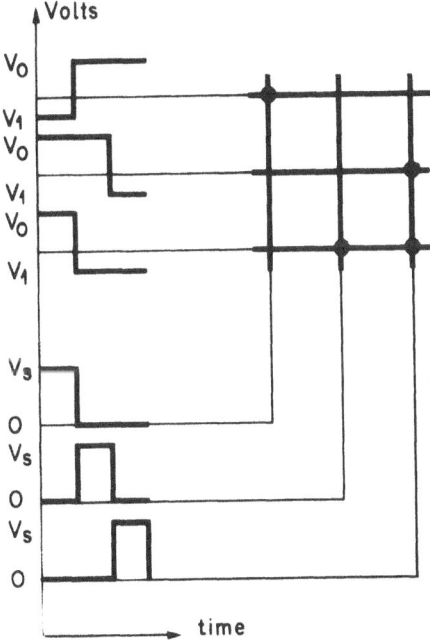

Fig. 2. Schematic illustration of dynamic (time-multiplexed) matrix addressing.

data source format (e.g. television), it may be desirable to interchange the functions of rows and columns. Point-at-a-time addressing with data presented sequentially to each row during each column strobe interval is also possible, but the reduction of effective element-selection time is generally deleterious to display performance.

With time-multiplexed or dynamic matrix addressing the drive voltages change with time, so the transient response characteristics of the display device are an important consideration in analyzing matrix performance. In fact it is convenient to divide non-emissive displays into four categories according to transient response as shown by Table 1. When the risetime $t_\uparrow$ and fall time $t_\downarrow$ are both short compared with the drive pulse duration (and therefore also with the frame repetition period T), the electrooptic response follows the instantaneous drive voltage waveform. When the fall time is at least as long as the period, the display effect produced by one pulse persists until it is refreshed by the next pulse. However, if the risetime is also long (specifically with respect to the pulsewidth), then the steady-state response is determined by the integral over a full period of some function of the whole drive waveform. Finally, memory behavior is obtained if a given displayed pattern will be changed or ignored after an interval shorter than the fall time.

The analysis of multiplexed display performance which is appropriate to each of these categories of transient response is illustrated by specific examples in the next section. Although a single device can in principle be operated

## TABLE 1

Display Categories According to Transient Response

| Display Mode | Response Times | Example |
|---|---|---|
| Instantaneous | $T \gg t_{\downarrow}, t_{\uparrow}$ | Non-remanent PLZT ceramic |
| Persistence (slow scan) | $t_{\downarrow} \gtrsim T > t_{\uparrow}$ | Cholesteric phase change |
| Integrating (fast scan) | $t_{\downarrow} > t_{\uparrow} \gtrsim T$ | Twisted nematic |
| Memory | $t_{\downarrow} \to \infty$ | Cholesteric texture change |

in more than one category by changing the drive frequency, the examples show the most practical mode for each device chosen. After the discussion of the techniques of multiplexing, the numerous technologies in the field of non-emissive electrooptic displays are surveyed in the final section to assess their applicability to matrix displays.

## 2. DETERMINATION OF MULTIPLEXING CAPACITY

### 2.1. Displays with Instantaneous Response

Slim-loop PLZT ceramics with 9/65/35 composition exhibit switching times in the microsecond range for electrooptic effects having potential utility for displays.[9] The application of an electric field induces a reversible phase transition to a ferroelectric state evidenced by the appearance of birefringence or scattering, depending on grain size. Upon removal of the field, the material reverts immediately to a cubic phase which is optically isotropic. Both transitions are sufficiently fast that the device response effectively follows the instantaneous voltage waveforms found in most display applications.

The optical performance of a flicker-free display is determined by the average contrast ratio. If the N columns of a matrix with instantaneous response are scanned sequentially in a period T, then a selected element achieves its peak contrast ratio $CR_{pk}$ only during the strobe interval T/N, having unity contrast ratio for the remainder of the scan period. Consequently the average contrast ratio $CR_{av}$ is considerably lower than $CR_{pk}$:

$$CR_{av} = \frac{CR_{pk}}{N} + \frac{N-1}{N}. \tag{1}$$

If a contrast ratio of 5 to 7, which is typical for newsprint, is taken as the lower limit for acceptable performance, then even a simple calculator display with N = 8 demands a peak contrast ratio around 40. Although the peak contrast ratio of an emissive display can be increased almost at will by raising the drive current, the contrast ratio of a non-emissive display is limited by saturation of the electrooptic effect. Improbably high saturation contrast ratios are sometimes reported for particular measurement conditions (collimated illumination, normal incidence, very narrow detector acceptance angle, monochromatic light, dark polarizers with nearly perfect extinction), but contrast ratios under conditions appropriate to direct viewing of practical displays seldom exceed 10 for any non-emissive device. Instantaneous response therefore virtually disqualifies a non-emissive device from consideration for dynamic matrix addressing of flicker-free displays with N > 2.

A corollary of this conclusion is that a non-emissive device must exhibit a threshold if it is to be used for flicker-free display of unrestricted patterns by matrix selection. It has been suggested[7] that the threshold requirement can be circumvented by stacking two device arrays, one defining rows and the other columns, optically in series. The composite display is opaque everywhere except at the intersections of clear rows and a clear column. Such a device has been used as a page composer with column-sequential matrix addressing for multiple-exposure holography.[10] However in a flicker-free display, the requirement that one column turns off before the next turns on implies instantaneous response, and then low average contrast prevails.

If more than two levels of selection are used to reduce the lead count by the methods outlined at the end of Section 1.2, then point-by-point selection replaces line-at-a-time selection and average contrast becomes substantially worse:

$$CR_{av} \cong 1 + \frac{CR_{pk}}{M}, \qquad (2)$$

where the total number of display elements $M \gg N$. Hence threshold and slow decay are both requirements for directly viewed non-emissive matrix displays, in practice if not in principle.

## 2.2. Displays with Persistence

If a cholesteric layer[11] initially in the focal-conic texture has a positive dielectric anisotropy, then the application of an electric field larger than a critical value causes a relatively rapid transition to the homeotropic nematic state. The layer reverts to the cholesteric state when this field is removed. However nucleation and growth of the focal-conic texture are hindered by a bias field slightly below the critical value, so the relaxation time can be greatly prolonged. Display operation in the slow-scan mode is indicated: the nematic state produced by a single pulse persists under the influence of the bias until refreshed. The performance of such a display is analyzed by considering independently the response during the selection interval and during the interval between successive strobes.

Continuous refreshing. The first addressing scheme[12] used for a phase-change matrix is the familiar "three-to-one" technique which is shown schematically in Fig. 3. Data are carried by the polarity of row signals with am-

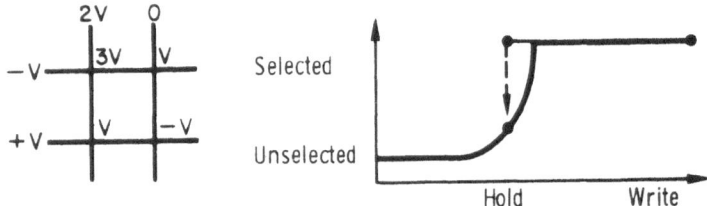

Fig. 3. Schematic explanation of 3-to-1 addressing of cholesteric phase transition matrix.

plitude V, while the columns are strobed sequentially with 2 V. Therefore the selected elements are driven with a net voltage of 3 V and all other elements see ± V. Since the phase-change effect is insensitive to polarity, the latter voltage can serve as a uniform bias to prolong the decay time. The bias level is chosen as a compromise between longer decay times for selected elements and weaker scattering of unselected elements as V approaches the critical voltage $V_c$.

In operation, the selected elements of the first column are driven nematic with a risetime limited by the overvoltage which corresponds to 3 V. Unselected elements remain focal-conic at the voltage V. After a dwell time at least equal to the risetime for 3 V, the strobe moves on to address the other columns in sequence. The bias voltage then tends to hold the written states of the elements in the first column: the unselected elements remain focal-conic and the selected elements relax slowly from the nematic state.

For a flicker-free display, the strobe must progress through all N columns of the matrix and return to refresh the first column before this relaxation becomes appreciable. This time for maximum tolerable relaxation can be termed $t_\downarrow^*$ to distinguish it from the considerable larger $t_\downarrow$ which is conventionally measured to 90% relaxation. Then the frame repetition period T must satisfy the following relationships:

$$t_\downarrow^*(V) \geqslant T = N\, t_\uparrow(3V).\tag{3}$$

Hence

$$N \leqslant \frac{t_\downarrow^*(V)}{t_\uparrow(3V)}.\tag{4}$$

In one experiment[13] it was found that moderate contrast was obtained with V = 35 V, for which $t_\downarrow \gtrsim 10$ s, $t_\downarrow^* \cong 4$ s and $t_\uparrow = 75$ ms. The implied limiting value is N ~ 50; the prototype used N = 28. Several seconds were required to write a new frame of information with this display.

Clear/write operation. Some improvement in writing speed and multi-plexing capacity can be obtained, at the cost of higher operating voltages and some flicker, by the use of the "two-to-one" addressing scheme shown in Fig. 4. Here the matrix is initially cleared to the nematic state by applying

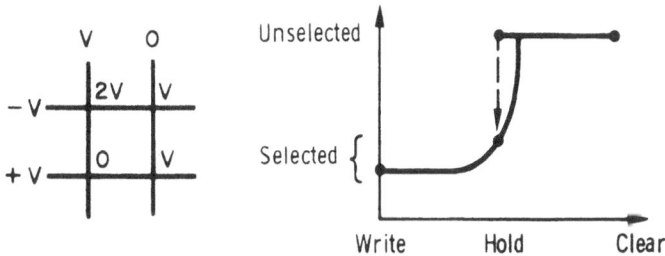

Fig. 4.  Schematic explanation of 2-to-1 matrix addressing of phase transition display.

$V$ to all columns and $-V$ to all rows. Since the drive is then only 2 $V$, this transition is relatively slow but it occurs simultaneously for all elements and is needed only once each frame. (This clear operation could in principle be accelerated by using a larger clear pulse.) Information is then written into the display in column sequence. Unselected elements in a strobed column are again driven toward the nematic cleared state by 2 $V$, while selected ele-ments relax rapidly to the focal-conic state when the bias voltage is re-moved. Columns not being strobed are held in their previously written states by the bias voltage $V$. The resulting display has contrast reversed with re-spect to the previous scheme: scattering characters appear on a clear back-ground.

The clear/write cycle must be repeated before the nematic background relaxes too far into the focal-conic state, so the frame period must again be less than the decay time at the bias voltage $V$. An upper limit for the multi-plexing capacity is

$$N < \frac{t_\downarrow^*(V) - T_{min}}{t_\downarrow(0)}. \tag{5}$$

The cholesteric-to-nematic transition time in the denominator of (4) is here replaced by the nematic-to-cholesteric relaxation time in zero applied field. The latter transition can be significantly faster, especially if a short pitch material is used,[14] so the multiplexing capacity can be improved. The term $T_{min}$ in (5) refers to the minimum acceptable viewing time for the last col-umn, before the frame is cleared for the next write cycle. This term is re-moved if, instead of clearing all columns at the start of the frame, each col-umn is cleared immediately before it is written by a column pulse $V_{clear}$ which is large enough that

$$t_\downarrow(V_{clear} - V) \leqslant t_\uparrow(0). \tag{6}$$

Because of the difference in scattering intensity at the holding voltage and during writing at 0 V, a bright line appears to sweep across the display during writing. The frame-clear pulse is, of course, also clearly perceptible. Therefore flicker-free operation is impossible with this addressing technique. Since some flicker is inevitable, a more liberal definition of $t^*_\downarrow$ may be tolerated for a larger multiplexing capacity. In a practical application of this technique,[15] a cholesteric mixture with $V_c = 85$ V was used to obtain $t_\downarrow(0) = 13$ ms. With N = 56, the time to write a full frame was thus 728 ms. Clear time at 2 V = 160 V was 30 ms.

## 2.3. Displays with Integrating Response

Nematic liquid crystal displays commonly use layer thicknesses of around 10 $\mu$m for which decay times of a few hundred milliseconds are typical. For a flicker-free display the drive pulse must recur with a period T short enough that the response to the previous pulse has not decayed appreciably, typically within 50 ms. If N lines are to be scanned, the dwell time available to drive each line is T/N. But the resulting pulsewidth is then considerably shorter than the risetime for a single pulse a few times larger than the threshold voltage. The display responds to the cumulative effect of repeated drive pulses, the steady-state response being related to an integrated function of the drive waveform. With the exception of the flexoelectric effect[16] the electrooptic phenomena in liquid crystals are insensitive to the polarity of applied voltage so the appropriate function is the rms (root-mean-square) value of the drive waveform.[17,18] Over a broad range of frequencies including those generally of interest for dynamic matrix addressing, then, the steady-state average response of a nematic liquid crystal display to a given addressing waveform will be the same as that to a sinusoidal or square-wave excitation having the same rms value.

By considering the drive waveforms of Fig. 2 over a full frame period and using the definition of rms voltage, the rms value of the potential difference across an excited matrix element can be calculated in general form as

$$\hat{V}_{on} = \sqrt{\frac{1}{T}\left[(V_s - V_1)^2 \frac{T}{N} + V_1^2(n-1)\frac{T}{N} + V_0^2(N-n)\frac{T}{N}\right]}. \qquad (7)$$

Likewise the rms voltage seen by an unselected element is

$$\hat{V}_{off} = \sqrt{\frac{1}{T}\left[(V_s - V_0)^2 \frac{T}{N} + V_1^2 \frac{nT}{N} + V_0^2(N-n-1)\frac{T}{N}\right]}. \qquad (8)$$

Here N is the number of columns scanned in the frame period T, and n is the number of excited elements in the same row as the element considered. We seek to maximize the contrast ratio between selected and unselected elements, so we are concerned with the ratio $\hat{V}_{on}/\hat{V}_{off}$. This selection ratio can be written as

$$\left(\frac{\hat{V}_{on}}{\hat{V}_{off}}\right)^2 = 1 + \frac{2V_s(V_0 - V_1)}{V_s^2 - 2V_sV_0 + NV_0^2 + n(V_1^2 - V_0^2)}. \tag{9}$$

Restricted patterns. To find the extrema of this function, we set the derivatives with respect to the three voltage levels to zero, solve simultaneously for $V_0/V_1$ and $V_s/V_1$, and substitute to obtain

$$\left(\frac{\hat{V}_{on}}{\hat{V}_{off}}\right)^2_{max} = \frac{N-n}{N-n-1}\left(1 \pm \sqrt{\frac{N-n}{n(N-n)}}\right). \tag{10}$$

These solutions are plotted in Fig. 5 as functions of n for N = 8. Clearly the selection ratio can become large if n → N. Furthermore we recognize that

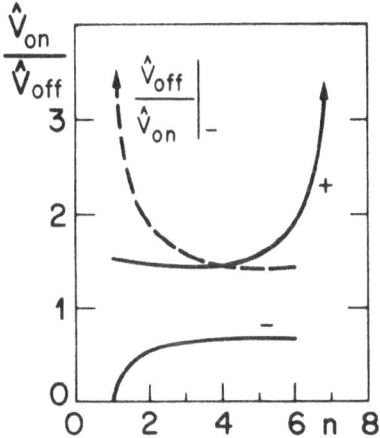

Fig. 5. Extrema of $\hat{V}_{on}/\hat{V}_{off}$ as functions of the number n of selected elements per row in a display with N = 8 columns

logical inversion of the input data interchanges the roles of "off" and "on" elements so the broken curve of $\hat{V}_{off}/\hat{V}_{on}$, which plots the reciprocal of the lower solution, is also relevant and shows that the selection ratio also becomes large as n → 0. Since the selection ratio varies with n, contrast will be uniform over the display and optimum selection will obtain only if n has the same value on all rows. This constitutes a restriction on the range of patterns that can be displayed in this manner.

An illustrative example may clarify this addressing technique.[19] Consider a matrix with rms response which is driven with $V_s = V_1 = V$ and $V_0 = 0$ to display patterns such that $n = 1$ for all rows. As shown in Fig. 6,

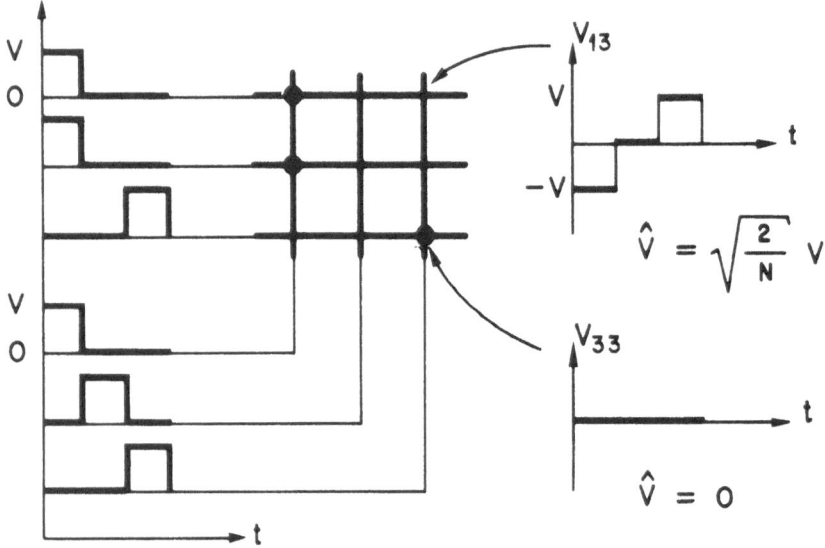

Fig. 6. Optimum addressing of rms-responding matrix for the display of restricted patterns having exactly one element selected in each row.

the strobe and data pulses coincide to give a potential difference of zero on the selected element in each row. For each unselected element the single strobe pulse and the single data pulse do not overlap so the rms drive voltage is $V(2/N)^{1/2}$. This infinite selection ratio produces a display with selected elements turned off against a uniform background of elements turned on to full contrast. Such a system could be used as an oscillograph to display single-valued functions, with size and resolution limited only by peak voltage and fabrication constraints.

General patterns. The uniform display of unrestricted patterns requires that the effective voltage on selected elements be independent of n. Rewriting (7) in the form

$$N\hat{V}_{on}^2 = V_s^2 - 2V_1 V_s + V_0^2 N + n(V_1^2 - V_0^2) \qquad (11)$$

makes it clear that the dependence on n can be removed by setting $V_1^2 = V_0^2$. For a non-trivial result, $V_0 = -V_1 = V_D$. Then the data to be displayed are carried by pulses of constant amplitude $V_D$ whose polarities relative to the strobe pulse convey the picture information. Making these substitutions in (9) and maximizing with respect to $V_D/V_s$, we obtain after Alt and Pleshko[20] the optimum addressing conditions for the display of unrestricted patterns on a matrix with rms response:

$$V_s = \sqrt{N}\ V_D \text{ for which } \frac{\hat{V}_{on}}{\hat{V}_{off}} = \sqrt{\frac{\sqrt{N}+1}{\sqrt{N}-1}}. \qquad (12)$$

This important result shows that increasing the number of scanned lines places increasingly stringent demands on the threshold characteristic of the display device, as indicated schematically in Fig. 7. For successful operation with N = 32, the display contrast must switch through its full operational

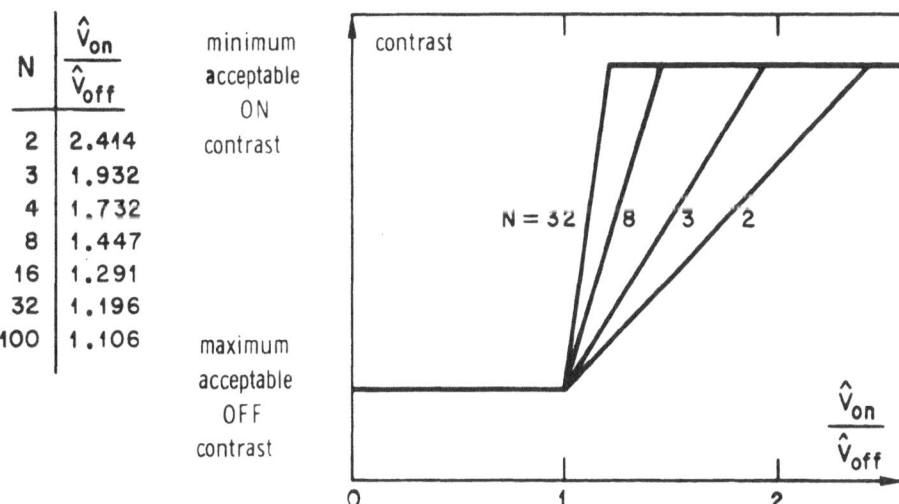

| N | $\dfrac{\hat{V}_{on}}{\hat{V}_{off}}$ |
|---|---|
| 2 | 2.414 |
| 3 | 1.932 |
| 4 | 1.732 |
| 8 | 1.447 |
| 16 | 1.291 |
| 32 | 1.196 |
| 100 | 1.106 |

Fig. 7. Increasing the number N of scanned lines reduces the rms selection ratio $\hat{V}_{on}/\hat{V}_{off}$ so a steeper threshold curve is required.

range with only a 20% change in applied rms voltage. Since the measurement of device contrast as a function of rms voltage indicates the minimum selection ratio required, it is the steepness of this characteristic curve which determines the maximum size of matrix in which that device can be satisfactorily applied. Matrix performance depends on the number of scanned lines N rather than on the total number of elements M, so it is preferable to strobe the smaller dimension of a non-square matrix. Where the steepness of the contrast curve is insufficient for operation with $N = (M)^{1/2}$, then M elements must be arranged as a non-square matrix to reduce N at the cost of additional addressing connections.

Multiplexing capacity. The application of these results to the determination of the multiplexing capacity of a real display is not always straightforward. Figure 8 demonstrates that the threshold characteristic for a twisted nematic device varies with the viewing angle and with the amount of tilt in the quiescent alignment.[21] A large tilt bias occurs when surface alignment is accomplished by vacuum evaporation at grazing incidence.[22] Small tilt bias can be achieved by unidirectional rubbing,[23] which unfortunately is not compatible with high-temperature glass sealing. Cells with small tilt bias have steeper threshold curves and less spread with viewing angle, so small tilt bias is desirable for multiplexing. (Zero tilt can be obtained[22] but is unacceptable because of the occurrence of regions with reverse twist and tilt.[23])

274 A. R. KMETZ

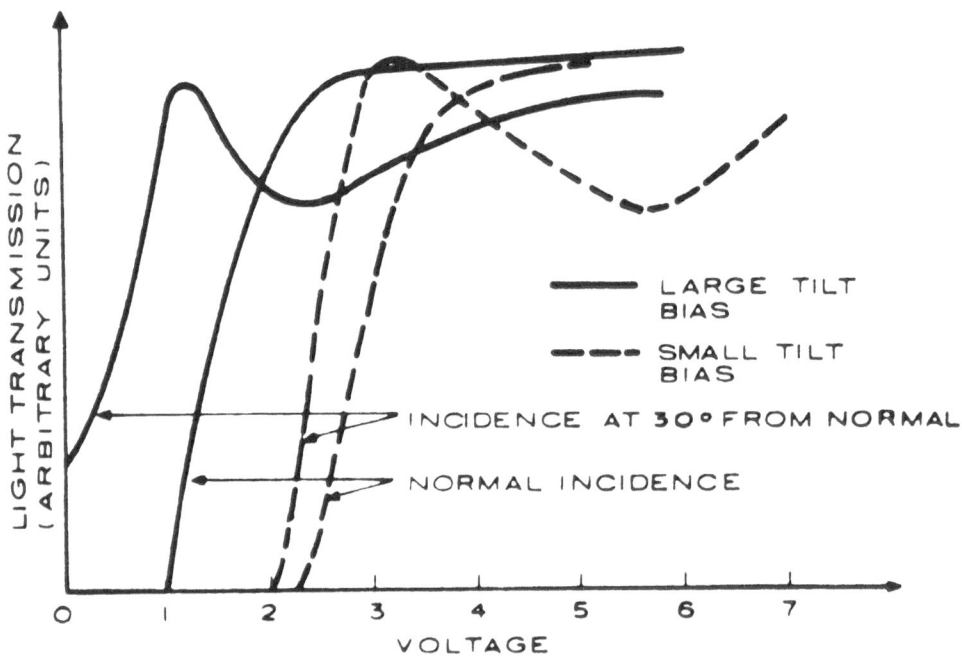

Fig. 8. Effects of alignment tilt bias and of viewing angle on the threshold characteristics of a twisted nematic device.[21]

Variation of threshold characteristics with temperature can also be important.[24]

Multiplexing capacity of a device can be determined only after the required contrast ratios, field of view and temperature range have been specified for a particular display application. Assuming that threshold decreases
with temperature, the voltage for which the maximum tolerable contrast between unselected elements and the background is reached at maximum temperature anywhere in the field of view is taken as $\hat{V}_{off}$. Similarly the minimum voltage necessary to produce the required contrast between selected
and unselected elements at minimum temperature everywhere in the field of
view is taken as $\hat{V}_{on}$. These voltages determine the minimum selection ratio
required by the device itself in the given application, but this ratio must be
increased to account for deviations from nominal drive voltage levels.[20]
Then the actual multiplexing capacity can be calculated as the largest integer
value of N which can be substituted into (12) to yield at least this selection
ratio.

There is clearly a trade-off between the number of scanned lines and
field of view. The contrast curve for a twisted nematic cell at normal incidence indicates that eight to ten lines could be multiplexed in a projection
display. For direct viewing over a relatively wide field, multiplexing is

limited to three or four lines in a reflective display, where mirror symmetry compensates for some of the viewing-angle dependence, and to even fewer lines for a transmissive display with diffuse rear illumination.

Since multiplexed performance is a complicated function of many variables, it is highly desirable to carry out direct visual evaluations. However it is not always necessary to fabricate display prototypes with complex photolithography and to construct sophisticated addressing electronics for this purpose. So long as the assumption of rms response remains valid, the simple arrangement of Fig. 9 provides an accurate simulation of multiplexed opera-

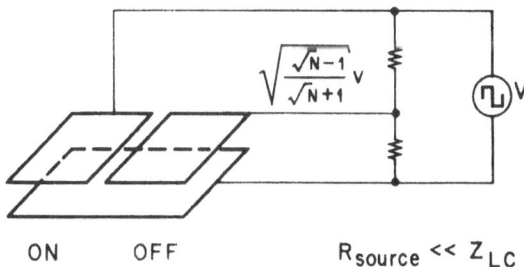

$$\sqrt{\frac{\sqrt{N}-1}{\sqrt{N}+1}}\,V \qquad\qquad V$$

ON        OFF                $R_{source} \ll Z_{LC}$

Fig. 9. Simple arrangement for visual evaluation of multiplexing in device with rms response.

tion. It is feasible, for instance, by visually comparing selected and unselected elements against the background under the given viewing conditions, to determine experimentally the optimum generator voltage from which $V_s$ and $V_D$ can be derived by the foregoing formulas. Likewise the optimum multiplexing capacities of different cells can be directly compared with little investment in time or equipment.

Power consumption. Power consumption must be computed from the actual multiplexing drive waveforms, rather than simply from their rms values, because the display elements have significant reactance. Energy stored in the device capacitance is lost every time the drive voltage changes. Since the number of voltage transitions seen by a given element depends on the selection of other elements in the same row, power consumption varies with the pattern displayed. Total dissipation of a square matrix, normalized by the dissipation of an equivalent display with individually driven elements, is plotted as a function of N in Fig. 10. For these curves, polarity reversal to assure zero average voltage is assumed to occur at the data frame repetition rate; reversal during each strobe interval would result in higher power consumption. The approximation RC = T is chosen for twisted nematics, and source resistance is neglected. Direct-drive power is calculated at three times the threshold voltage.

The increase in the multiplexed power ratio with increasing N in Fig. 10 for the fully selected raster is caused by the increase in peak strobe voltage $V_s$ with N. The power ratio for the checkerboard is higher because the number of voltage transitions per frame is larger. In both cases the capacitive ef-

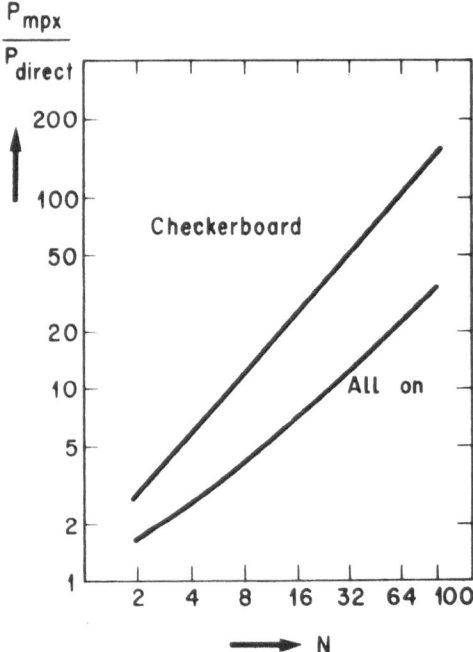

Fig. 10. Ratio of power dissipated by a multiplexed display to dissipation with direct addressing as a function of matrix size for a limiting-resolution checkerboard pattern and for a fully selected raster.

fect dominates. For the pattern with all elements turned off, zero voltage is applied for direct addressing; all elements of the multiplexed display see a waveform with rms value $\hat{V}_{off} \neq 0$ and thus dissipate power. For all cases, a multiplexed display consumes more power than an equivalent display with direct addressing, the ratio increasing with N.

Frequency limitations. To conclude the discussion of this category of device response, it should be noted that the range of frame frequencies over which rms analysis holds is limited. The average optical response is determined simply by the rms value of the applied voltage only when the frequency is high enough that the frame period is short compared to the decay time. For lower frequencies contrast will be smaller than that predicted by rms analysis and must be calculated from a convolution of the drive waveform with the device impulse response.

Several factors can affect the high-frequency limit. The threshold voltage for dynamic scattering diverges at the space-charge relaxation frequency, which is typically a few kilohertz. Consequently the Fourier components which approach this frequency are less effective in producing an optical response. In this case the effective selection ratio can be calculated by using a weighted sum of Fourier components in the foregoing analysis. Neglecting the inhibiting effect of components above the cutoff frequency, Fig. 11 shows that the effective selection ratio for a given number of scanned lines is reduced, and the maximum achievable N is limited, by a finite ratio of cutoff frequency to frame frequency.[20]

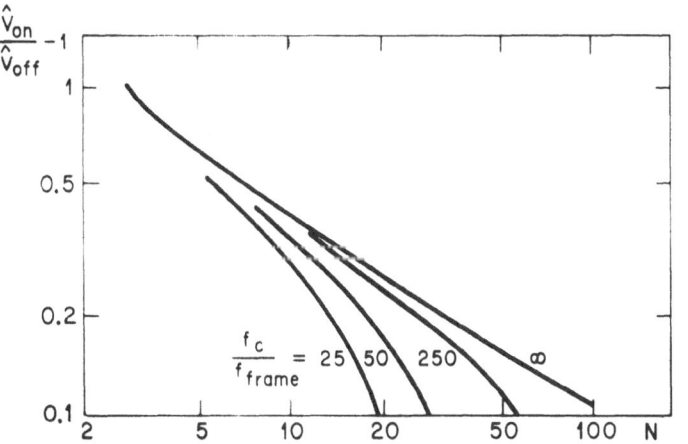

Fig. 11. Limitation of multiplexing capacity by a cutoff frequency.[20]

A similar effect occurs in materials with positive dielectric anisotropy because of the intrinsic relaxation of $\epsilon_\parallel$. For the materials commonly used in twisted nematic devices simple rms analysis is usually adequate because this cutoff frequency occurs only around 1 MHz, but materials are known[25] with relaxation frequencies of a few kilohertz. With all materials, the element capacitance combines with the resistance of the transparent electrodes to form a dispersive transmission line which attenuates high frequencies. This cutoff frequency typically lies above 50 kHz in simple twisted nematic watch cells, but it can cause non-uniform response at much lower frequencies in a complex matrix with long constricted leads.

### 2.4. Memory Displays

The cholesteric texture change exemplifies display effects with intrinsic memory.[11] Electrically induced turbulence at selected locations in a layer of long-pitch cholesteric material with negative dielectric anisotropy transforms the initial clear planar texture to a scattering focal-conic texture. The write time and threshold voltage are similar to the dynamic scattering effect since the same electrohydrodynamic mechanism is operative. Once written, the scattering state persists without refreshing for weeks or months.[26] The display can be erased by applying sufficient voltage at a frequency above the space-charge relaxation frequency. Here the hydrodynamic effect is suppressed and the planar texture is restored by dielectric reorientation in the field. Because of the relatively small negative dielectric anisotropies of typical materials and the difficulty with which dislocations are anihilated by a field, voltages on the order of 100 V are required to clear a typical display in about 100 ms.

Since the decay time of a memory display is by definition effectively infinite, matrix addressing is not limited by refresh requirements as it is in the other categories of device response. The number of display elements is thus determined in principle only by the constraints of mechanical and electronic feasibility and cost. However matrix addressing prohibits drive voltages in excess of three times the threshold which, for most non-emissive devices, implies relatively slow writing. Consequently the time to write a full frame of information may constitute the practical limit on the size of a memory display for a given application.

Experimental prototypes have verified that quite large texture-change displays can be operated with straightforward matrix addressing. In one example,[27] a 240-line display took about half a minute for column-sequential writing of a full frame. Almost two minutes were needed for frame-simultaneous erasure at 50 V. Because of the strong thickness dependence of these transients, it was predicted that decreasing the cholesteric layer thickness from 30 $\mu$m to 10 $\mu$m would reduce both transition times to a few seconds.

Clearly the texture-change effect can be used in relatively large matrix displays if the displayed information need be changed only infrequently. As with any memory display, energy consumptions calculated independently for the write and erase operations can be combined to find total power dissipation only when the rate of change of information is known for a particular application.

## 3. TECHNOLOGY SURVEY

Having described the principal modes of operation and developed the techniques for analysis of multiplexing capability for specific examples of each of the categories of device response, we conclude by surveying the field of non-emissive electrooptic devices. (Electromechanical[28,29] and magneto-optic[30-32] displays are not considered.) For each display technology, the present status of dynamic matrix addressing is given, the potential for future improvement is estimated and, in some cases, promising directions for development are indicated.

### 3.1. Dynamic Scattering and Texture Change

A state-of-the-art dynamic scattering display[33] is multiplexed with four lines scanned (N = 4). Straightforward matrix addressing is used, as described in Section 2.3 for devices with rms response. Strobe and data amplitudes of 12 V and 6 V are typical. Good legibility is obtained over a somewhat restricted field of view by the use of a flip-up vane for contrast enhancement in reflection.

Two-frequency multiplexing[34,35] at first seemed to be a breakthrough. An ac bias signal $V_{hf}$ above the space-charge relaxation frequency $f_c$ is superimposed on the usual low-frequency drive to inhibit the hydrodynamic instability. The onset of scattering then occurs at a higher value $V_t$ of the low-frequency voltage which is related to the ordinary threshold $V_{to}$ by:[34]

$$V_t{}^2 = V_{to}^2 + \gamma V_{hf}^2 . \tag{13}$$

The parameter $\gamma$ is a combination of material constants which increases with the magnitude of the negative dielectric anisotropy. By raising the threshold without appreciably affecting the steepness of the contrast curve, the multiplexing capacity calculated by the methods of Section 2.3 is increased.

However, the principal Fourier components of the low-frequency drive must lie below $f_c$, so $f_c$ must be increased in proportion to the number of lines scanned. Practical limits on doping levels restrict $f_c$ to a few kilohertz. Raising $V_{hf}$ to increase the threshold also reduces the decay time, so the frame rate must be increased to retain flicker-free operation. These effects combine to restrict the bandwidth available for the low-frequency addressing waveforms, effectively limiting the number of scanned lines to about ten. A prototype system with such a display having N = 8 has been shown.[36] Because $\gamma$ varies only weakly with temperature, a useful operating temperature range was achieved despite the strong variation of $f_c$ by choosing the high frequency to be above $f_c$ for the highest required temperature.[35] Choosing 20 kHz for the N = 8 display yields a power consumption around 90 mW/cm$^2$, which is comparable to LED displays!

Alt and Pleshko[20] used a 40×40 matrix to show that the bandwidth limitation can be circumvented and a lower $f_c$ can be used if selection is performed instead by switching the high-frequency signal superimposed on a low-frequency bias. In order that the high-frequency drive have an rms value large enough to inhibit scattering despite its short dwell time, a 70 V signal had to be used; its frequency was 40 kHz, presumably to prevent chevron-mode scattering. Voltage and power considerations therefore make this alternative unattractive unless N is again small.

The failure of two-frequency addressing to significantly enhance the multiplexing capacity of dynamic scattering displays in practical applications can be seen in the recent report[37] of a calculator product using this approach for N = 3. Although the development of materials with more negative dielectric anisotropy might benefit two-frequency schemes, it seems preferable to seek to improve the steepness of the device contrast curve through better understanding of the transition from Williams domains to scattering turbulence and of the relationship between scattering intensity and ionic dopant species.

Much of this discussion applies as well to cholesteric texture-change displays since the writing mechanism is similar to dynamic scattering. Section 2.4 showed that large texture-change storage displays (N $\geqslant$ 240) are feasible by simple matrix addressing. Neither significant technical developments nor marketable applications have appeared in the last few years, so the technology can be considered stagnant. Some improvement in attainable contrast ratio and in writing speed can be achieved by two-frequency addressing and evolution in dynamic scattering materials would likewise benefit texture-change displays. Nevertheless the display performance expectable from this inherently slow scattering effect is insufficient to motivate its use in a practical display system.

## 3.2. Twisted Nematics

The present multiplexing limit for reflective twisted nematic displays under open viewing conditions was shown in Section 2.3 to be three or four scanned lines. Although this is no more than with dynamic scattering, the voltage and power requirements are significantly lower and user preference for absorbing over scattering displays is evident. In fact the twisted nematic display is the clear leader among all contenders for the recognized markets for non-emissive displays, viz. watches, calculators, instruments.

The discovery of nematic materials whose dielectric anisotropies change sign at relatively low frequencies raised the possibility of improving the multiplexing capacity of twisted nematics by two-frequency addressing.[25] However both $f_c$ and $\gamma$ were found to vary strongly with temperature, so power consumption is again excessive and temperature compensation is no longer simple. Domains of reverse tilt are another undesirable consequence of two-frequency drive.[38] This addressing scheme does not seem useful for twisted nematics.

Improving the relative steepness of the contrast-voltage characteristic by altering material parameters also seems unpromising. Although it is easy to adjust the dielectric anisotropy of a mixture, the contrast curve scales with the threshold voltage so the multiplexing capacity is unchanged to first order. Perturbation of the field by the distorted configuration is a second-order effect which slightly favors low-threshold materials.[39] The relative steepness could be enhanced[40] by increasing $K_{11}/K_{33}$ from present typical values around 0.8, but it is unlikely that the ratio can exceed unity. Even if chemists learn how to control this parameter, the available range for improvement is therefore small.

Since alteration of the bulk properties does not offer significant improvements in multiplexing, one is led to consider the effect of relaxing the usual assumption of rigid surface coupling. Rapini and Papoular[41] calculated the reduction in threshold for homeotropic cells which results from weak anchoring. In a recent extension of their analysis,[42] we find that the steepness of the contrast curve generally increases faster than the threshold decreases, so the multiplexing capacity tends to improve with weak anchoring. The improvement is smaller and less certain with twisted nematics than with homeotropic or planar cells. The computed curves for twist cell transmission in Fig. 12 imply that multiplexing capacity would be doubled if the anchoring energy could be reduced to $10^{-3}$ of present typical values. However this improvement at normal incidence may be offset by a larger variation of optical threshold with angle. Furthermore such a cell would have more stringent tolerances on thickness and somewhat slower turn-off. Nevertheless loose

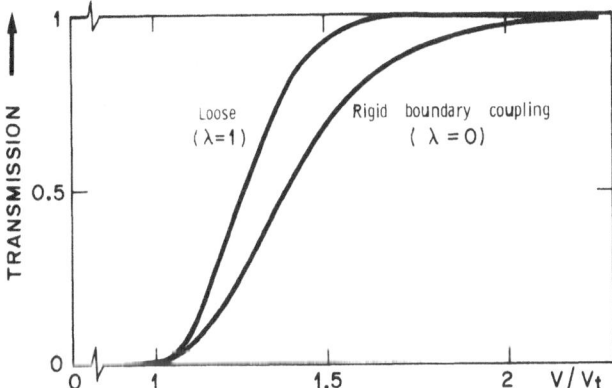

Fig. 12.   Computed twist cell transmission versus voltage normalized by optical threshold for boundary coupling strengths differing by a factor $10^3$.

boundary coupling remains an interesting possibility for improving the multiplexing capacity of twist cells. The required weak-anchoring surfaces apparently can be found,[43] but then grooving of the substrate will be inadequate to support the twisted alignment so a new mechanism for azimuthal orientation must be sought.

### 3.3. Electrochromics

Superior appearance is the main advantage of electrochromic displays over twisted nematics: their contrast and brightness can be higher, and their field of view is virtually unrestricted. However their power consumption is proportional to the switching rate and lifetime is counted as a maximum number of cycles, so applications are favored where infrequently changing data permit operation in a storage mode. In fact, electrochromic displays have been called storage batteries with a visible state of charge.[44] This metaphor provides an insight to the obstacles to matrix addressing of electrochromics.

Figure 13 shows schematically the write operation in a symmetrical electrochromic display cell. The element is "off" when the colored electrochromic deposit is on the bottom surface where it is hidden from view. This deposit is necessarily accompanied by a battery emf. When the writing voltage is applied, current flows to discharge this internal battery and recharge it in the opposite polarity. This process involves removal of the colored layer on the bottom substrate and deposition of a colored layer on the top substrate. When the upper deposit is clearly visible, the cell is considered "on" and the writing source is removed to prevent degradation due to overcharging. The display remains in its written state as long as no current flows.

Connecting together the "on" and "off" elements in Fig. 13 would cause a current flow tending to discharge both batteries, thereby destroying the

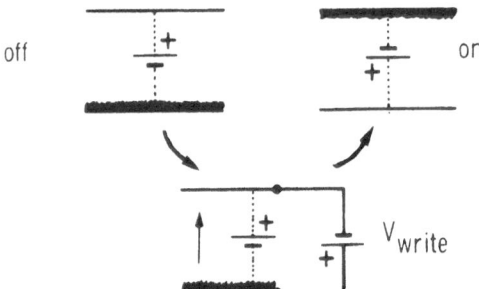

Fig. 13.  Schematic illustration of writing operation, showing internal battery emf associated with electrochromic deposit.

stored display information. In a matrix, many display elements are interconnected. Moreover the matrix address lines are connected to low impedance sources which also permit current flow. Information written into such a matrix therefore dissipates by locally non-uniform crosstalk at a rate governed by internal resistance. Clearly an electrochromic matrix display is not capable of true storage.[45]

The current through an element of an electrochromic matrix will be zero only if its internal battery emf is balanced by an opposing external voltage. If the external voltage increases, the electrochromic battery will be further charged. If the external voltage decreases, a discharge current evidenced by a change in coloration will flow. Although reaction potentials measured against a reference electrode are important for understanding the electrochemical mechanisms and are relevant to initial operation of virgin cells, it is clear that a practical electrooptic threshold does not exist.[46]

Lacking both memory and true threshold, an electrochromic matrix might still be operated by utilizing a difference in writing rates for full-select and half-select drive voltages and by refreshing the display by periodic clearing and rewriting. However the voltage dependence of the writing rate is small,[47] the effects of repeated half-selects are cumulative and the required refresh interval depends on the information displayed. Even without considering such problems as lifetime under non-uniform charging conditions and voltage drops in transparent electrodes due to high current densities,[45] it appears that electrochromic displays are fundamentally unsuited for matrix addressing.

### 3.4. Electrophoretic and Dipole Suspension Displays

Individually driven electrophoretic elements have demonstrated optical performance superior to any other reflective electrooptic display.[48] Their brightness, contrast, angular field of view and potential color choice are comparable to a printed page. Driving signals in the 30 - 100 V range and response times of 0.1 - 1 s are typical; reducing the voltage degrades both switching speed and ultimate contrast. The scientific basis for this technology is poorly established, and the number of relevant publications is small. Even less is known about matrix addressing of electrophoretics.

Electrophoretic displays as first reported had memory but no useful threshold: the pronounced "knee" found when brightness was measured as a function of voltage under single-pulse conditions was identified as a transit time effect.[49] Displays with a true threshold have recently been achieved[50] through an additive to the suspension fluid. This is thought to strengthen the adsorption forces so that particles held against the cell walls require a measurable electric field to break free before they can participate in electrophoretic migration. Matrix addressing of an electrophoretic storage display using this modified suspension has been demonstrated by a prototype with N = 32. The time needed for line-at-a-time writing of the full frame was 40 s and contrast was relatively poor, but feasibility was established.

The insertion of a grid between the electrodes of the usual electrophoretic cell has been reported[48] to permit display operation controlled by 1 V logic signals. Matrix addressing with this technique has not been investigated, but one can imagine a structure similar to that of multiplexed vacuum fluorescent displays. Only a single unswitched supply would then be required for the acceleration potential ($\gtrsim 30$ V). Fabrication of such a device would be difficult, especially since crosstalk and slow response are to be expected unless the separation between the grid and the segmented electrode is small.

Compared with the electrophoretic migration of charged pigment particles, the orientation of suspended dipole particles by an electric field[51] appears poorly suited to display application, despite some recent progress with materials.[52] In particular the lack of threshold, at least in practical cell geometries,[53] precludes matrix addressing of dipole suspension devices.

### 3.5. Tunable Birefringence

The voltage dependence of the optical retardation of an initially homeotropic cell containing a nematic liquid crystal with negative dielectric anisotropy is illustrated in Fig. 14. When viewed between crossed polarizers in white light, the cell is black until threshold is reached and then progresses as indicated through several orders of interference colors with increasing voltage. For conventional devices, rms response obtains[18] and multiplexing performance is determined by the methods of Section 2.3. However, unlike the twisted nematic device, there is no need here to operate between threshold and saturation of the electrooptic effect.[54] In particular, first-order yellow and red are easily distinguished by an observer, yet they are separated by a voltage difference which is small compared to the threshold. This implies a relatively large multiplexing capacity: such two-color matrix displays have been made with $N \cong 100$. However their field of view is very narrow, re-

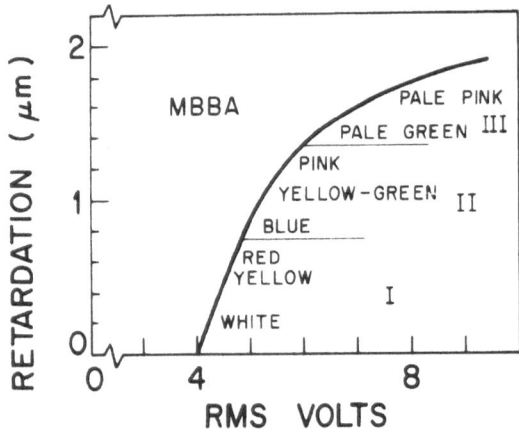

Fig. 14.  Voltage dependence of retardation of MBBA in a homeotropic cell 12.5 μm thick, showing three orders of interference colors seen with crossed polarizers in white light.

flective operation is impractical, thickness tolerances are very tight and strobe voltages are unattractively high. We calculate that weak boundary coupling could improve the multiplexing capacity by about a factor of four while lowering the voltage requirement.[42]

Because both display states correspond to weak deformations near the Freedericksz threshold, the response times are long. Improvement in speed by an order of magnitude has been obtained by using a material with positive dielectric anisotropy and initially planar alignment.[55] Then the first order retardation is reached only near saturation of the deformation, so the display states have voltages which are well separated and far above threshold. Transitions are therefore faster but the multiplexing capacity is lower. An N = 40 matrix using monochromatic light was reported with rise and fall times of 5 and 20 ms.

To circumvent the limitations of multiplexing with rms response, the tunable birefringence effect has recently been operated instead as a refreshed display with persistence.[56,57] A narrow pulse of large amplitude produces a peak response which is allowed to decay almost completely before the pulse is repeated. Very thin cells are needed if the natural decay time is to be short enough that this frame repetition rate be above the flicker-fusion frequency. It is found that the average contrast is a strongly nonlinear function of the pulse height, so a low-amplitude row signal added to or subtracted from the large column strobe is sufficient to modulate the response. Maximum average contrast ratio decreases[56] with the number of lines scanned: matrix displays with N = 24 and N = 32 have been reported with frame-write times on the order of 10 ms and strobe amplitudes of 25 - 65 $V_{rms}$. Field of view is better (24° to 45°) with these thin cells, but it seems unlikely that uniform thicknesses of 2 to 5 μm are attainable in mass production.

## 3.6. Ferroelectrics

Many ferroelectric materials are known which exhibit a variety of optical effects of potential utility for displays.[58] However monolithic fabrication is a necessity for a practical matrix display, so only those materials need be considered which are available in defect-free slices at least 1 cm$^2$ in area. Since transverse electrooptic effects necessitate complex electrode structures, high voltages and poor display filling-factors, we will consider only longitudinal effects for matrix displays. Finally the conclusions of Section 2.1 lead us to consider only memory effects. Three ferroelectric materials remain as potential matrix display candidates.

PLZT ceramics are by far the best in terms of availability as large-area samples of high quality and low cost. The 7/65/35 composition has satisfactory memory capability but driving voltages are high (30 - 200 V). As with all ferroelectrics, no true threshold exists; nevertheless the resistance to disturbance of the written states by repeated half-select pulses is adequate for N ~ 100.[59] Several display effects are possible.[60] The polarization scattering and fringe-field effects provide inadequate contrast for display application. Good contrast has been obtained in the laboratory with electrically controlled birefringence in strain-biased plates, but no practical means of fabrication is known, nonuniformity across the array is inherently a problem and the field of view is narrow.

Bismuth titanate[61] can be grown in platelets of useful size. Matrix addressing is accomplished with about 35 V signals, producing switching times around 25 $\mu$s. Epitaxial growth of bismuth titanate may make larger arrays possible at the cost of higher operating voltages. Efficient optical readout is a problem with this material. Adequate contrast has been obtained over about a 30° range of viewing angles by the differential-retardation method, but very exacting mechanical fabrication is required and display brightness is poor.

Gadolinium molybdate can be grown as very large, optical-quality crystals which are easily cut and polished by ordinary methods. Efficient optical readout with adequate field of view can be obtained by several methods.[62] Domain wall motion can take place in relatively low field (200 V/cm),[63] but wall velocity is low and useful switching requires about 1 ms. The large strains associated with domain walls in this ferroelectric material cause crystal cracking if orthogonal walls intersect, so the delineation of the two-dimensional arrays of isolated reversed domains required for a matrix display is a problem in a single crystal. A possible solution involves the controlled nucleation and translation of lens-shaped domains, similar to the shift-register operation of magnetic bubble memories.[64]

## 3.7. Cholesteric Phase Transition

It was shown in Section 2.2 that matrix displays having N ~50 are possible with the phase-change effect. Reasonably short frame-write times, appropri-

ate for example to small terminal readouts, can be obtained by repetitive
clear-write operation with rather high drive voltages (80 V). The dependence
of threshold voltage on cell thickness is unfortunate since thickness variations are more difficult to control in the larger cells required for direct viewing of such complex messages. Despite adequate legibility, the appearance of
the display is not very pleasing: the general disadvantages of a scattering display are compounded by the visible "flashing" due to the cyclic refreshing.

Greubel[65] discovered that a homeotropic surfactant can make the phase
transition bistable. If such a cell were maintained at a bias voltage above a
lower critical field after being driven nematic, the homeotropic state theoretically should persist indefinitely. True memory operation would then be
possible, without the periodic refreshing which restricts matrix size and
causes flicker. However defects in the homeotropic texture act as nucleation
sites from which the focal-conic texture grows. The elements of a simple
matrix are separated by narrow field-free regions which remain focal-conic,
so relaxation to the focal-conic state begins at the edges of a written element
and propagates inward. Thus refreshing is still necessary in practice; an
N = 64 prototype has been shown.

White and Taylor[66] proposed changing the display effect of the cholesteric phase transition from scattering to absorption by the addition of a pleochroic dye to the cholesteric material. Optical performance superior to
twisted nematic devices is theoretically possible, but no guest-host system
having sufficiently high order parameter and stability is yet known. Like the
bistability effect, this approach promises significant improvement if major
obstacles can be overcome.

## 3.8. Conclusions

Many of the non-emissive electrooptic display technologies have demonstrated impressive performance when each element is individually addressed,
but few seem likely to find widespread application in matrix displays and
none is even potentially a candidate for a flat TV. Tunable birefringence and
ferroelectric displays have relatively large multiplexing capacities but suffer
from poor appearance, especially in reflection, relatively high voltage and
difficult fabrication. Dynamic scattering and the cholesteric texture change
effect are obsolescent. Lack of a useful threshold makes electrochromic, dipole suspension and elastomeric[67] displays unsuitable for matrix addressing.
The twisted nematic device is presently the most important passive display,
offering appearance, speed and multiplexing capacity adequate for calculators, watches and instruments. No breakthrough in the multiplexed performance of twisted nematics is to be expected, but even small improvements
could make a significant difference in the competition with emissive displays. The cholesteric phase transition seems unlikely to be viable as a scattering display, but the development of a successful guest-host system could
challenge twisted nematics; improvement of multiplexing through exploitation of bistability could open up alphanumeric applications such as small

terminals· and office machines. Electrophoretic displays will probably never be fast enough for these applications, but their optical performance and less critical cell dimensions suggest that they may ultimately be developed for alterable signs like airport flight information displays.

## ACKNOWLEDGEMENT

Stimulating discussions and critical reading of the manuscript by T. J. Scheffer and by J. Nehring, who was especially helpful in clarifying the categories of display response in Section 1.3, are gratefully acknowledged.

## REFERENCES

1  North American Rockwell Microelectronics, "Lloyd's ACCUMATIC 100" (1972)

2  F. Kinji, German patent appl. OS 2 163 634 (1971)

3  A. Sobel, IEEE Trans. Electron Devices ED 18 (1971) 797

4  C. H. Gooch, R. Bottomley, J. J. Law and H. A. Tarry, J. Phys. E 6 (1973) 485

5  F. Mottier and J. Nehring, Brown Boveri Research Center, private communciation, 1975

6  S. Fukumoto, German patent appl. OS 2 403 172 (1974)

7  G. W. Taylor, Proc. IEEE 58 (1970) 1812

8  S. Sherr, IEEE Conference on Display Devices, 72 CH0707-0-ED (1972) 32

9  K. H. Härdtl, this book

10  A. Kumada, Ferroelectrics 3 (1972) 115

11  E. P. Raynes, this book

12  J. J. Wysocki, J. H. Becker, G. A. Dir, R. Madrid, J. E. Adams, W. E. Haas, L. B. Leder, B. Mechlowitz and F. D. Saeva, S.I.D. Int. Symp. Digest 2 (1971) 130

13  T. Ohtsuka, M. Tsukamoto and M. Tsuchiya, Jap. J. Appl. Phys. 12 (1973) 371

14  E. Jakeman and E. P. Raynes, Phys. Lett. 39A (1972) 69

15  M. Tsukamoto and T. Ohtsuka, Jap. J. Appl. Phys. 13 (1974) 1665

16  W. Helfrich, Appl. Phys. Lett. 24 (1974) 451

17  P. J. Wild, S.I.D. Int. Symp. Digest 3 (1972) 62

18  A. R. Kmetz, IEEE Trans. Electron Devices ED 20 (1973) 954

19  F. Ueda, H. Arai and H. Amagasaki, German patent appl. OS 2 414 608 (1974)

20  P. M. Alt and P. Pleshko, IEEE Trans. Electron Devices, ED 21 (1974) 146

21  L. Goodman and D. Meyerhofer, S.I.D. Int. Symp. Digest 6 (1975) 76

22  E. Guyon, P. Pieranski and M. Boix, Lett. in Appl. & Eng. Science 1 (1973) 19

23  E. P. Raynes, Rev. Phys. Appl. 10 (1975) 117

24  F. J. Kahn, this book

25  H. K. Bücher, R. T. Klingbiel and J. P. VanMeter, Appl. Phys. Lett. 25 (1974) 186

26  J. P. Hulin, Appl. Phys. Lett. 21 (1972) 455

27  H. Takata, O. Kogure and M. Murase, IEEE Trans. Electron Devices ED 20 (1973) 990

28  W. R. Aiken, S.I.D. Int. Symp. Digest 3 (1972) 108

29  R. N. Thomas, J. Guldberg, H. C. Mathanson, P. R. Malmberg and A. S. Jensen, S.I.D. Int. Symp. Digest 6 (1975) 28

30  D. E. Lacklison, G. B. Scott, R. F. Pearson and J. L. Page, IEEE Trans. Magn.
    MAG 11 (1975) 1118

31  R. E. Glusick and M. Fine, IEEE Conf. on Display Devices 70 C 55-ED (1970) 112

32  L. T. Romankiw, M. M. G. Slusarczuk and D. A. Thompson, IEEE Trans. Magn.
    MAG 11 (1975) 25

33  Sharp EL-8010 Pocket Calculator

34  P. J. Wild and J. Nehring, Appl. Phys. Lett. 19 (1971) 335

35  C. R. Stein and R. A. Kashnow, Appl. Phys. Lett. 19 (1971) 343

36  C. R. Stein, S.I.D. Int. Symp. Digest 4 (1973) 38

37  E. T. Fitzgibbons and R. G. Carlson, S.I.D. Int. Symp. Digest 5 (1974) 90

38  E. P. Raynes and I. A. Shanks, Electron. Lett. 10 (1974) 114

39  F. Gharadjedaghi, doctoral thesis Institute Nat. Polytech. Grenoble (1975)

40  D. W. Berreman, this book

41  A. Rapini and M. Papoular, J. Phys. 30 (1969) C4-54

42  J. Nehring, A. R. Kmetz and T. J. Scheffer, J. Appl. Phys. 47 (March 1976) to be
    published

43  G. Ryschenkow and M. Kleman, J. Chem. Phys. 64 (1976) 404

44  H. R. Zeller, this book

45  I. F. Chang and W. E. Howard, IEEE Conf. Display Devices and Systems
    74 CH 0892-0 ED (1974) 148

46  J. H. McGee, W. E. Kramer and H. N. Hersh, S.I.D. Int. Symp. Digest 6 (1975) 50

47  J. Bruinink, this book

48  A. L. Dalisa and R. A. Delano, S.I.D. Int. Symp. Digest 5 (1974) 88

49  I. Ota, J. Ohnishi and M. Yoshiyama, Proc. IEEE 61 (1973) 832

50  I. Ota, T. Sato, S. Tanaka, T. Yamagami and H. Takeda, presented at LASER 75,
    Munich (June 1975)

51  A. M. Marks, Appl. Optics 8 (1969) 1397

52  A. Davis and I. M. Thomas, S.I.D. Int. Symp. Digest 6 (1975) 88

53  A. M. Marks, Proc. S.I.D. 11 (1970) 2

54  M. Hareng, G. Assouline and E. Leiba, Proc. IEEE 60 (1972) 914

55  K. Uehara, H. Mada and S. Kobayashi, IEEE Trans. Electron Devices ED 22 (1975)
    804

56  G. Labrunie, J. Robert and J. Borel, Appl. Optics 13 (1974) 1355

57  P. Carosi, P. Maltese and C. M. Ottavi, IEEE Trans. Electron Devices ED 22 (1975)
    801

58  H. Schmid and J. Schwarzmüller, presented at IEEE Symposium on Applications of
    Ferroelectrics, (June 1975) Albuquerque, N.M.

59  H. W. Roberts, Appl. Optics 11 (1972) 397

60  M. D. Drake, Appl. Optics 13 (1974) 347

61  G. Marie and J. Donjon, Proc. IEEE 61 (1973) 942

62  A. R. Kmetz, IEEE Trans. Electron Devices ED 18 (1971) 756

63  J. R. Barkley, L. H. Brixner, E. M. Hogan and R. K. Waring, Jr., IEEE Symp. on
    Ferroelectrics 3 (1972) 191

64  W. Bindloss, E. I. Du Pont de Nemours and Co., Wilmington, Del., private communi-
    cation, 1971

65  W. Greubel, Appl. Phys. Lett. 25 (1974) 5

66  D. L. White and G. N. Taylor, J. Appl. Phys. 45 (1974) 4718

67  Y. Asano, Proc. IEEE 64 (1976) 189

SHORT COMMUNICATION

## TEMPERATURE DEPENDENCE OF MULTIPLEXED TWISTED

## NEMATIC LIQUID CRYSTAL DISPLAYS

F. J. KAHN and R. A. BURMEISTER, Jr.

Hewlett-Packard Company, Palo Alto, California, USA

## 1. INTRODUCTION

Operating temperature range can make the difference between a laboratory curiosity and a practical display device. For a limited range of instrument applications, operating temperature ranges as narrow as +5 to +45° C may be acceptable. However, operating ranges of −10 to +65° C and wider are desired by potential users. Reaching these goals with twisted nematic (TN) liquid crystal displays (LCDs) is not an easy task — particularly if the LCD is to be multiplexed. Nevertheless, we have recently reported on performance of multiplexed TN-LCDs with satisfactory operating properties between 0 and 50° C[1] and have other laboratory displays with improved performance which will operate satisfactorily from −6 to +66° C.

The factors limiting operating temperature range, in addition to the nematic temperature range of the liquid crystal, are transient response at low temperatures and temperature dependence of the threshold voltage. For reasons of fast transient response as well as reliability, we have concentrated on biphenyl liquid crystals. In this paper, we will present the temperature dependence of transient response and voltage threshold for typical biphenyl liquid crystals and indicate the origins of these temperature dependences.

## 2. PERFORMANCE PARAMETERS FOR MULTIPLEXED TN-LCDs

Performance parameters including threshold sharpness $\rho$, threshold temperature coefficient $\delta$, and voltage figure of merit $M_v$ for a TN-LCD viewed in transmission between crossed polarizers are defined in Fig. 1. The smaller these values, the better the multiplexed display performance.

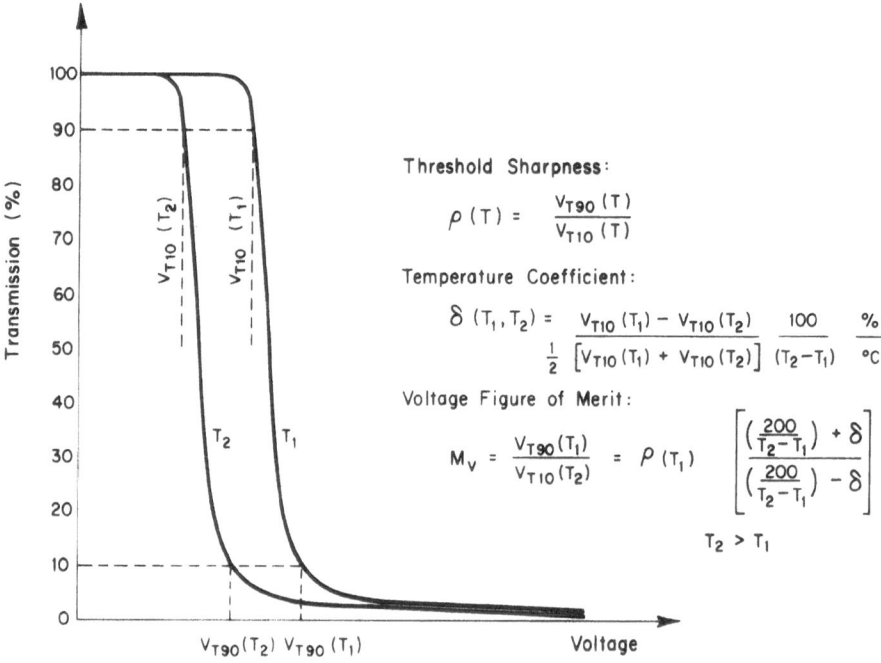

Fig. 1. Performance parameters for multiplexed twisted nematic displays.

The threshold sharpness coefficient $\rho(T)$ determines the maximum number of display elements (corresponding to a minimum duty cycle) which can be multiplexed at a given temperature with a 9:1 contrast ratio. The required $\rho_{max}$ for constant-contrast multiplexed operation in an optimized fast-scan[2] or rms voltage address mode is given in Table 1 as a function of duty cycle. Although other definitions of $\rho_{max}$ are possible, the results would be qualitatively similar. Threshold sharpness coefficients of about 1.40 are readily obtained. Thus ignoring voltage and display tolerances, viewing angle limitations, and other similarly important considerations, we see from Table 1 that multiplexing up to 10 lines with good contrast is possible without technological breakthroughs.

Temperature dependence of the threshold voltage will also influence the number of lines which can be multiplexed. Suppose, for example, the non-select voltage level $V_{NS}$ modifies the transmission of a display element by 10% at the lowest operating temperature $T = T_1$, i.e. the element is considered to be off. What threshold temperature coefficient $\delta$ would result in this element being on (90% transmission with $V_{NS}$ still applied) at the highest operating temperature $T = T_2$? Calling this value $\delta_{max}$, we find

$$\delta_{max}(T_1,T_2) = \left(\frac{\rho(T_2) - 1}{\rho(T_2) + 1}\right)\left(\frac{200}{T_2 - T_1}\right) . \tag{1}$$

in % per °C. Values for $\delta_{max}$ are listed in Table 1 as a function of duty cycle. We see that as duty cycle and $\rho_{max}$ decrease, the acceptable threshold temperature dependence also decreases. Other definitions for $\delta_{max}$ are possible, but they would give qualitatively similar results. For a variety of liquid crystal mixtures of device interest, we have measured $\delta(0°, 50°)$ values in the range 0.35 %/°C to 1.6 %/°C. Thus, temperature dependence of the threshold voltage is large enough to be an important practical consideration in multiplexed TN-LCD applications. For practical multiplexed displays, values for $\delta$ much lower than $\delta_{max}$ are desired.

TABLE 1

Upper Limits on Performance Parameters for Multiplexing
TN-LCDs as a Function of Duty Cycle

| Duty Cycle (%) | Number of Multiplexed Rows | $\rho_{max}$ | $\delta_{max}$ (%/°C) | $M_{vmax}$ |
|---|---|---|---|---|
| 25 | 4 | 1.73 | 1.07 | 3.00 |
| 12.5 | 8 | 1.41 | 0.74 | 2.11 |
| 8.3 | 12 | 1.35 | 0.60 | 1.83 |
| 6.3 | 16 | 1.29 | 0.51 | 1.66 |
| 4.0 | 25 | 1.22 | 0.40 | 1.49 |
| 2.0 | 50 | 1.15 | 0.28 | 1.32 |
| 1.0 | 100 | 1.11 | 0.21 | 1.23 |

In practice, materials which minimize $\delta$ do not necessarily minimize $\rho$. A convenient figure of merit for a multiplexed display which accounts for both $\rho$ and $\delta$ is $M_v$, the ratio between the minimum voltage for which the display is fully on at all operating temperatures, $V_{T90}(T_1)$, to the maximum voltage for which the display is fully off at all operating temperatures, $V_{T10}(T_2)$, where $T_1$ and $T_2$ are respectively the minimum and maximum operating temperatures of the display. Values for $M_v$ may be calculated from $\rho$ and $\delta$ using the expression in Fig. 1. Values for $M_{vmax}$ are listed in Table 1 using $\rho_{max}$ and $\delta_{max}$. It is easily shown that for $\delta = \delta_{max}$

$$M_{vmax} = \rho(T_1)\rho(T_2) . \tag{2}$$

Thus if $\rho(T)$ is temperature independent,

$$M_{vmax} = \rho_{max}^2 .$$ (3)

For typical displays, we find $M_v$ values between 2 and 3. Thus, 8 to 12 lines ($\sim 10\%$ duty cycle) seems to be the multiplexing limit for reflective TN-LCDs with present technology. Future developments resulting in $\rho \leqslant 1.29$ and $\delta \leqslant 0.51$ would enable multiplexing 16 lines. More lines could be multiplexed in transmission.[2,3]

So far, we have considered only steady-state response. Transient response is also an important consideration. We have compared response times of displays operated in a multiplexed mode,[1] i.e. switching between the voltages $V_{NS}$ (non-select) and $V_S$ (select), to those for operation in a direct-drive mode, i.e. switching between 0 V and 2.7 $V_{T10}(T)$. For multiplexing (MPX) with a 25% duty cycle we find that, at a given temperature, both turn-on and turn-off times are 1.5 to 2 times larger than for non-multiplexed (NMPX) operation. Temperature dependences of transient response times are very similar in all cases. An important parameter for a multiplexed display is $t_{OFF}(\text{MPX})$ at the lowest operating temperature, i.e. generally the longest transient response time. Typical user specifications for $t_{OFF}$ at the lowest operating temperature $T_1$ are as follows:

$$t_{OFF}(T_1) \leqslant \quad 1 \text{ s required}$$

$$\leqslant 100 \text{ ms desired.}$$

With biphenyl materials, such as BDH E-7, displays with $t_{OFF}(\text{MPX},0°C)$ < 1 s are readily fabricated using 10 $\mu$m thick LC layers. Thinner layers will give proportionately faster response. For convenience of analysis, we will restrict our discussion in the following to $t_{OFF}(\text{NMPX})$ which as noted above has the same T-dependence as $t_{OFF}(\text{MPX})$.

## 3. TEMPERATURE DEPENDENCE OF TRANSIENT RESPONSE

Following Jakeman and Raynes,[4] we take

$$t_{OFF} = \frac{\gamma_1 L^2}{K_J \pi^2}$$ (4)

where

$\gamma_1$     rotational viscosity
$K_J$     an equivalent Franck elastic constant
$L$     LC layer thickness

for small LC deformation by an applied field. Berreman[5] has shown that the threshold for molecular reorientation is given by

$$V_T = 2\pi^{3/2} \left( \frac{K}{\epsilon_\| - \epsilon_\perp} \right)^{1/2} \tag{5}$$

where

$V_T$     threshold voltage for 90° twist cell

$\epsilon_\|, \epsilon_\perp$     LC dielectric constants

$K = K_{11} + \frac{1}{4}(K_{33} - 2K_{22}) + 2K_{22}(L/P)$

$K_{11}, K_{22}, K_{33}$     Frank elastic constants

$P$     LC pitch.

We make the identification $K_J = K$ and assume that equation (4) applies even to large deformations. While (4) has not been rigorously derived (a closed-form solution to the boundary value problem does not exist), it is at least dimensionally correct. Following Gruler[6] and Saupe,[7] we define a set of reduced elastic constants

$$C_{ii} = \frac{K_{ii} V^{7/3}}{S^2} \tag{6}$$

where V is the molar volume and S the order parameter. The reduced constants $C_{ii}$ take into account the theoretically expected dependences of elastic constants on molar volume and order parameter and should therefore be independent of temperature in the absence of changes in short-range order.

Before we proceed, one additional modification of (4) is necessary. Rotational viscosity $\gamma_1$ is not as convenient to measure as shear viscosity $\eta$. The latter can be determined in bulk samples relatively independent of sample boundary conditions using a rotating cone plate viscometer. We will assume $\eta(T) \propto \gamma_1(T)$. Although this is not rigorously true, our results indicate that this is a useful assumption at least for cyanobiphenyl materials. Thus, (4) becomes

$$t_{OFF} \propto \frac{\eta L^2 V^{7/3}}{C S^2} \tag{7}$$

where     $C = C_{11} + (C_{33} - 2C_{22})/4 + 2C_{22}(L/P)$.

Since the coefficient of volume expansion is typically[8] $\sim 10^{-3}/°C$, the changes in $L^2$ and $V^{7/3}$ are negligibly small compared to the T-dependence of $\eta$ and $S^2$. Thus in the absence of changes in short-range order, we expect

$$\frac{t_{OFF}}{\eta}(T) \propto \frac{1}{S^2}. \tag{8}$$

Data for $t_{OFF}(T)$ and $\eta(T)$ are plotted in Fig. 2 for BDH E-3 (nematic range $-2$ to $+54°$ C). Referring to Fig. 2 and (8), we observe that the temperature dependence of transient response is primarily due to the viscosity $\gamma_1$ and secondarily due to the equivalent elastic constant K which gives the $1/S^2$ dependence of (8). An interesting aspect of the effect of K(T) is the leveling-off of $t_{OFF}(T)$ at high temperatures near $T_{NI}$. The data clearly show that this leveling is due to K(T) and not to $\gamma_1(T)$. Note that $\eta(T)$ and $t_{OFF}(T)$ have similar temperature dependences for $T \ll T_{NI}$, thereby confirming the validity of using $\eta$ in place of $\gamma_1$. A similar temperature dependence of $t_{OFF}(T)$ has recently been reported independently by Tarry[9] for three other cyanobiphenyl mixtures.

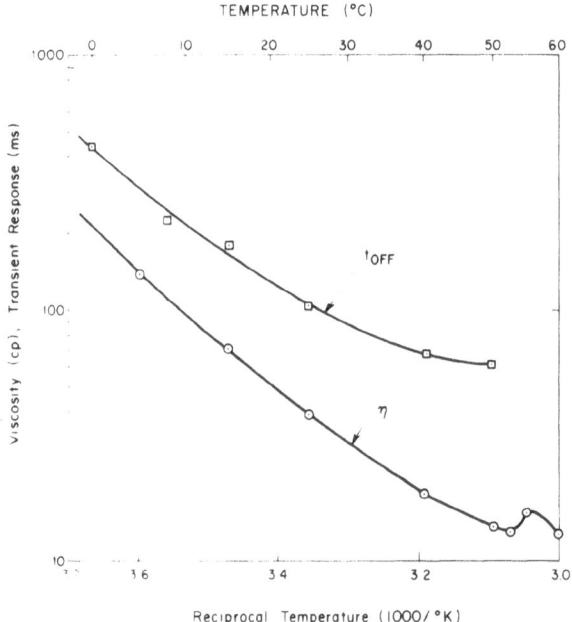

Fig. 2. Temperature dependence of transient response time ($t_{OFF}$) and shear viscosity ($\eta$) of BDH E-3 biphenyl liquid crystal mixture (see text).

## 4. TEMPERATURE DEPENDENCE OF THRESHOLD VOLTAGE

Temperature dependence of the optical threshold voltage $V_{T10}$ is given in Fig. 3 for a $90°$ twist cell containing BDH E-3 liquid crystal with 0.3% by weight cholesteryl nonanoate added. This datum is for a low tilt ($\sim 5°$) cell. However, we have found that the temperature dependence of $V_{T10}$ is approximately independent of molecular tilt at the substrate boundary.

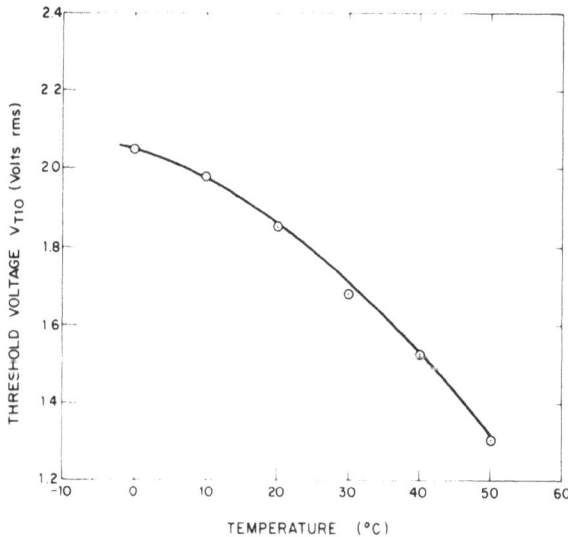

Fig. 3. Temperature dependence of threshold voltage for BDH E-3 liquid crystal (0.3% by weight cholesteryl nonanoate added).

An analytical expression for the temperature dependence of the voltage threshold is obtained by combining (5) and (6) and using the relation

$$(\epsilon_{\parallel} - \epsilon_{\perp}) = (\frac{A}{T} + B)S .\qquad(9)$$

A and B are constants giving the contributions of permanent and induced dipoles to the dielectric anisotropy, respectively.[10] In the absence of changes in short-range order, A and B will be relatively temperature-independent. If we assume that the optical threshold $V_{T10}$ is proportional to the deformation threshold $V_T$, then $V_{T10}$ is given by

$$V_{T10} \propto S^{1/2}(\frac{A}{T}+B)^{1/2}V^{-7/6}[C_{11}+(1/4)(C_{33}-2C_{22})+2C_{22}(L/P)]^{1/2}\qquad(10)$$

Equation (10) shows that the principal temperature dependence of $V_{T10}$ will occur near the nematic-isotropic transition temperature $T_{NI}$ due to the rapid decrease in S near $T_{NI}$. Additional T-dependence will be found in materials having changes in short-range order, e.g. stiffening of the reduced elastic constants at low temperatures as a result of smectic tendencies and formation of cybotactic groups[6] or changes in the dielectric tensor associated with dipole-dipole interactions.[10]

Thus neglecting possible changes in short-range order and molar volume, the lowest values of threshold temperature coefficient $\delta(T_1,T_2)$ for a given operating temperature range will be obtained for materials with the highest

$T_{NI}$. Since short-range order changes are most pronounced near the melting point $T_{KN}$, it might be expected that materials with lower $T_{KN}$ values would tend to have lower $\delta(T_1, T_2)$ when $T_{NI}$ values are equal. However, this trend tends to be obscured by the mixture-to-mixture differences in short-range order which do not seem to correlate with $T_{KN}$. Figure 4 shows measured values of $\delta(0°, 50°C)$ for 17 different liquid crystal mixtures. Most of the mixtures are biphenyls, but also included are at least one ester, Schiff base and azoxy mixture. Points corresponding to BDH E-3, E-7 and E-9 biphenyl mixtures are specifically identified. The solid curve has been fit by eye to define the trend of the data. The data of Fig. 4 confirm that $\delta(0°, 50°)$ tends to decrease with increasing $T_{NI}$. Variations from the trend lines can be qualitatively explained by the effect of the specific mixture compositions on short-range order. However, those explanations go beyond the intended scope of this paper.

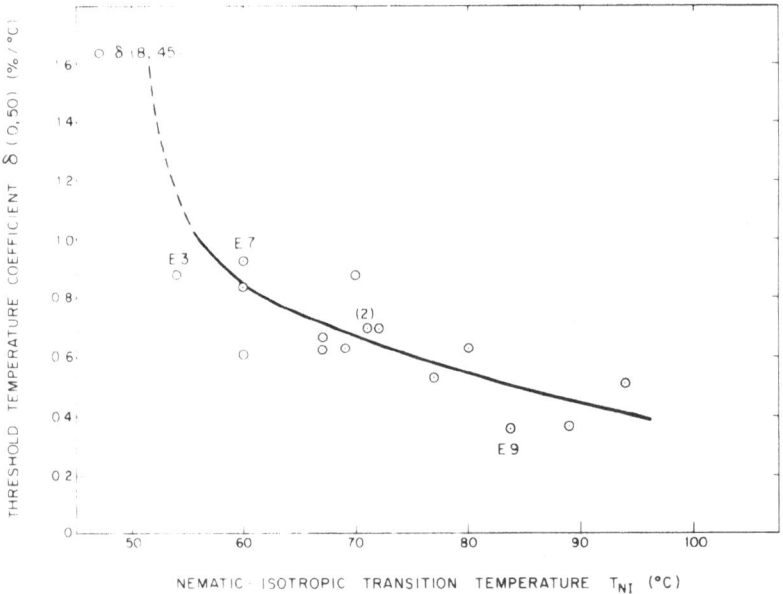

Fig. 4. Threshold temperature coefficient $\delta(0°, 50°)$ versus $T_{NI}$ for 17 liquid crystal mixtures, including BDH E-3, E-7 and E-9 biphenyl mixtures.

## 5. CONCLUSIONS

We have defined performance parameters for multiplexed TN-LCDs, namely threshold sharpness $\rho$, threshold temperature coefficient $\delta$ and voltage figure of merit $M_v$. The smaller these values the better the multiplexed performance. Parameter values of $\rho \sim 1.4$, $\delta \sim 0.35$ %/°C and $M_v \sim 2.1$ are readily obtained. These values would permit multiplexing 8 to 12 lines in reflection using an optimized fast-scan address mode assuming that satisfactory device and voltage tolerances could be maintained reproducibly. With biphenyl materials, such as BDH E-7, multiplexed transient response times less than 1 second can readily be obtained at $0°$ C using 10 $\mu$m thick LC layers. The temperature dependence of the transient response of TN-LCDs has been shown to be primarily due to T-dependence of viscosity and to a lesser extent due to T-dependence of the equivalent elastic constant which in turn results largely from T-dependence of the order parameter. Finally, the temperature dependence of $V_{T10}$ has been shown to be primarily due to changes in order parameter and to a lesser extent due to changes in short-range order affecting the reduced elastic constants or dielectric constants. Changes in molar volume could also have an effect. Data for 17 mixtures demonstrate that $\delta$ decreases with increasing $T_{NI}$.

## ACKNOWLEDGMENTS

We gratefully acknowledge numerous stimulating technical discussions with Dr. Paul E. Greene, electrooptical measurements by George S. LaBelle, viscosity measurements by Dr. Hsia Choong and Gene Koch and cell processing by Don Bradbury and Erwin Littau. We also thank W. H. de Jeu for his comments on temperature dependence of dielectric constants as well as for a preprint of his review article.[10]

## REFERENCES

1  F. J. Kahn and R. A. Burmeister, S.I.D. Int. Symp. Digest 6 (1975) 86

2  J. E. Bigelow, R. A. Kashnow and C. R. Stein, IEEE Trans. Electron Devices, ED-22 (1975) 22

3  C. J. Gerritsma, this book

4  E. Jakeman and E. P. Raynes, Phys. Lett. 39A (1972) 69

5  D. W. Berreman, Appl. Phys. Lett. 25 (1974) 12

6  H. Gruler, Z. Naturforsch. 28a (1973) 474

7  A. Saupe, Z. Naturforsch. 15a (1960) 810

8  M. J. Press and A. S. Arrott, Phys. Rev. A8 (1973) 1549

9  H. A. Tarry, Electron. Lett. 11 (1975) 339

10  See for example, W. H. de Jeu, to be published in Orsay Liquid Crystal Group (eds.), 'Advances in Liquid Crystal Research', supplement of Solid State Physics Series

SHORT COMMUNICATION

## SCANNING LIMITATIONS OF TWISTED NEMATIC

## DISPLAY DEVICES

C. J. GERRITSMA and P. van ZANTEN

Philips Research Laboratories, Eindhoven, Netherlands

Alt and Pleshko[1] have shown that the maximum number of scanned lines $N_{max}$, in the case of multiplexing rms-responding liquid crystal devices, is directly dependent on the optical response vs. voltage curve of the device. In their analysis the rms-value of the on- and off-voltage is:

$$\hat{V}_{on} = \hat{V}_{th} + \hat{\Delta} ,$$

$$\hat{V}_{off} = \hat{V}_{th} .$$

In these relations $\hat{V}_{th}$ is the optical threshold voltage and $\hat{\Delta}$ is the increment of voltage above $\hat{V}_{th}$ for which the contrast of the device is just acceptable. According to Alt and Pleshko, $N_{max}$ is given by the following relation:

$$N_{max} = \left[ \frac{(1 + P)^2 + 1}{(1 + P)^2 - 1} \right]^2 \qquad (1)$$

where $P = \hat{\Delta}/\hat{V}_{th}$. When the results of this analysis are applied to dynamic scattering or field effect displays, it must be emphasized that the optical response of liquid crystal devices depends on illumination and viewing angle. This implies that the response-curve changes with operating conditions, and consequently $N_{max}$ depends on the mode of operation (e.g. transmissive or reflective).

Let us first examine the situation when the display device is used in a transmissive mode under normally incident light (e.g. projection onto a diffuser). In this mode the optical performance is independent of viewing angle. The corresponding transmission vs. voltage curve is shown in Fig. 1. For this mode we assume $\hat{V}_{off} = \hat{V}_{th}$, and equation (1) can be used directly to determine $N_{max}$. According to (1), $N_{max}$ increases with decreasing P.

A completely different situation occurs when the display is directly viewed (without diffuser) in diffuse light instead of "parallel" light. For transmission at normal viewing angle, we again have the condition

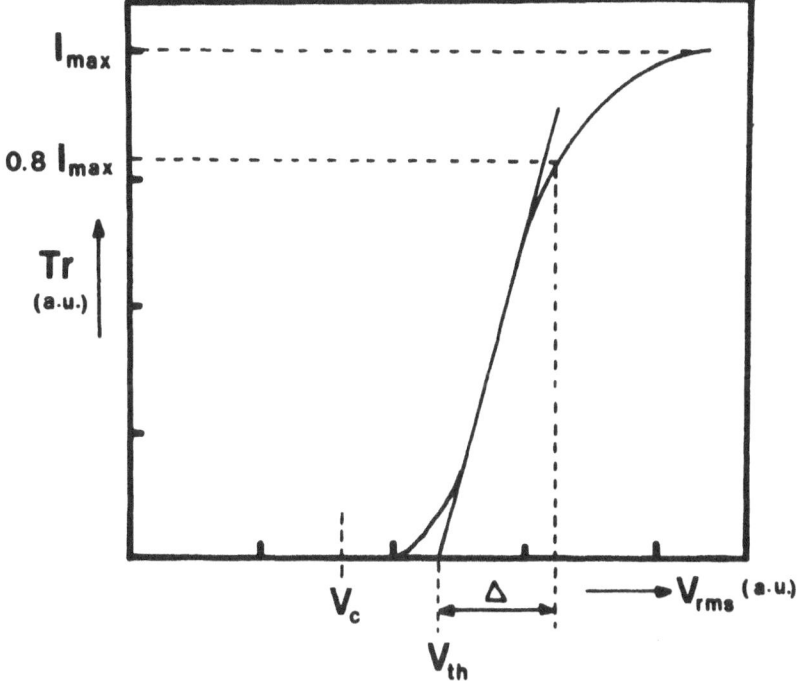

Fig. 1.  Transmission vs. voltage for a twisted nematic layer between parallel polarizers under normally incident illumination.

$$\hat{V}_{on} = \hat{V}_{th} + \hat{\Delta} \ . \tag{2}$$

However, it is well known[2] that for twisted nematic displays the threshold voltage $\hat{V}_{th}$, at which changes in transmission for normally incident light are first observed, is much higher than the critical voltage $\hat{V}_c$, at which reorientation of the director begins (Freedericksz transition). In addition the transmission is strongly angle-dependent[3] at voltages above $\hat{V}_{th}$. There is thus a different requirement for $\hat{V}_{off}$ since all the off-elements now have to appear "off" at every viewing angle between zero and $\pi/2$. This requirement is satisfied if $\hat{V}_{off} = \hat{V}_c$.

For this situation we introduce a second performance parameter $Q$, defined by $Q = \hat{V}_c/\hat{V}_{th}$ ($Q \leqslant 1$). The off-voltage then is given by

$$\hat{V}_{off} = Q \, \hat{V}_{th} \ . \tag{3}$$

From the expressions (2) and (3), $N_{max}$ becomes

$$N_{max} = \left[ \frac{(1 + P)^2 + Q^2}{(1 + P)^2 - Q^2} \right]^2 \ . \tag{4}$$

The dependence of strobing voltage $V_s$ and the information-pulse voltage $V_D$ on the performance parameters P and Q is given by (5) and (6):

$$\hat{V}_D = \frac{1}{2}\hat{V}_{th}\left[(1+P)^2 + Q^2\right]^{1/2} \tag{5}$$

$$\hat{V}_s = \frac{1}{2}\hat{V}_{th}\left\{\frac{[(1+P)^2 + Q^2]^3}{[(1+P)^2 - Q^2]^2}\right\}^{1/2}. \tag{6}$$

Equations (4), (5) and (6) correspond to equations (16), (17) and (18) in Ref. 1.

Comparison between (1) and (4) shows that $N_{max}$ decreases with Q at constant P. The calculated values of $N_{max}$ as a function of P are given in Fig. 2 for different values of Q.

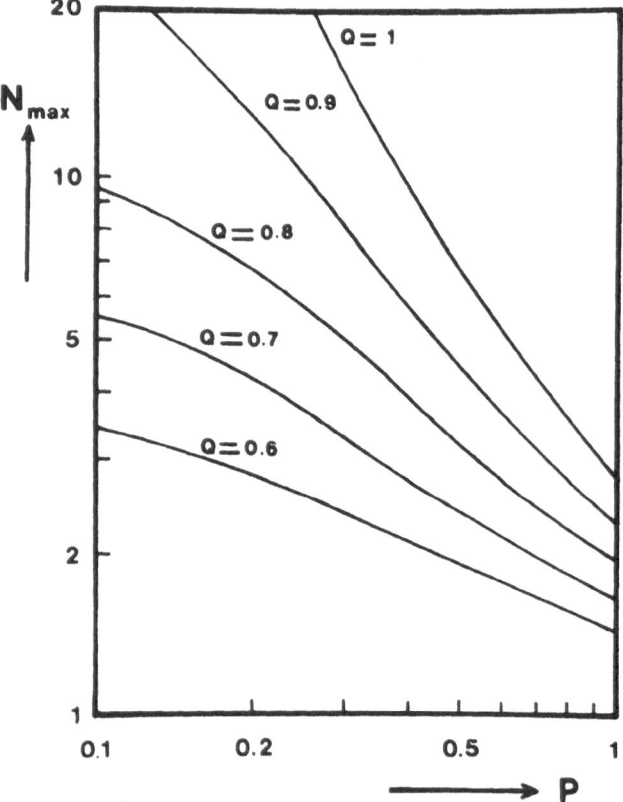

Fig. 2. Maximum scanning capability as a function of the device performance parameters P and Q.

To compare the two different situations (normally incident light and diffuse light), we measured P and Q under standard conditions for a number

## TABLE 1

### Performance Characteristics of Twisted Nematic Devices for Different Materials (see text)

| Material | P | Q | Normal incidence | | Diffuse light | |
|---|---|---|---|---|---|---|
| | | | $N_{max}$ | $V_s/V_D$ | $N_{max}$ | $V_s/V_D$ |
| FEI | 0.45 | 0.74 | 8 | 2.8 | 3 | 1.7 |
| BDH | 0.25 | 0.53 | 21 | 4.6 | 2 | 1.4 |

of commercially available materials. The transmission vs. voltage curves for twisted nematic layers between parallel polarizers were measured with a He-Ne laser beam at normal incidence. The voltage $\hat{\Delta}$ was chosen to correspond to 80% transmission of the total light intensity (Fig. 1). For the various materials examined, P-values between 0.25 and 0.45 were measured. The Q-values were between 0.53 and 0.80. It is interesting to note that substances with small P-values also had small Q-values. Therefore the positive influence on $N_{max}$ by a small P-value is suppressed by the negative influence of a small Q-value.

As an example we present the performance characteristics of two materials, namely Eastman Kodak product 11900 Field-Effect Mixture I (FEI) and a mixture of biphenyls supplied by BDH Chemicals Ltd. (BDH). The composition of the latter mixture was 56% 4'-n-pentyl-4-cyanobiphenyl and 44% 4'-n-heptyl-4-cyanobiphenyl. The relevant data for these substances (P and Q) are given in Table 1. For normally incident light, $N_{max}$ from (1) is equal to 21 for BDH and 8 for FEI. The difference in $N_{max}$ is due to the difference in P. In the case of diffuse light, the values of $N_{max}$ from (4) are much lower, and $N_{max}$ for BDH is even less than for FEI. In Table 1 we also list the $V_s/V_D$-ratios at which the corresponding values for $N_{max}$ are realized.

The agreement between theoretical analysis and experimental data is good. We found that satisfactory matrix addressing of twisted nematic devices is possible in the reflective mode with no more than about 4 scanning lines. This number is higher than the calculated value (2 or 3), because we restricted ourselves to viewing angles up to 60° from the normal instead of 90°. Furthermore, the index of refraction of the liquid crystal is higher than that of air, so the angle of incidence inside the liquid crystal is less than the external angle and the situation is in fact more favorable.

## REFERENCES

1    P. M. Alt and P. Pleshko, IEEE Trans. Electron Devices ED 21 (1974) 146
2    C. J. Gerritsma, W. H. de Jeu and P. van Zanten, Phys. Lett. 36A (1971) 389
3    C. Z. van Doorn and J. L. A. M. Heldens, Phys. Lett. 47A (1974) 135

## DISCUSSION

W. G. Freer (Rank)
What was the surface tilt angle in your cells?

C. J. Gerritsma
It was on the order of two to five degrees.

P. Wild (Brown Boveri)
Couldn't you obtain the same results without introducing the parameter Q simply by using the experimentally determined threshold at the worst-case viewing angle in the conventional calculation?

C. J. Gerritsma
I wanted to emphasize that unselected elements are sure to be off regardless of viewing angle if their voltage does not exceed the Freedericksz threshold.

A. R. Kmetz (Brown Boveri)
From a physical point of view, the relevance of the Freedericksz threshold in practical cells having tilted alignment is questionable; from an engineering standpoint, the a priori choice of a conservative operating threshold is unnecessarily restrictive for applications requiring fields of view intermediate between the extremes you considered.

# INTEGRATED ELECTROOPTIC DISPLAYS

T. P. BRODY

Westinghouse Research Laboratories, Pittsburgh, Pennsylvania, USA

## SUMMARY

By an integrated display we understand a subsystem in which a significant or major portion of the display electronics is assembled or fabricated on a common substrate which itself forms the display surface. As the number of display elements increases and approaches that of a cathode-ray tube, the extension of the concepts and techniques of large-scale integration to the display field becomes mandatory, for both technical and economic reasons. Matrix addressing is employed in most attempts to build higher resolution displays. Passive matrix addressing schemes however present many problems, particularly with display materials having a slow response to electrical excitation. This paper reviews the work so far carried out on integrated non-emissive displays, with particular emphasis on recent work involving the use of "active matrices". Such matrices contain gain-producing, switching and/or memory elements at every mesh point, and represent the most ambitious thrust to date toward full display integration. An attempt is made to establish the scope and limitations of the various schemes employed, and the direction of probable future developments are indicated.

## 1. INTRODUCTION

It appears that displays are undergoing an evolution similar to that of integrated electronic circuits, progressing from small, discrete-component assemblies to large, integrated displays containing many picture elements. The driving force towards LSI technology in the semiconductor industry has been the major cost reduction per gate (or function) possible through integrating an increasing number of components on a chip. Likewise, in the field of solid-state displays (in which we include, for convenience, all electronically addressed displays, whether the display material itself is solid, liquid or gas), the driving force is that of potential cost reduction per picture element — a visual near-equivalent of the logic gate — through the application of the concepts or techniques of large-scale integration.

A major difference between the two fields (logic and display) is, however, the factor of size. Logic systems handle abstractions, namely units of information, and nature has determined that these units require very minute amounts of matter and energy for their storage and processing. A bit of information can, after all, be stored in an assembly of a few atoms, and requires just a few kT's worth of energy for its survival. Hence, logic elements can be miniaturized, and this, indeed, has been the most powerful single reason for the dramatic cost reduction and increased computing power brought about by large-scale IC's. In contrast, displays consist of highly concrete physical elements, which, by their size and energy output must directly influence the human eye, a highly specified receiver. Thus, displays cannot be meaningfully miniaturized, and must also handle comparatively large amounts of energy, whether that energy is generated by the display itself (active or light-emitting displays) or is external to the display and is passively modulated by it (passive or non-emissive displays).

Thus, although there are major conceptual similarities between the two fields of endeavor, there are also major physical differences, and there is no prima facie reason to believe that the specific materials, structures and technologies which have been successful in semiconductor LSI can be directly applied to the problems of large-area, multi-element displays. On the other hand, concepts, principles of circuit design and organization, and many tools, such as computer-aided design, could fruitfully be borrowed from the semiconductor field.

In this paper, we review some of the main approaches to the integration of non-emissive displays. No consideration is given to non-electronically (e.g. optically or electron-beam) addressed displays or to projection systems. While the paper is a review, it has been written from a rather specific point of view. This point of view can be simply stated: the primary problem to be solved in constructing a large, multi-element display is that of signal distribution to the screen, and those techniques which hold the promise of solving this problem for a wide range of different display materials are to be preferred to solutions which are less general or are specific to only one type of display. In other words, we seek a measure of universality which is analogous to that of silicon IC technology for different types of electronic circuits.

In a cathode-ray tube — still by far the most elegant and cost-effective display systems — the signal distribution is performed by the rapid movement across the screen of a single, low-inertia beam of electrons. A solid-state equivalent of the steerable beam is normally some form of matrix addressing, although other, analog-type scanning schemes (e.g. through traveling acoustic waves) have been explored.

Matrix addressing of solid-state displays has been extensively used in the past, both with active and passive display media. Usually two sets of parallel busbars, orthogonal to each other, are employed with the display material sandwiched between them. These matrices can be termed "passive", since they are simply conductors of signals and power. Gain-producing, non-linear or switching and memory functions have to be provided by external circuits in combination with the electrical properties of the display medium.

There are certain inherent limitations[1,2] in the use of passive matrices. These limitations relate partly to the onerous electrical performance requirements placed on the display materials and partly to the need for rather elaborate external driving circuits. The requirements placed on passive display materials include:

— Need for large non-linearities
— Uniformity of thresholds
— High speed of response
— Local storage capability, consistent with gray-scale response.

The problems related to the external driving circuits we may list as:

— Current and/or voltage requirements, particularly for fast operation
-- Complexity
— Expense
-- Mounting and interfacing with matrix (size disparity).

For these reasons, and also for the more generalized reasons mentioned earlier, significant efforts are currently being devoted to the development of "active" matrices for display addressing. Such matrices contain switching and/or memory elements at every intersection point and can, in principle, be completely integrated with the display medium. The purpose of an active matrix is to separate out the electronic functions to be performed on the display, leaving the display medium with only the optical requirements. By extension, the active matrix will ideally also incorporate those scanning, decoding and driving functions which have to be carried out at the edges of the display panel, thus eliminating the large number of external connections that otherwise have to be made to a large matrix.

Clearly, if such large-area active matrices can be developed and fabricated, they would provide a general solution to the solid-state display addressing problem, in a manner independent of the specific properties of the display medium they are designed to activate. It is our thesis that such a goal is attainable, and that the achievement of this goal is a necessary condition for the eventual technical and economic viability of solid-state displays with a performance comparable to that of cathode-ray tubes. For intermediate types of display goals, such as alphanumeric displays with a few tens to a few hundred characters, vectorgraphic displays, monochrome and on-off (i.e. not gray-scale) displays, tradeoffs will exist at any given time along the evolutionary curve.

As is inevitable in a developing technology, boundaries and definitions are not sharp and precise. Many of the larger displays so far reported, for example, use passive addressing schemes combined with some form of time-division multiplexing. If such a display is constructed in a fashion so that integrated circuit chips or discrete devices are directly attached to the substrates carrying the conductors, such a display could well be considered "integrated" by our definition. On the other hand, by usual electronics terminology, we would speak of such an assembly as a hybrid circuit, to distinguish it from a truly integrated circuit which is manufactured, not assembled. Much of the current display work is moving in this (hybrid) direction, and compatible drive and scanning IC's are being developed for this purpose. However, such displays more appropriately fall into the category discussed in the paper by A. R. Kmetz. We therefore touch upon them only briefly.

Another fuzzy boundary is represented by the use of materials with non-linear electrical properties, which can compensate for the absence of sharp thresholds and local memory from the display material. Such materials (e.g. ferroelectrics) bring in many attractive features, and in combination with an otherwise passive matrix, represent a reasonable halfway-house between active-matrix displays and time-multiplexed passive-matrix systems. We discuss this work in somewhat greater detail.

The main part of the discussion concerns the two major thrusts towards truly integrated displays, namely those using, respectively, silicon and thin film technology. A review covering this topic has recently been written,[3] and although it emphasizes TV-type applications, there is inevitably an overlap between the present discussion and this very recent publication.

## 2. MATERIALS FOR INTEGRATED NON-EMISSIVE DISPLAYS

Much work in the past has been done on a wide variety of light-modulating display materials, but only liquid crystals have so far been employed in the types of integrated displays which are of main interest in this review. On the other hand, as mentioned in the Introduction, there is a converging trend of development, the aim of which is to provide large-scale integration of display drive electronics in a form which is sufficiently universal to interface with a variety of different display materials for different applications. Since this convergence of materials and integrated drive circuits is still in the future for materials other than liquid crystals, the discussion of such combinations is necessarily hypothetical. Nonetheless, the capability for these combinations already exists, and it is predictably just a matter of time before they will make their appearance. We therefore briefly review the more important current candidate materials for integrated displays. More extensive discussions will be found in the materials-oriented papers of this volume.

In discussing these materials, we have to anticipate to some extent the conclusions emerging from our section on integration techniques. We may summarize the essential points by stating that techniques now exist which allow us to build active matrix drives for all those display materials which possess the property of modulating (or emitting) light in response to some form of electrical excitation, provided only that these materials can be applied in the form of a thin, extended film or layer over (or under) the surface of the matrix. By this criterion then the following passive materials qualify as candidates for integration:

- Liquid crystals of all types
- Electrochromics
- Electrophoretics
- Ferroelectrics
- Colloidal suspensions
- Fluorescent dyes.

The first three groups are currently prime candidates for near-future applications, but the others are also good matches. Chance rather than any inherent logic will determine the actual time-evolution of such integrated displays, but it is appropriate to list some of the advantages and disadvantages of these classes of materials, as we see them today.

First of all, however, we must establish the display objectives we seek, in order to measure up our materials against these. Integration of displays, just as that of logic systems, becomes important in proportion to the number of elements which they contain. While single-digit alphanumerics, watch and instrument readout panels, and multi-digit displays for calculators could also benefit from integration, the real motivation is the drive toward displays with capabilities approaching and eventually surpassing that of cathode-ray tubes. Between these extremes there is a large field of heterogeneous applications, such as computer terminals, graphic panels, bank terminals, vector-graphic displays for military and other purposes, office systems and probably many others, requiring various sizes, resolutions, speeds of response, digital or analog, monochrome or several colors. We may therefore define the shorter term goal of integrated solid-state displays as that of technically and economically satisfying the various emerging needs for displays of performance intermediate between that of assemblies of a few alphanumerics and that of the CRT. In many applications, the small bulk and weight, lower power consumption and potentially lower total system cost are added attractions of a solid-state display unit. From our point of view, interest centers on the ease of combination of the materials listed with the driving matrix, chemical, electrical and processing compatibility, and, of course, range of performance capability when integrated.

## 2.1. Liquid Crystals

The variety of liquid-crystal (LC) materials and electrooptic effects is so large that even a mere listing of them is impractical within the scope of this review. Both major classes of nematic electrooptic effects, namely dynamic-scattering and field-effect types, are well compatible with active matrix drives. Field-effect operation is probably preferable, as the high impedance and very low current requirements form a better electrical match to the matrix. The voltage range of operation of most liquid-crystal materials is fully compatible with both silicon and thin-film matrix devices, while the higher voltage capability of the thin-film matrix also permits fast operation well above the material threshold and the use of the cholesteric-nematic phase transition[4] which requires higher voltages.

All normally used liquid-crystal materials are inactive chemically with the metal and inorganic insulator surfaces with which they make contact in an integrated display. Processing compatibility is excellent; the materials can be simply applied by normal techniques over the surface of the finished and passivated matrix. All the standard surface alignment and preparation techniques can be carried out on the matrix surfaces. Both reflective and transmissive displays can be built with thin film active matrices. With silicon matrices, reflective displays can be constructed, but transmissive displays are more of a challenge. Such displays, however, have been postulated.[5]

With respect to overall performance, liquid crystals in integrated displays share some of the limitations of non-integrated displays, such as temperature range of operation, sluggishness at low temperatures, unsatisfactory viewing angle and limited ability to display color. On the other hand, limitations relating to the absence or non-uniformity of thresholds, slow response and consequent multiplexing and contrast problems are essentially removed by the active matrix, and this, of course, is the main reason for their development. Size and resolution of the integrated display is governed entirely by the state of the art in active matrices and not by any inherent limitation of the liquid-crystal materials. Life limitations of integrated LC displays are expected to be broadly similar to those of non-integrated displays. It should be remarked, however, that ac operation, which is preferable from the standpoint of long life, is easily provided through the active matrix.

## 2.2. Electrochromic Materials

Electrochromism is the color change induced in a material by an electric field — or, more accurately, by a change in the charge state of certain material species.[6] The most important effects are those of color-center effects in solids, and electrochemical (usually redox) reactions in solids and liquids. Recent work has been mainly in the latter field. Again, the range of materials and effects requires a separate review, and here we restrict our discussion to the same form which was followed with liquid crystals.

From the point of view of building an integrated electrochromic (EC) display, these materials present rather different electrical requirements, in that they are voltage- rather than field-driven, and their action depends on a charge transfer, which represents quite substantial currents if the transfer is to be done fast. For these reasons, EC materials are less desirable candidates for integration. EC devices normally involve highly conductive liquid electrolytes which will react electrochemically with most metals. It will be necessary to use noble-metal electrodes for the output terminals of the active matrix, or to protect the terminals with a layer of evaporated carbon or sputtered $In_2O_3$. Otherwise, there appears to be no chemical or processing incompatibility, and an assembly such as that for the integrated LC displays can readily be visualized, such that the active matrix simply replaces one of the display electrodes.

Concerning display performance, the situation is again quite similar to that of LC's. Temperature range, visual appearance and maintenance properties will not be influenced by integration. It is recognized that the color and light scattering properties of EC displays are superior to those of LC displays while high contrast and particularly, high contrast combined with long life (many cycles), are more difficult to attain in EC's with the current state of the art. When driven by an active matrix of suitable design (see Section 4), EC-based displays could show a dramatic improvement in terms of the number of display elements, and hence overall resolution, possible. Since the speed of response of some EC materials can be above the reciprocal flicker-fusion frequency of the eye,[7] moving, gray-level video imagery through such materials is an entirely reasonable possibility. Multicolor EC displays would be quite difficult to construct with liquid materials, even though the colors are available. With solid EC films, colors other than blue are probably difficult to obtain. EC displays generally have a memory which may reduce or eliminate the need for an external refresh memory for fixed (i.e. non-moving) displays.

Since no active matrix-driven EC displays have so far been built, the above discussion is necessarily speculative. However, there is no good reason why an attempt to build one through thin-film-matrix techniques should not be successful.

### 2.3. Electrophoretic Suspensions

Far less work has been reported on electrophoretic materials (EP) than on the previous two classes of materials. Two reasons may account for this, namely 1) the rather high voltages required to drive EP displays, which puts them beyond the voltage conveniently available from IC's, and 2) the difficult problem of stability (tendency of the particles to separate and settle).

From the point of view of integrated displays, EP materials look very attractive. Although the voltages reported as necessary for fast response are high ($> 100$ $V_{p-p}$), the mechanism (drift of charged particles across the gap) is field-driven, and thinner cells require lower voltages. In any case, these voltages can be easily supplied by a thin-film active matrix. The materials are good insulators, and thus the impedance match with field effect transistors in the matrix is good. Chemical and processing compatibility is similar to that of LC's, and integrated EP displays could be assembled by the same techniques as the previously discussed materials.

As regards display performance, EP's have excellent optical characteristics, probably the best of all passive materials in terms of contrast as a function of viewing angle.[8] This is illustrated in Fig. 1. Their temperature range of performance so far reported is only fair, but there seems to be no reason

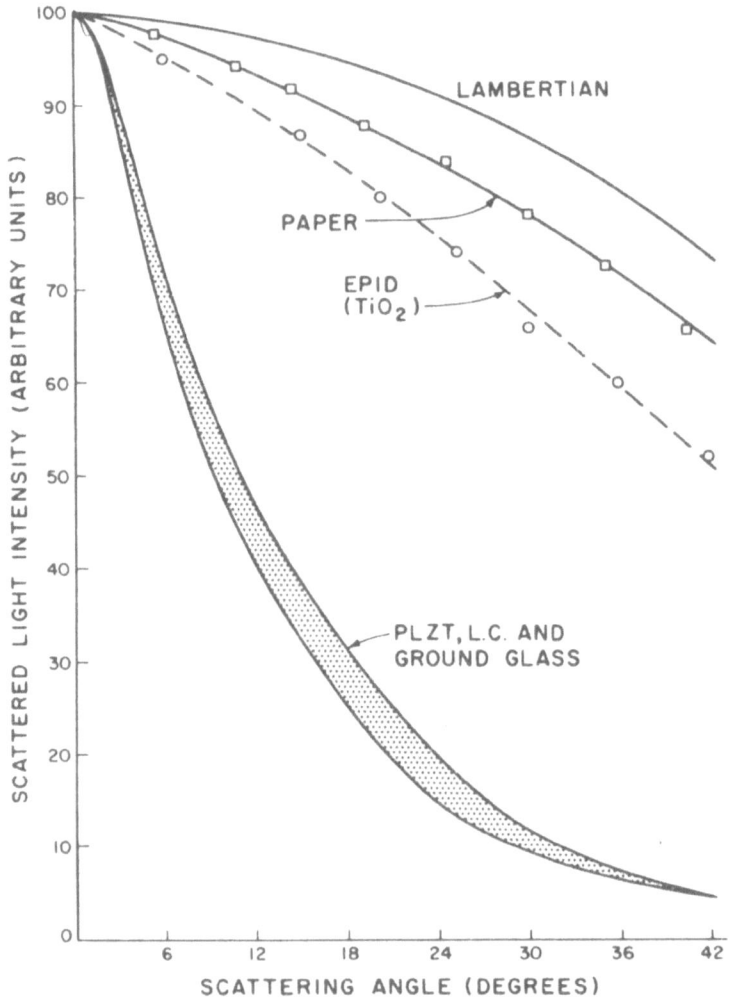

Fig. 1. Scattering profiles of several passive display media.[8]

why it should not improve, just as that of LC's did. Since the speed of response is controlled by viscous drag, it appears that a slower response at
lower temperatures will be unavoidable. The problem of settling and the
matching of densities of particles and suspension over a wide range of temperatures seems to be a difficult one.

The absence of sharp thresholds and comparatively "soft" contrast vs.
voltage characteristics of EP displays,[9] (Fig. 2) while a major limitation on
their multiplexing capability, does not detract from their value for integrated
displays. Indeed, a softer characteristic makes for easier gray-scale control.

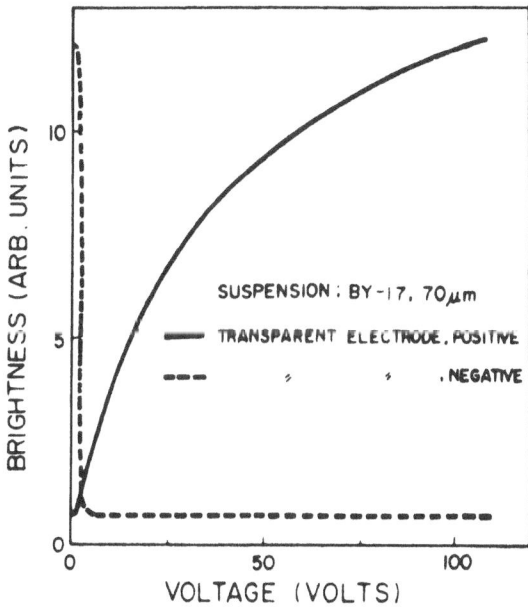

Fig. 2.  Brightness-voltage characteristic of EP suspension.[9]

The speed of response of EP displays so far reported is marginal for a moving display, but near enough to what is needed to make the effort at integration worthwhile. The very low power requirements (a few mW/cm$^2$ at TV frame rates) and the inherent memory are very attractive features for an integrated display, as is the range of available colors. The problem of building a multicolor EP display is similar to that of an EC display, namely that the materials are liquid and hence quite difficult to keep separated. Overall however, despite the present limitations, we consider EP materials highly promising for integrated multi-element displays.

## 2.4. Ferroelectrics

Ferroelectric single crystals and films represent another class of materials widely explored for display and related (e.g. optical storage) applications. They are the only solid materials in our list, and have the dual property of optical performance and a high degree of electrical non-linearity which has led to their use as ancillary layers in other types of displays, e.g. LC. The literature on display applications of ferroelectrics is quite extensive, and is reviewed by K. H. Härdtl in this book. Ferroelectric (FE) materials are capable of several different modes of electrooptic behavior such as field-induced scattering, optical attenuation and birefringence. The materials can be used in direct viewing, transmission or as projection displays.

Chemical compatibility is good, and indeed, both silicon/ and thin-film/FE combinations have been reported. Ferroelectric films can be deposited on silicon by sputtering, but not on thin-film matrices, since the deposition temperatures needed are too high. On the other hand, thin-film arrays can be, and have been, deposited on FE films and ceramics. Much work has been reported on FE ceramics, mostly PLZT (lanthanum-doped lead zirconate-titanate). These ceramics are available over fairly large areas (at present, several cm) and can be polished, so that active thin-film matrices could well be deposited on their surface.

Ferroelectrics have several attractive properties, such as high electrical and electrooptical nonlinearity, switching thresholds, high speed of response and both bistable and gray-scale memory, which in combination indicates a high potential for integrated display applications. Electrically, ferroelectrics are insulators and require fairly high voltages (~150 V or greater) for their operation. These requirements exceed the voltage capability of silicon matrix arrays but could be supplied by thin-film matrices. The high voltages needed particularly for the prepoling operation, which often involves applying fields in the plane of the device, are clearly a deterrent to the development of larger, electrically addressed devices. No power-consumption measurements of FE displays have been reported, but from the reported switching voltages and device dimensions, we estimate an energy of the order of 1 mJ/cm$^2$ per switching cycle.

The available viewing angle is also not satisfactory; indeed at angles of more than 20° from normal, FE ceramic displays have the lowest contrast merit factor (defined as contrast ratio times intensity as a function of scattering angle). Figure 3 illustrates this point.[10] For some types of FE displays, such as the longitudinally electrically poled types, there is, in addition, an undesirable contrast reversal as a function of viewing angle, similar to liquid crystals. Multicolor effects have not been reported and may be difficult to obtain.

## 2.5. Dipole Suspensions

This class of materials is probably not as well known as those so far discussed. The materials are tiny needles, of suitable dimensions to constitute dipoles at optical wavelengths, suspended in transparent fluids.[11] The needles can be metals or orgainic crystals such as herapathite (iodoquinine sulfate). In a field-free suspension, the dipole orientation is random and the liquid is opaque. Under the influence of electric fields, the dipoles align parallel to the field, causing a substantial change in light absorption. The particles actually behave like broadband electromagnetic antennae and will interact with light in a manner derivable from antenna theory.

These materials appear to have very attractive properties for an integrated display. They are field-driven, very low-current, low-voltage systems. With some suspensions, ionic species tend to form, leading to electrolysis if there is direct contact to conducting electrodes. In these cases, insulating

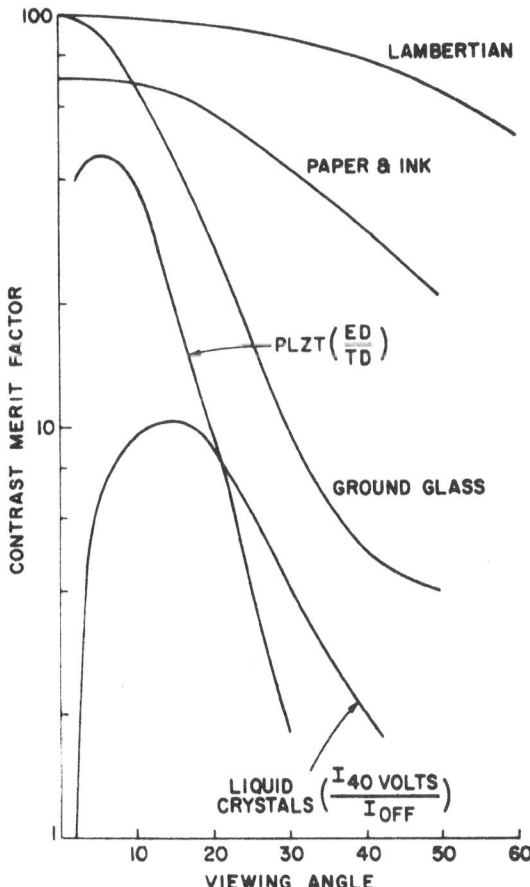

Fig. 3. Contrast merit factor of several passive display media.[10]

electrodes and ac drive are needed. Otherwise, there appears to be good electrical, chemical and processing compatibility with active matrix arrays.

Very little has been reported on these materials in actual display applications. A recent report by Davis and Thomas[12] indicates some stability problems in the suspensions. There could also be problems with the long-term stability of the organic crystals against dissolution in the fluid. The metal dipoles should be stable, but may be quite difficult to produce uniformly with the required dimensions. More work will be required in the preparation and study of these materials before their potential for integrated displays can be determined. The response times reported by Marks[11] are quite short (a few ms), while Davis and Thomas feel that they need to be improved. The same applies to the contrast obtainable. Colors would appear to be difficult or impossible to obtain in this medium, but gray scale is available through amplitude or pulse-width modulation.

## 2.6. Fluorescent Dyes

This medium provides an electrically controlled degree of UV-stimulated fluorescence, and hence qualifies as a passive display.[13] The typical material is a rare-earth chelate, dissolved in an electrolytic solvent. Absorbed UV produces Stokes fluorescence, with a color depending on the rare-earth metal ion. The fluorescent emission can be quenched by the injection of charge, and the process is reversible, constituting a reduction to a non-fluorescent state followed by oxidation back to the emissive state. Not much work has been reported so far on such materials, so their future potential is hard to estimate.

Significant currents ($10 \text{ mA/cm}^2$) have to be supplied, and the response speed is a function of the current (Fig. 4). Speeds of writing adequate for video imagery have been reported, though the recovery time was not short enough for moving images. Recovery speed is limited by diffusion mechanisms, in that the discharge has to take place at the electrodes. The operating voltages are low (2 - 3 V), so this material is quite similar to EC liquids in its electrical characteristics. Electrode reactions and molecular degradation similar to that of dynamic scattering LC's might be expected, but little information is available on life.

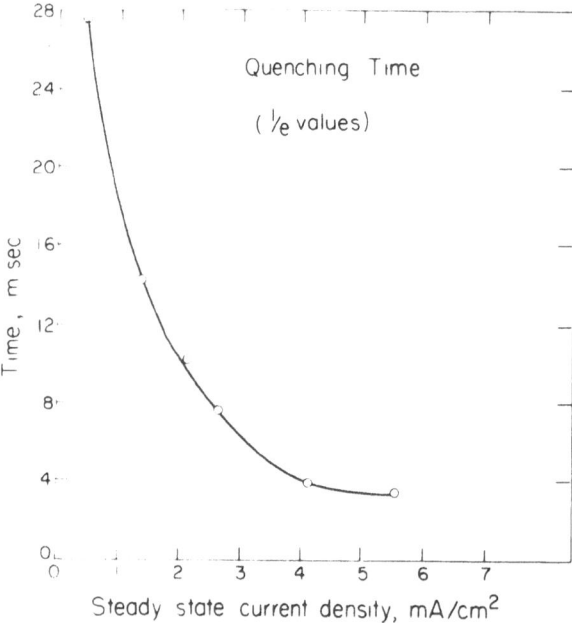

Fig. 4. Fluorescence quenching time as a function of steady-state current delivered to the cell electrodes.[13]

In terms of optical performance, fluorescent dyes are quite attractive. The brightness is governed by the incident UV, and 80 fL has been reported.

The emission is isotropic and can be seen at any viewing angle. The liquids are transparent, which permits additional modes of operation. Contrast ratios up to 25:1 have been reported, and gray scale is possible by amplitude or duty-cycle variation. Red and green have so far been reported as available colors. The need for an UV pump source is a disadvantage, of course, but on the whole, these materials also lend themselves well to integration with active matrices, and merit further exploration.

## 3. APPROACHES TO INTEGRATION

In roughly ascending order of approach to larger displays and full integration, we may distinguish five stages of evolution. (Insofar as all these "stages" coexist at present, the evolutionary terminology is inappropriate. However, as indicating an evolution of concepts, it is not altogether misleading.) They can be described as follows:

1) Passive arrays or matrices driven by discrete components or IC's mounted separately
2) Passive arrays or matrices with drivers mounted on the display panel
3) Matrices containing non-linear layers
4) Silicon active matrices
5) Thin film active matrices.

The distinction between 1) and 2) is perhaps a little forced, but does broadly correspond to the conventional distinction in solid-state electronics between discrete assemblies, e.g. on circuit boards, and hybrid circuits. Both stages, however, still use passive addressing, and therefore, of necessity, some form of time-division multiplexing (unless each display segment is directly addressed in parallel). These display systems are reviewed separately by A. R. Kmetz, and are included in this review mainly for the sake of logical consistency.

The modification introduced by the use of non-linear materials might be thought of as a qualitative rather than a quantitative step towards integration. By this we mean that certain global properties of the display materials, such as non-linearity, presence or absence of an electrooptical threshold, speed of response and retentivity are improved and supplemented by the non-linear layers, irrespective of the size and resolution of the display. These improvements in turn permit some significant savings in external circuitry and also permit the construction of displays with a larger number of elements.

Finally, in 4) and 5) we reach the active matrices which, in combination with the display materials, represent the only displays which can be accurately described as integrated. We first deal with the basic technologies involved in each class of displays and then review, through some examples, the state of accomplishments in each class.

### 3.1. Passive Arrays or Matrices with Discrete Components

This type of display is currently the most widespread. It consists of the display panel itself, which is usually an assembly of two glass plates with the display material sandwiched between them, and the drive circuits assembled by conventional techniques on circuit boards. The display panel carries on its edges a set of contact fingers, which are linked by conductive paths to the display segments. The conductive paths are usually photoetched in a transparent tin-oxide layer covering the glass. A typical cell construction for an LC display is shown in Fig. 5. Connection between circuit and panel is made by multistrand cable or wire, soldered to edge connectors which mate with the fingers on the panel. The overall assembly will, of course, be governed by the particular display application, and no general principles can be derived from the heterogeneous material so far published.

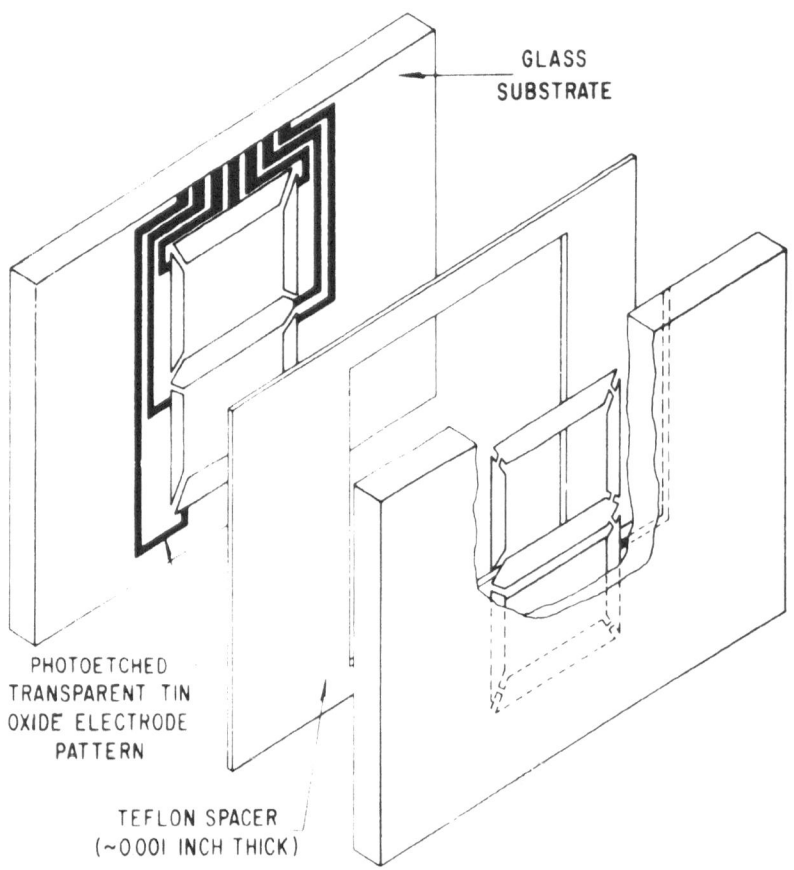

GLASS
SUBSTRATE

PHOTOETCHED
TRANSPARENT TIN
OXIDE ELECTRODE
PATTERN

TEFLON SPACER
(~0.001 INCH THICK)

Fig. 5. Construction of a typical LC display cell.[36]

We may include in this category the development of special integrated circuits which contain display driver circuits in addition to the computing or other functions for which the circuit was designed. An impressive example of such a multipurpose chip is the calculator reported by Fitzgibbons and Carlson.[14] Figure 6 shows the MOS LSI chip they developed. Nakada et al.[15] reported a similar development. A block diagram of their system is shown in Fig. 7. A 40×40-element LC passive matrix display has been reported by Uehara et al.[37] They use a two-voltage, high-frequency addressing scheme, but the circuits are not described in the paper. A test pattern on their display is shown in Fig. 8.

### 3.2. Passive Matrix with Integrated Drivers

We have found few references to such assemblies, which leads us to question whether the technical or economic benefits gained can be really significant.

Fig. 6. MOS-LSI calculator/display-driver chip.[14]

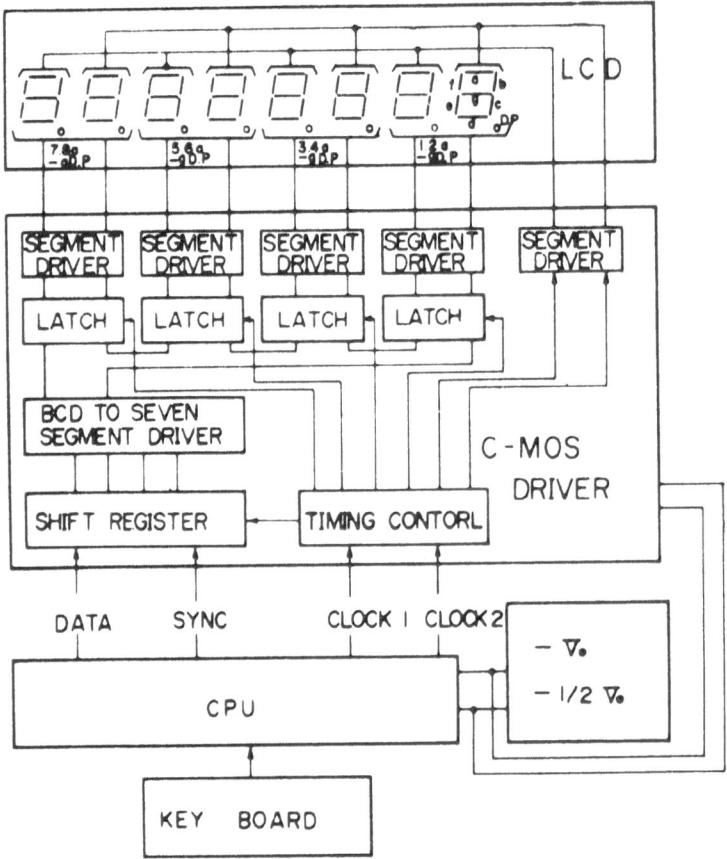

Fig. 7   Block diagram of LC display system.[15]

Effectively, the glass panel which carries the display segment electrodes is extended to serve as an assembly board for a hybrid circuit. The best known work in this category is probably that reported by Gerritsma and Lorteye.[16] Figure 9 shows their liquid-crystal display assembly. It consists of a 9-digit by 7-segment display, with each segment directly connected to one stage of a 63-bit shift register. Three serially connected IC shift-register chips are used, and these IC's are attached to the glass plate by reflow soldering. Only four external connections to this display assembly are needed. Either packaged or bare chips can be used; in the latter case, we have a close equivalent to a "conventional" thin-film hybrid circuit.

### 3.3. Matrices with Non-Linear Layers

The concept of using a highly non-linear material in combination with a passive matrix, to assist in the problem areas of inadequate thresholds,

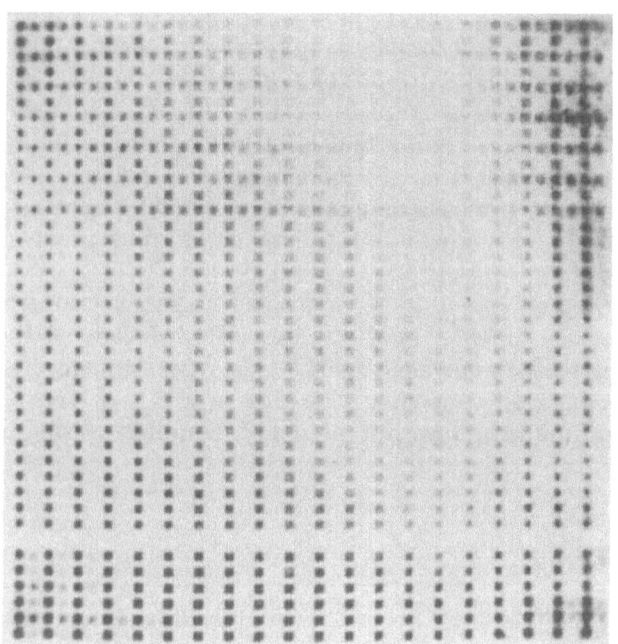

Fig. 8. Test pattern displayed on 40X40 element LC matrix.[37]

Fig. 9. Photograph of 9-digit hybrid LC display assembly.[16]

crosstalk and lack of memory, goes back almost 20 years.[18] The original
concept was applied to electroluminescent panels, and a fair amount of development was done on such panels in the late 50's and early 60's. More
recently, the concept has been applied to liquid-crystal displays,[19] and some
very interesting results have been reported[20] using PLZT ceramic layers.

The construction of such a display is illustrated by Fig. 10. The x and y
matrix electrodes are applied to opposite sides of the ferroelectric wafer,
which effectively forms the back plate of the display. The ferroelectric
material here performs several important electrical functions. Its hysteretic
characteristic provides local storage, its non-linear capacitance, in series
with the LC capacitance, provides an excellent threshold, and it also allows
rapid writing of data, independent of the speed of response of the LC
material.

Figure 10 shows a cross-section of the display, together with an approximate equivalent circuit. Figure 11 shows an exploded view of the entire
assembly. Referring to the equivalent circuit (Fig. 10), the capacitance $C_{xy}$
is made small to reduce driver loading. The ratio $C_{yL}/C_{xL}$ is large to reduce
the drive voltage during the addressing period. A complete display frame is

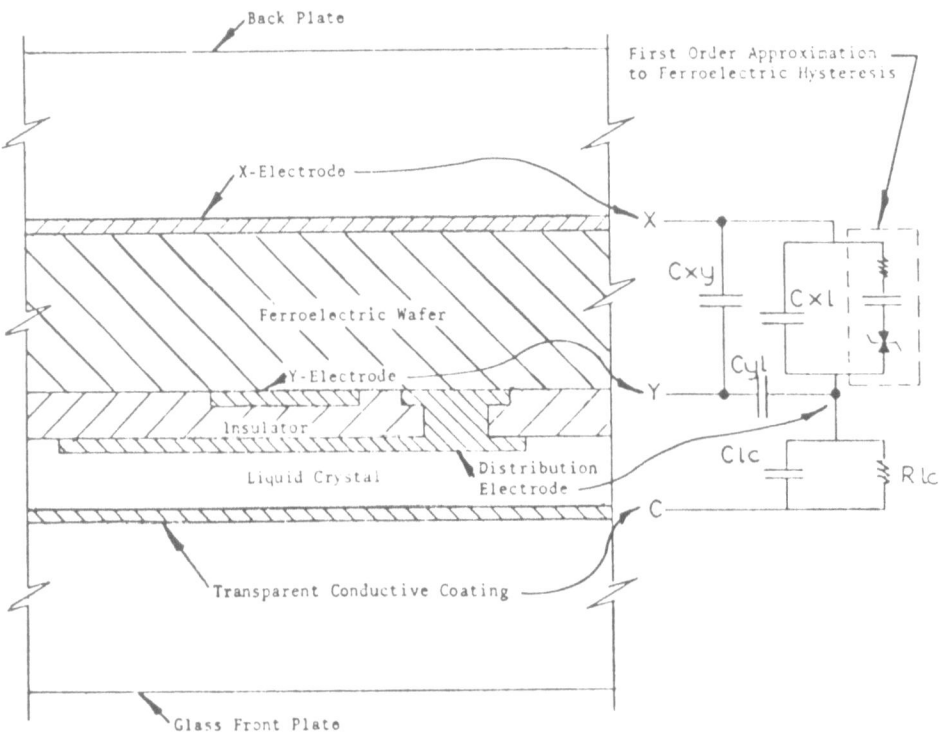

Fig. 10. Cross-section of FE-LC display, with approximate equivalent circuit.[20]

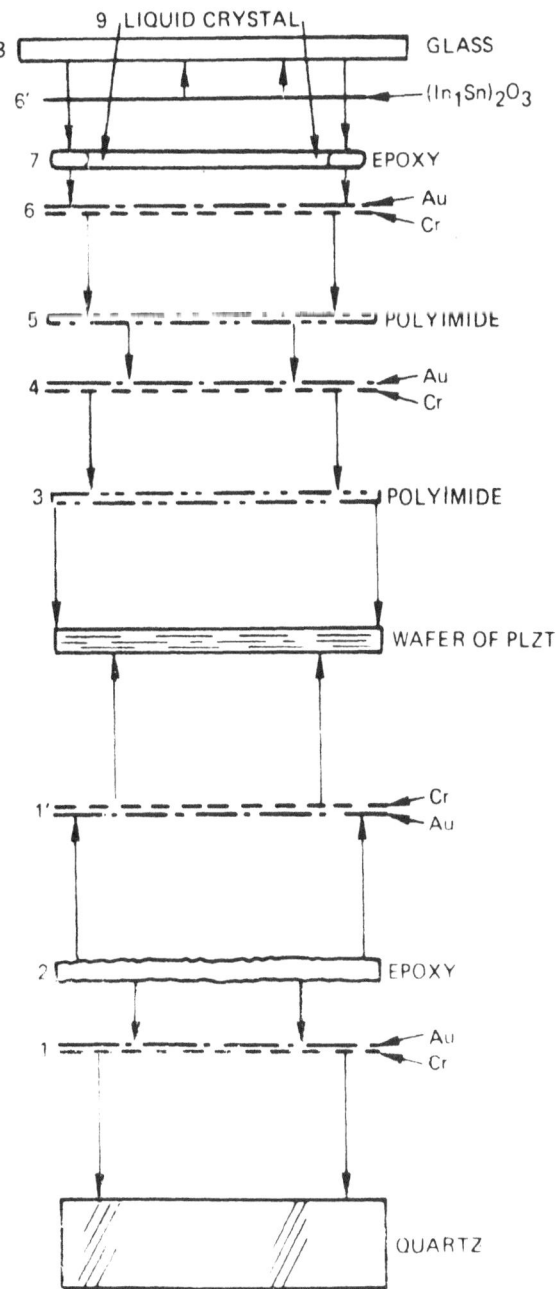

Fig. 11. Assembly sequence of FE-LC display.[20]

stored in the FE by line-at-a-time writing. It is then read out in parallel, a whole frame at a time. During the readout, $C_{yL}$ is switched in parallel with $C_{LC}$ and $R_{LC}$. The time constant of this combination is larger than the read pulse (a few ms) and smaller than the frame time. During the readout pulse, the stored information is erased and the FE is reset to accept the next frame. Tannas and York[20] have built 1-inch-square displays with 32 and 64 lines per inch. Line patterns were displayed with frame repetition rates of 20 to 90 frames/s. The response speed of the dynamic scattering LC material was about 50 ms. Contrast ratios up to 20:1 and limited gray scale capability were observed. The driving voltage was 120 V.

In this work, as in that of Grabmaier et al.,[19] the ferroelectric layer was used purely as an electronic component of the display. A different type of matrix-addressed FE display, in which the optical properties of the ferroelectric are also utilized, was recently reported by Wu et al.[21] Their display is based on the electrooptical switching of epitaxially grown films of bismuth titanate ($Bi_4Ti_3O_{12}$) on (110) spinel substrates. The operation of this display is based on switching fields applied in the plane of the FE film, and is somewhat complex. Basically, regions of the film between interdigitated electrodes are first poled. A display cell consisting of a center electrode and two outer electrodes (on the same side of the film) can then be switched by a smaller field opposite to the poling field. This smaller field switches only the c-axis polarization components, which have lower coercive force. This switching produces an identical tilt of the optical indicatrix in the two regions of opposite electrical polarity, thus producing the same change of optical transmission everywhere in the addressed cell. On arrays with a single set of interdigitated electrodes, contrast ratios up to 7:1 were measured, using crossed polarizers and narrow band illumination. A response time of 5 $\mu$s was measured at a field of 60 kV/cm. The actual voltages used were 450 V for poling and 225 V for switching. Gray scale effects, due to partial switching have also been observed. A matrix-addressible array of 5-mm-square, containing 30×26 elements was also constructed (Fig. 12). Only preliminary data have so far been published, indicating that the contrast ratio was somewhat lower and some cross-talk problems were present.

### 3.4. Silicon Active Matrices

The impact of silicon LSI technology on computing and information processing in general has been overwhelming, and hence the attempt to extend this technology to the display field is understandable. Although we have argued in the Introduction that the main strength of silicon processing lies in miniaturization, some very impressive results have been obtained with MOSFET matrices[22,25] which suggest considerable potential for a range of display applications.

An active matrix, as defined in the Introduction, is an electronic circuit coextensive with the display surface. It is, however, a rather peculiar circuit which has a very large number of output terminals, i.e. one for each picture

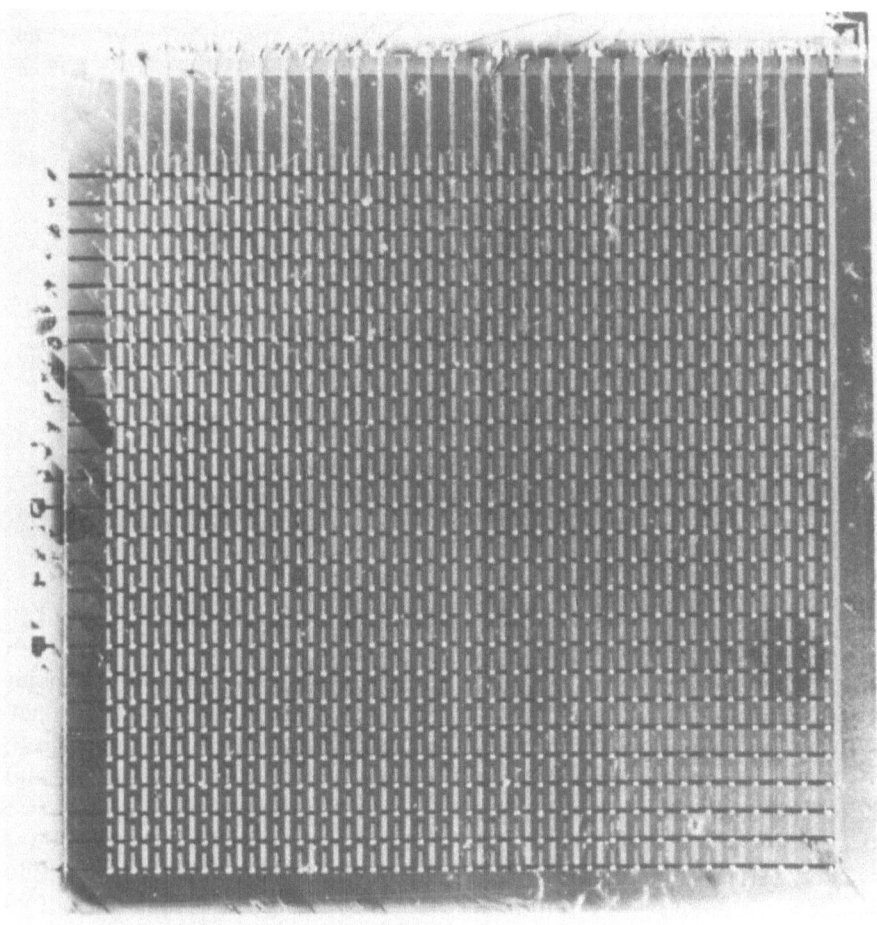

Fig. 12. Bismuth titanate matrix-addressed display.[21]

element. Again, the idea of using such matrices for display-addressing is not new, and many conceptual versions of circuits have been described in the past. A very detailed discussion, with particular application to LC displays, has been given by Lechner.[23] It is only during the last two years, however, that solid-state technology has begun to supply practical solutions. The basic advance made during this period was the breaking of a psychological barrier. Now that large active matrices have been made by two different technologies, the goal, not long ago considered utopian, appears to be attainable by fairly straightforward measures.

Obviously, the best chance for constructing an active matrix lies in choosing the simplest possible design. For liquid crystal displays (the only ones so far constructed with a silicon matrix), the unit cell can be extremely simple, consisting of a single logic transistor and a storage capacitor, as shown in Fig. 13. Such a cell constitutes a sample-and-hold circuit, which

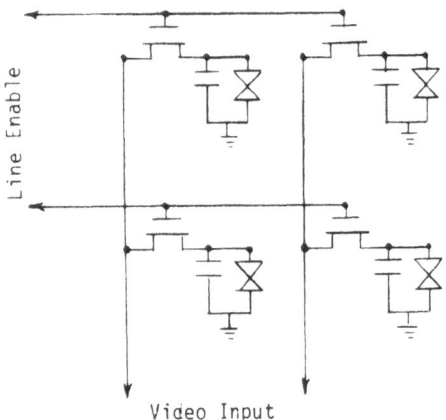

Fig. 13. Elementary circuit for LC display cell.[25]

permits rapid writing and subsequent slow transfer of excitation to the LC. The requirements on the performance of such a circuit element have been previously analyzed.[24] Briefly, the ON-impedance of the switching transistor must be low enough to permit charging of the storage (and parasitic) capacitance during a line period, while the OFF-impedance must be high enough to keep the element capacitance at its charged voltage for a refresh (frame) period. These conditions define a switching ratio for the transistor: this ratio has to exceed the number of lines on the display by about a factor of 10 for satisfactory operation. ON/OFF ratios of the order of $10^5$ to $10^6$ are easily obtainable with MOSFET's, and hence, the line-multiplexing capability of an active silicon matrix comfortably exceeds any practical display requirements.

Line-at-a-time addressing has been adopted as an optimum by most workers on matrix-driven displays. As display data inputs normally form a serial data stream, serial-to-parallel conversion and line storage has to be provided by external circuits. These circuits, however, could be integrated with the matrix at a later stage of development.

The Hughes group has so far reported on two different embodiments of a 1-inch-square, $100 \times 100$ element LC matrix display. In the first version,[22] the matrix was fabricated on a silicon wafer by conventional IC techniques, i.e. photomasking, etching and diffusion. The matrix lines terminate in pads, and the external circuits (row-scanner, serial-to-parallel converter) are attached by wire bonding to the pads. The display is assembled by sandwiching the LC (here of the dynamic-scattering type) between the silicon chip and a glass cover. Details of the assembly process have not been published.

More recently, a display of the same size and resolution has been made on a silicon-on-sapphire (SOS) substrate.[25] Here a p-channel, self-aligned-gate technology has been used, employing ion-implantation techniques. Details of the processing will be found in Ref. 25b. The advantages claimed for the SOS approach include:

— Possibility of operating in transmission as well as in reflection
— High element density possible, due to the insulating substrate which eliminates crosstalk
— Simplified processing
— No light sensitivity, as carrier lifetime in the SOS material is low.

The display otherwise was similar in construction to the earlier one. It was exercised with video inputs and the following results have been reported:

|  |  |
|---|---|
| Contrast Ratio | 18:1 |
| Response Times | 100 ms rise |
|  | 160 ms decay |
|  | (40 ms decay with reduced contrast) |
| Gray Shades | 6 |
| Ambient Operation | no fading under 10,000 fL illumination. |

Figure 14 shows a still picture of a video image displayed on the Hughes array.

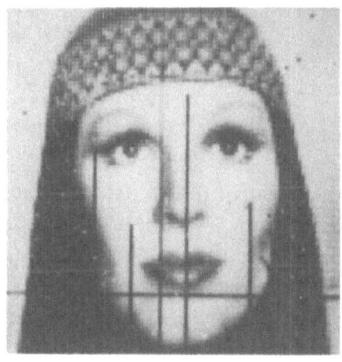

Fig. 14. Photograph of gray-scale video image displayed on the 1-inch-square silicon LC display.[25]

Another, rather more ambitious attempt to construct a silicon-matrix-based display was made by Kmetz et al.[26] Their objective was to build a 3-inch-square LC display, with 30 lines/inch resolution. The display was to be assembled from four 1.7-inch-square wafers, each including addressing and drive circuits as well as the matrix. Each matrix point consisted of a bistable flip-flop and enable devices (6 MOSFET's per cell) constituting a random-addressible memory (RAM) bit. Unfortunately, this task turned out to be too ambitious, and required too big an advance in silicon processing over very large areas. In consequence, no operating circuits or displays were obtained. Figure 15 shows the complete assembly of the four quadrants.

Fig. 15. Assembly of 3-inch-square LC display matrix from four quadrants.[26]

An earlier, important development properly belongs in this section. This is the work reported by Borel,[17] which concerns the fabrication of single 5×7 dot matrix LC characters, where the display is formed on top of a 35 bit MOS shift register. The construction of the display is quite similar to that of the Hughes matrix, but it is not matrix-addressed. Data corresponding to a given character are entered serially into the shift register and are then stored there for subsequent display. This is an important development, and the first reported instance of a truly monolithic display unit. Unfortunately, we have not found further references to this work since the original 1971 reports.

### 3.5 Thin-Film Active Matrices

The use of thin-film transistor technology for building scanners for displays was suggested by Weimer.[27] Brody and Page,[28] and Lechner et al.[23] discussed the possibility of building active matrix displays. Thin-film transistors (TFT's) were first reported in the early 60's, but were soon completely overshadowed by the rapidly developing MOSFET technology. The early application objectives of the two devices were quite similar, and the successful solution of the MOS surface stability problems, coupled with the better theoretical understanding of single crystal devices and much larger industry effort devoted to these devices, resulted in the gradual abandonment of work on the thin-film device. (A somewhat more extended account of TFT history can be found in Ref. 3.)

Recent work on TFT's has sought to capitalize on the relative advantages of thin-film techniques. Among these are the ability to form circuits on insulating, non-crystalline substrates, the speed, simplicity and uniformity of the fabrication steps and, above all, the ability to deposit an unlimited variety of materials over extended areas with a high degree of uniformity and with good control of the film properties. It is the latter feature (large area uniformity) which is particularly attractive for display applications. Thin-film techniques appeared to be the only possible approach to fabricating circuits of the size needed for integrated displays. Modular approaches were considered but did not seem practical, except perhaps for low resolution, very large area displays

Work on thin-film active matrices has so far been reported by three groups: Westinghouse, Aerojet General and Hughes. The Aerojet work[29] has been devoted to light-emitting displays and will not be covered here. Lipton and Koda (Hughes) have reported on a small (2×3 element) dynamic scattering LC display driven by a thin-film-transistor matrix.[30] The circuit principle is similar to that of the silicon matrix described earlier. The construction of their panel is shown in Fig. 16. The area definition of the various regions was done by photoetching techniques. Only rudimentary performance data have been reported, and this work was not continued.

The Westinghouse group has reported on both passive and light-emitting displays driven by active matrices. Only the former falls within the scope of this volume, but the philosophy of approach and most fabrication details are common to both. This is so by intent rather than by accident, since the point of departure of our work has been to try to develop a display-integration technology with the largest possible universality.

Our group has reported on the construction and performance of a 6-inch-square, 20 lines/inch twisted-nematic LC display[24,31] and on the development of an active matrix whose cells are capable of long-term gray-scale memory.[32] The electrical design of the matrix elements used was the simplest possible, as shown in Fig. 17. A storage capacitor was not used, since the field-effect LC element itself provided sufficient capacitance. The geometrical design of the element is illustrated in Fig. 18. It can be seen that most of the area is taken up by the (transparent) drain pad, permitting an almost completely transparent construction.

Area definition of the thin-film circuit was done through metal stencil masks, rather than photo-etching, for several reasons: 1) An all-vacuum technology, without the need for repeated pump-downs and chemical non-vacuum operations, is preferred for reasons both of throughput and of over-

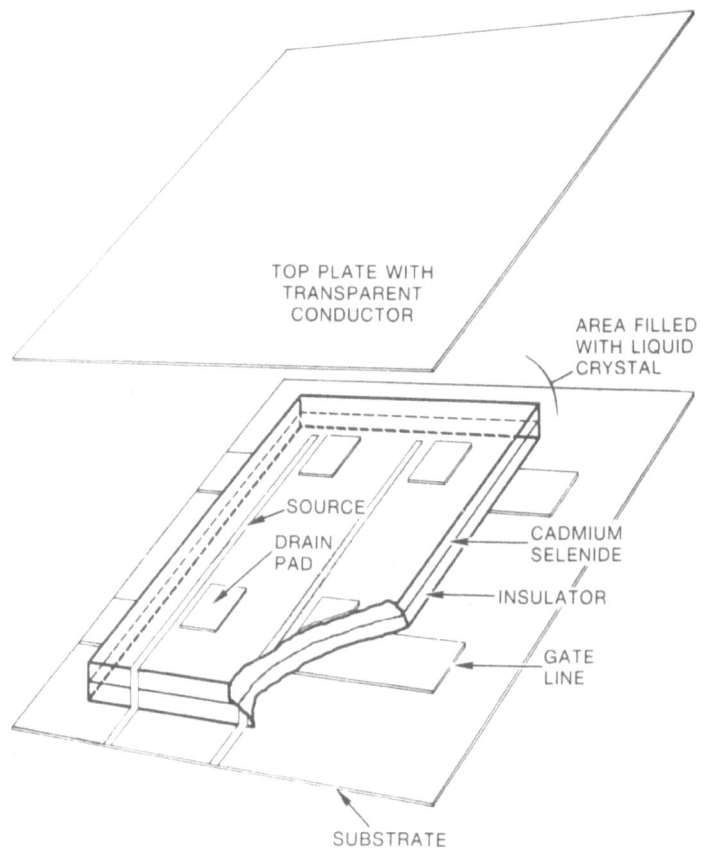

Fig. 16.  Schematic of thin film transistor matrix LC array construction.[30]

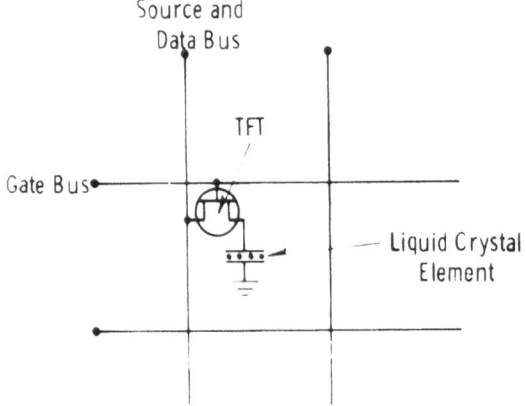

Fig. 17.  Matrix elemental cell circuit design of TFT-LC panel.[24]

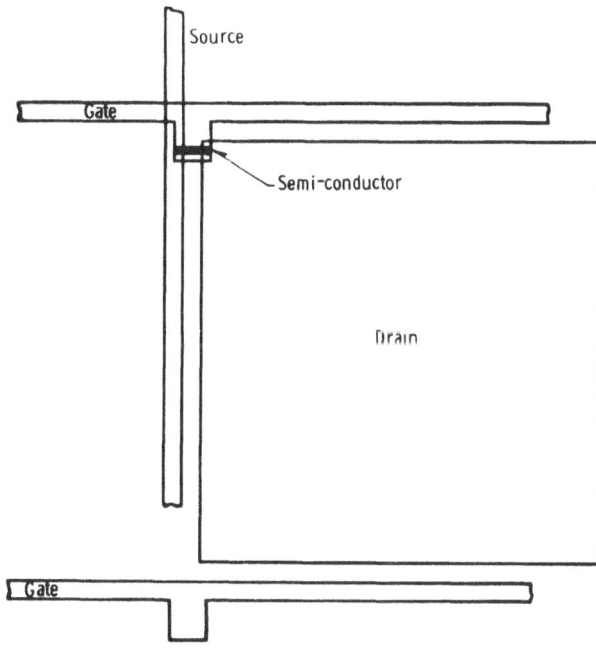

Fig. 18. Geometrical layout of TF-matrix circuit element.[24]

all cleanliness. 2) Photoetching technology at the resolutions and over the large areas required is not well developed, and would be expected to have a low yield. 3) The variable-aperture metal stencil mask employed gives great versatility and permits changes in the detailed layout without the need for new masks.

After a unit-cell design and layout have been established, the thin-film matrix is fabricated in a continuous sequence of vacuum depositions, layer by layer. All materials used are stored in separate baffled evaporation sources. Each material (metal, semiconductor or insulator) is deposited through the mask apertures appropriate to that material layer. The entire sequence of mask and material changes could be (but is not at present) controlled by a programmable controller. Structural details of the 6-inch-square LC matrix circuit are shown in Fig. 19 and Fig. 20. Figures 21 and 22 show characteristics of typical transistors in the matrix.

The finished matrix circuit is passivated with a protective coating, then removed from the vacuum chamber for assembly. To form a twisted-nematic LC display, the matrix surface and the NESA glass cover have to be treated with an orienting layer (e.g. by oblique evaporation of SiO.[33]) The display then can be assembled by normal LC-display assembly and sealing techniques. Figure 23 shows the performance of one half of the 6-inch-square panel (only one half could be exercised at one time when this picture was taken). The square patches are crude spacers, not defects.

Fig. 19. Finished 6-inch-square TFT matrix for LC display.[24]

Thin-film memory matrices have not so far been applied to passive displays, but since they could very conveniently be so applied, a brief mention
of them seems appropriate. The memory-TFT's are close relatives of silicon
memory transistors. By virtue of the double-layer gate-insulator construction (Fig. 24), these transistors are capable of storing a charge on the floating
gate for long periods of  time.[34] By controlling the amount of injected
charge (e.g. through pulse-width modulation), the transistor can be put into

Fig. 20. Detail from Fig. 19, showing geometry of matrix unit cell.

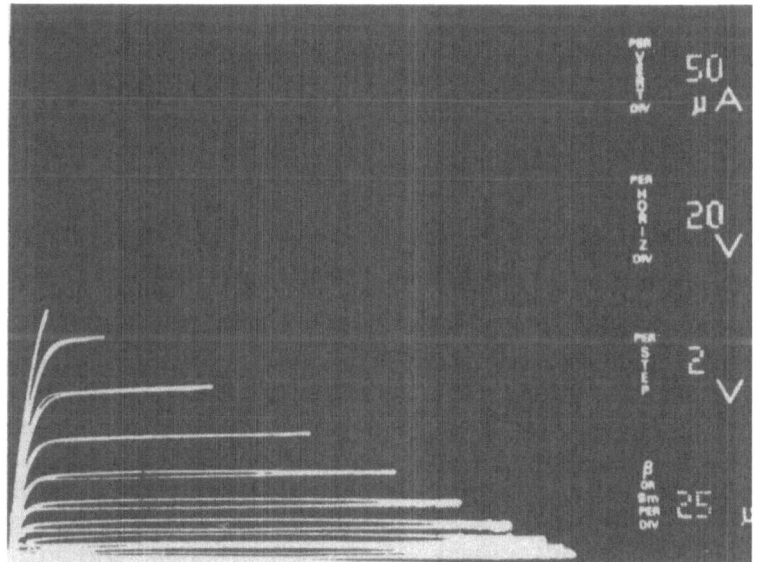

Fig. 21. Transistor characteristics from 6-inch-square matrix.[3]

Fig. 22.  Gallery of transistor characteristics, randomly selected from corresponding regions of 6-inch-square matrix.[3]

a continuum of memory states. Fully ON and OFF states are shown in Fig. 25. A 1-inch-square, 40×40 matrix of such transistors is shown in Fig. 26. The memory TFT's can be written into and erased at comparatively high rates, although not yet at the 15 kHz rate needed to address them for line-at-a-time TV.

## 4. CONCLUSIONS

In the historical development of the science and art of electronics, materials research has invariably led, and provided stimulus for, device and systems development. Without pure and perfect germanium and silicon crystals, transistors and digital computers would not have existed, even as a concept. Without synthetic ruby, no masers or lasers. Without ferrites, no high-speed memories. And certainly, without III-V compounds and liquid crystals, no digital watches and displays. Solid-state electronics is, quite simply, a recurring case of new materials looking for new applications. Contrary to the popular belief that "necessity is the mother of invention", we can say with far more justice that it has been the case of "invention being the mother of

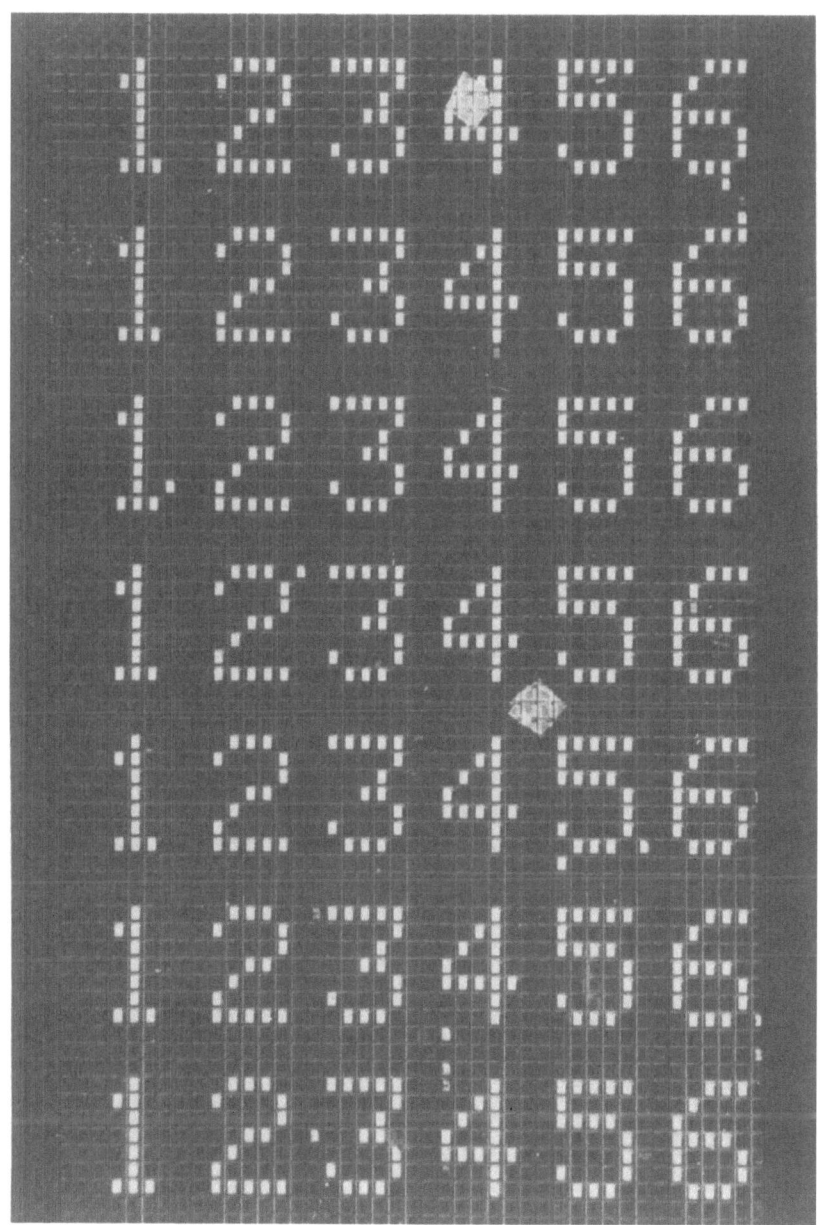

Fig. 23. Performance of (one half of) the 6-inch-square, 20 lines/inch LC display from Fig. 19.

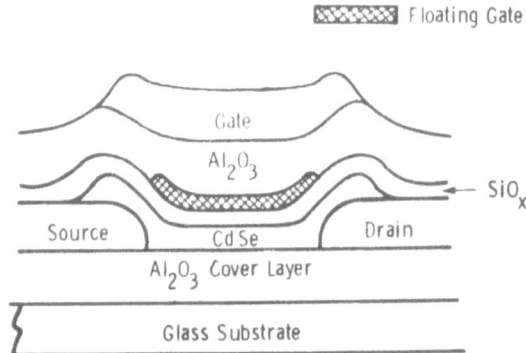

Fig. 24.  Cross section of thin-film floating-gate memory transistor.[34]

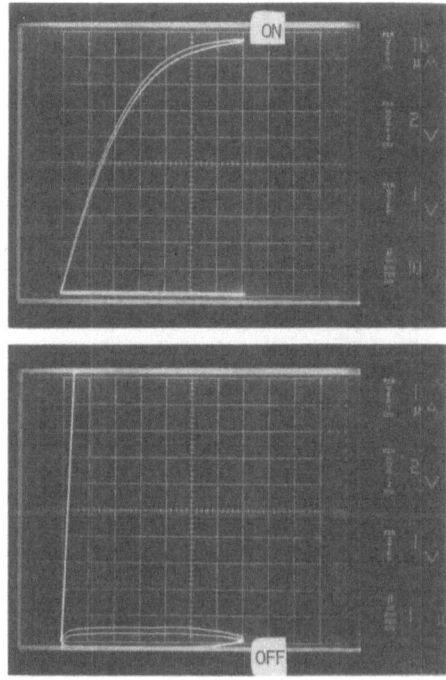

Fig. 25.  ON and OFF characteristics of TF memory transistor.[34]

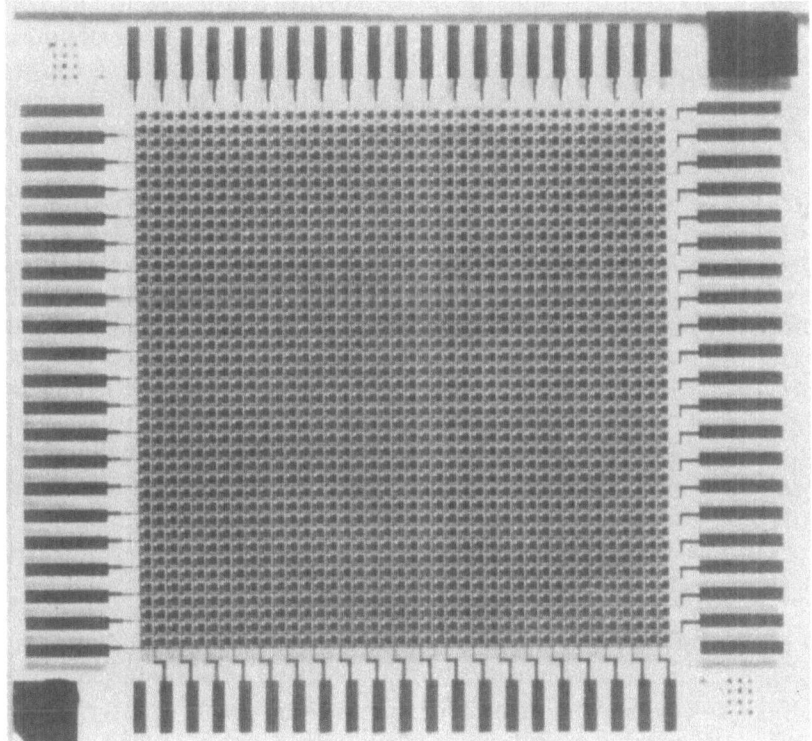

Fig. 26. Completed 1-inch-square, 40×40 element TF memory matrix.[32]

necessity"! In other words, the availability of new materials, or materials in new form, has inevitably led sooner or later to their practical exploitation in fields which otherwise would not even have been conceived as possible fields of endeavor.

Passive display materials have appeared on the scene fairly recently but are clearly destined to open up new territory by virtue of their novel properties, the most important of which are low power consumption, potentially low cost, simplicity of use, and absence of area restrictions. Liquid crystals are very much in the lead, due to the "bandwagon" effect consequent to their being the first in the field, but it is by no means certain that they will establish a dominant position as their visual performance is really quite marginal. Electrochromics and electrophoretics are much superior in this respect, and if their stability problems can be solved by the comparatively low level of effort available, they will provide very stiff competition to LC's.

Ferroelectrics have an area problem, require high voltages and are deficient in optical performance. These are drawbacks which will be difficult to overcome, and FE's are the least likely of the materials on our list to find widespread application. Fluorescent dyes and dipole suspensions have not been sufficiently explored to allow more than a guess about their future. The need for a UV pump lamp is a considerable practical disadvantage of the dyes, which will probably militate against further development. Dipole suspensions, on the other hand, are a very attractive medium from all points of view. The question there is whether they offer any real advantages over the principal contestants which would generate sufficient additional support for development. Very little current activity is apparent with these last two classes of materials.

The foregoing, perhaps too summary, discussion of future prospects of passive display materials has, at any rate, the merit of objectivity. The author is not committed to any of them and has certainly not made any final choices as to which, if any, of the materials is the most attractive for a higher performance, multi-element integrated display. However, when it comes to discussing the merits and prospects of the various assembly and integration schemes, he is forced to declare a point of view very different from that of the objective observer, namely that of a scientist or manager responsible for taking certain courses of action in display development. Consequently the following discussion contains some highly intuitive judgements and is formulated to stimulate a particular action decision.

We will start by trying to establish the scope of the various integration schemes discussed. By this we mean the apparent technological limits to the display size and resolution, as determined by the technique of addressing them, rather than by the inherent properties of the materials. We are assuming here that the remaining stability problems of each material can be successfully solved, and hence each of them will be usable in future displays. Since, with the current exception of ferroelectrics, they are amenable to being spread over almost arbitrarily large areas, they will not, by themselves set a limit to display size, or resolution.

Our first two categories (from the beginning of Section 3) represent displays where either each element is separately addressed or a simple multiplexing scheme is employed. It turns out that the performance limits of these two schemes are not far apart, even though they are governed by very different factors. In the case of separately addressed elements, the limiting factor is the number of output pins which a single LSI circuit can reasonably provide, or else the geometrical constraints of bringing out the etched connection paths from each display segment to the edge of the display or to some region where IC chips can be attached. The largest parallel-addressed display reported has 63 segments, and one would estimate, erring on the side of generosity, that 150 - 200 elements thus addressed will represent the upper limit attainable by such "brute force" techniques.

Matrix or matrix-like time-division multiplexing schemes should, in principle, yield displays with a far larger number of resolution elements, but here the material limitations discussed earlier quickly become the dominant factor. Liquid crystals are the only materials whose multiplexing capability has been well explored, and many ingeneous schemes have been proposed and demonstrated. Problems arise from the slow response of the material, the rapid reduction of contrast ratio as the number of lines is increased and the deterioration of the viewing angle. These problems are more fully discussed in the companion paper of A. R. Kmetz, and here we simply state that the practical limits to multiplexed LC displays are of the order of 500 - 1000 elements, consistent with the maintenance of acceptable visual performance. This picture could change, of course, if LC materials with significantly improved properties were to be discovered and developed. No work on simply multiplexed EC or EP materials has so far been reported; these materials are not expected to reach even the LC level of performance, due to additional deficiencies.[35] The multiplexing capability of fluorescent dyes and dipole suspensions is, at best, expected to compare with, or fall short of the LC performance.

All these estimates are highly vulnerable, unless we carefully define our "level of performance". For example, since both EC and EP materials have memory and so have certain cholesteric-nematic LC mixtures, one could visualize a much larger multiplexed matrix if writing and erasure can be allowed sufficient time at each point. In the case of EC or EP materials, other problems appear[35] which restrict even a slow operation to a small number of elements. With cholesteric-nematic mixtures, however, the number of elements might approach the limit given by the ratio of retentivity to risetime. We should therefore point out that our estimates relate to displays with fast update and erasure capability, approaching that of video frame-rates.

We now come to the case of ferroelectric and FE-assisted matrix displays. These are really two quite separate developments, related only through a material common to both, and should be discussed separately. Ferroelectric electrooptic displays containing a large number of matrix-addressed elements could in principle be developed. However, problems of material uniformity over large areas, high processing temperatures, the need for high voltages coupled with rather marginal optical performance, and the absence of any significant advantage of such a system over competitors suggest a low probability that such displays will be further developed.

The combination with LC layers looks more promising, and could lead to comparatively large displays. No major weaknesses are evident in this approach, but it is questionable whether the present amount of effort in this area can lead to practical systems. For a large matrix, the display will require a large number of interconnections with discrete or IC-chip high-voltage line drivers, which may set an economic limit to the display size. The need for a very thin, large area FE wafer may also present problems in assembly; sputtering may provide a solution. If the effort in this area can be maintained or increased, we see no reason why matrix-addressed FE-LC displays should not reach a capability of several hundred characters, $10^4$ or more resolution elements, and moving, gray-scale pictorial displays.

Finally, we must attempt to evaluate the scope, problems and promise of the two generalized efforts in large-scale display integration: that based on an extension of silicon IC technology and the thin-film approach. The scope of both these technologies is potentially immense. They possess a repertoire of device functions which is fully adequate to generate active matrices for most or all of the display materials discussed. Further, since the matrix itself represents only part of the total solution, these technologies can also provide the scanning, decoding, drive and peripheral memory functions and can integrate these with the matrix to generate a "fully decoded" integrated display. It appears to us that such fully integrated displays are both possible and will be necessary for any solid-state display which aspires to be competitive with the CRT.

Are these two technologies, silicon and thin-film, really in mutual competition, i.e. do they attack the same problems? We think not: each will find applications appropriate to its particular merits. There may very well be disputed land between them, but by and large there will be a division of territory. The obvious division is that based on size. Silicon technology, being single-crystal based, will continue to remain a high resolution, small-area process. The laborious progress year by year towards fractionally larger chip sizes in the semiconductor industry has been proceeding for too long to raise any expectation of a sudden breakthrough. Although display matrix circuits are less demanding in performance than high-speed logic or memory chips, the processing defects and substrate imperfections which limit acceptable yields to a chip of a given maximum size are the same for all silicon IC's. We predict, with some confidence, that yields on silicon matrices exceeding 1-inch on the side will not be economically acceptable for any practical applications for several years to come, and a 2-inch circuit is so far in the future that it may never be attempted.

Modularization, i.e. construction of larger displays from smaller "tiles" looks most unattractive for more than four modules, this being the largest number which can be fitted together into an operating whole without the need for electrical inter-module connections. A glance at Fig. 15 will verify this point. To make several hundred edge-to-edge connections, at a high resolution, without disturbing the continuity of the resultant display is a forbidding task, nor does any putative solution look economically plausible.

On the other hand, integrated wristwatch displays, even a wristwatch TV, helmet-mounted high resolution displays, and other small area, high resolution applications look entirely feasible through silicon matrices. Novel systems will also undoubtedly be developed if the original devices are successful. Liquid crystals are a good match. EC displays, which require higher currents at low voltages, appear a natural target for integration with bipolar IC's. EP materials and the fast cholesteric-nematic phase transition[4] require voltages which are too high for silicon IC's to supply.

In general, then, we predict a modest role for silicon active matrices in display integration, and see no prospect whatever of this technology achieving dominance in displays similar to its dominance of the digital logic field. That is not to say that there will not be much further progress in logic chips containing display driver outputs, in specialized IC drivers and in shift-register interfacing with otherwise passive display panels.

Thin-film technology, for reasons stated earlier and also in previous publications,[2,3,24] does appear to hold the promise of becoming the basis on which a very wide range of integrated displays can be built up. It alone has the capacity to generate active circuits over areas comparable to that of the CRT face. At the same time, it has enough geometrical resolution to form matrices with a line density which exceeds the resolution capability of the unaided eye. The high voltage capability of TFT's allows their interfacing with all display materials in current use or under development. However, it is less well suited to drive materials such as EC or fluorescent dyes which require substantial currents at low voltages. The problem is not so much that of the TFT's, which can be designed to provide quite large currents (approaching 1 mA per mil of channel width), but of the busbars. A 1000 Å thick busbar, 50 $\mu$ wide will have approximately 100 $\Omega$/cm impedance, far too high to carry currents of tens of mA. This is not an insurmountable problem, but militates against an early development of large-area integrated EC displays.

Problems in thin-film active matrix fabrication have been reviewed recently.[3] To summarize here, they relate to the need for 100 % perfection over large areas, the need for perfect deposition masks, thermal expansion and distortion of the masks, mask-cleaning and the adequate sensing and control of all deposition parameters. The art of fabricating thin-film active matrices is at present confined to a very small number of laboratories. It is to be hoped that once the true potential of this technology for integrated displays becomes more generally known, a cooperative, industry-wide effort will substantially accelerate progress. At the present time, the objectives are to prove the technology in applications which are beyond the scope of conventional techniques. However, there is no reason why thin-film techniques should not subsequently move in both directions: towards even larger, higher resolution and higher performance displays (both digital and gray-scale) and also towards smaller displays where the economics of fabrication and assembly may provide a competitive edge over present techniques.

## ACKNOWLEDGEMENTS

We wish to acknowledge the support of the US Air Force, US Army Electronics Command and the Office of Naval Research for part of the work reported here. Thanks are also due to the author's many collaborators at Westinghouse Research Laboratories for their important practical contributions.

## REFERENCES

1  A. Sobel, IEEE Trans. Electron Devices ED 18 (1971) 797
2  T. P. Brody, F. C. Luo, Z. P. Szepesi and D. H. Davies, IEEE Trans. Electron Devices ED 22 (1975) 739

3 T. P. Brody, IEEE Trans. Cons. Elec. CE 21 (1975) 260

4 J. J. Wysocki, J. H. Becker, G. A. Dir, R. Madrid, J. E. Adams, W. E. Haas, L. R. Leder, B. Mechlowitz and F. D. Saeva, Proc. S.I.D. 13 (1972) 114

5 L. T. Lipton, M. A. Meyer and D. O. Massetti, S.I.D. Int. Symp. Digest 6 (1975) 78

6 I. F. Chang, this book

7 C. J. Shoot, J. J. Ponjee, H. T. van Dam, R. A. van Doorn and P. T. Bolwijn, Appl. Phys. Lett. 23 (1974) 64

8 A. L. Dalisa and R. A. Delano, S.I.D. Int. Symp. Digest 5 (1974) 88

9 I. Ota, J. Ohnishi and M. Yoshiyama, Proc. IEEE 61 (1973) 832

10 A. L. Dalisa and R. J. Seymour, Proc. IEEE 61 (1973) 981

11 A. Marks, Appl. Optics 8 (1969) 1397 and paper presented at NEREM, Boston (Nov. 1973)

12 A. Davis and I. M. Thomas, S.I.D. Int. Symp. Digest 6 (1975) 88

13 D. P. Hamblen and J. R. Clarke, IEEE Trans. Electron Devices ED 20 (1973) 1028

14 E. T. Fitzgibbons and R. G. Carlson, S.I.D. Int. Symp. Digest 5 (1974) 90

15 K. Nakada, T. Ishibashi and K. Toriyama, IEEE Trans. Electron Devices ED 22 (1975) 725

16 C. J. Gerritsma and J. H. L. Lorteye, Proc. IEEE 61 (1973) 829

17 J. Borel and J. Robert, S.I.D. Int. Symp. Digest 2 (1971) 128

18 J. A. Rajchman, G. R. Briggs and A. W. Lo, Proc. IRE 46 (1958) 1808

19 J. G. Grabmaier, W. F. Greubel and H. H. Kruger, Mol. Cryst. Liq. Cryst. 15 (1971) 95

20 L. E. Tannas and P. K. York, IEEE Symp. on Applications of Ferroelectrics, Albuquerque, NM (June 1975) and S.I.D. Int. Symp. Digest 4 (1973) 178

21 S. Y. Wu, W. J. Takei and M. H. Francombe, IEEE Symp. on Applications of Ferroelectrics, Albuquerque, NM (June 1975)

22 M. N. Ernstoff, A. M. Leupp, M. J. Little and H. T. Peterson, IEEE Internat. Electron. Devices Meet. Tech. Digest 73 CH 0781-5 ED (1973) 548

23 B. J. Lechner, F. J. Marlowe, E. W. Nester, J. Tults, Proc. IEEE 59 (1971) 1566

24 T. P. Brody, J. A. Asars and G. D. Dixon, IEEE Trans. Electron Devices ED 20 (1973) 995

25 (a) L. T. Lipton, M. A. Meyer and D. O. Massetti, S.I.D. Int. Symp. Digest 6 (1975) 78. (b) M. A. Meyer, L. T. Lipton, G. H. Hershman and P. G. Hilton, 1975 Am. Electrochem. Soc. Meet., Toronto, Abstract 194

26 A. R. Kmetz, J. R. Pies and J. A. Lipman, Final Report on Contract DAAB07-72-C-0158, to US Army Electronics Command, Ft. Monmouth, NJ (Sept. 1974)

27 P. K. Weimer, in Wallmark and Johnson (eds.), "Field Effect Transistors", Prentice-Hall (1966)

28 T. P. Brody and D. J. Page, in 1969 Govt. Microcircuit Applications Conf. Digest of Papers

29 G. Kramer, IEEE Trans. Electron Devices ED 22 (1975) 733

30 L. T. Lipton and N. J. Koda, S.I.D. Int. Symp. Digest 4 (1973) 46

31 T. P. Brody, F. C. Luo, D. H. Davies, E. W. Greeneich, S.I.D. Int. Symp. Digest 5 (1974) 166 and WESCON 1974, paper 26.2

32  T. P. Brody, K. K. Yu and L. J. Sienkiewicz, S.I.D. Int. Symp. Digest 6 (1975) 82

33  J. L. Janning, Appl. Phys. Lett. 21 (1972) 173

34  K. K. Yu, T. P. Brody and P. C. Y. Chen, Proc. IEEE 63 (1975) 826

35  L. A. Goodman, IEEE Trans. Cons. Elec. CE 21 (1975) 247

36  A. R. Kmetz, IEEE Trans. Electron Devices ED 20 (1973) 954

37  K. Uehara, H. Mada and S. Kobayashi, S.I.D. Int. Symp. Digest 6 (1975) 80

## DISCUSSION

A. Schauer (Siemens)
Could you comment on the reproducibility and yield of your thin-film transistors?

T. P. Brody
The transistors are extremely reproducible; this is no problem at all. Yield is essentially a large-area cleanliness problem. A single speck of dirt in a 36 in² area can ruin the circuit. In the last few months, a typical yield from five consecutive runs for the 120×120 matrix has been two good devices and one acceptable device.

A. R. Kmetz (Brown Boveri)
Your work toward a discrete-element TV display and recent developments toward discrete-element CCD TV cameras raise the question of customer tolerance of point and line defects. Have any studies been made of this question?

T. P. Brody
CCD cameras now on the market have defect densities significantly higher than those in our displays, but customer reaction to these cameras is not yet known.

H. A. Dorey (Solartron/Schlumberger)
Silicon-target TV cameras can be programmed to interpolate through defects, which makes them much more tolerant of defects than would be a display where such post-processing is impossible.

T. P. Brody
I am very optimistic that before long we will have no defects.

SHORT COMMUNICATION


## PROGRESS TOWARD TFT-ADDRESSED TNLC

## FLAT-PANEL COLOR TELEVISION

A. G. FISCHER

Dept. of Electrical Engineering, University of Dortmund,
Dortmund, Fed. Rep. Germany

In the thin-film transistor (TFT) matrix for flat-panel displays shown in Fig. 1, the vacuum deposition of the X-Y part was performed until now using a "variable aperture mask".[1] Through it only one rectangular pad pattern could be deposited at a time, thus requiring numerous manual mask changes and shifts to deposit the complete matrix (even without the peripheral shift registers). This operation is prone to human errors and is difficult to automate.

We have now solved the problem of successively mating up to ten fixed-hole evaporation masks with the 15-cm-square glass substrate in a standard-sized 18-inch bell jar, as illustrated in Fig. 2 and 3. With only eight fixed-hole masks (Fig. 4), and with only three electron-beam evaporated materials (Ti-Au, $Al_2O_3$, CdSe), it is now possible conceptually to deposit the total matrix, including the shift registers, with double-gated TFT's (to attain higher off-on resistance ratios), within one hour and one pumpdown with full automation. (Incidentally, the same is possible for the more complex matrix needed to drive ac electroluminescent (EL), dynamic scattering liquid crystal and other electrooptic layers.[2] Previously, more than 30 manual changes of the variable aperture mask were required, with unavoidable human errors.[3]) This can be done using commercially available multi-layer deposition monitor systems with process computer and pneumatic and electric servomotors for mask moving, rotating of the crucible plate and for steering the electron beam.

It has recently been recognized by others that, in order to speed up the response of twisted nematic liquid crystal (TNLC) displays to TV rates and to improve the contrast,[4] a weak audio-frequency signal must be superimposed on the addressing dc signal, and an initial twist angle slightly less than 90° is required to eliminate reverse-twist ambiguities. We have found that the essential parts needed to construct the TNLC cell onto the TFT matrix can be vacuum-deposited during the same pumpdown during which the TFT matrix is made: by a simple drive, the substrate can be tilted for the deposition of the oblique Janning-type orienting film. After this, multiple spacers can be deposited onto opaque areas of the TFT matrix (e.g. the bus bars).

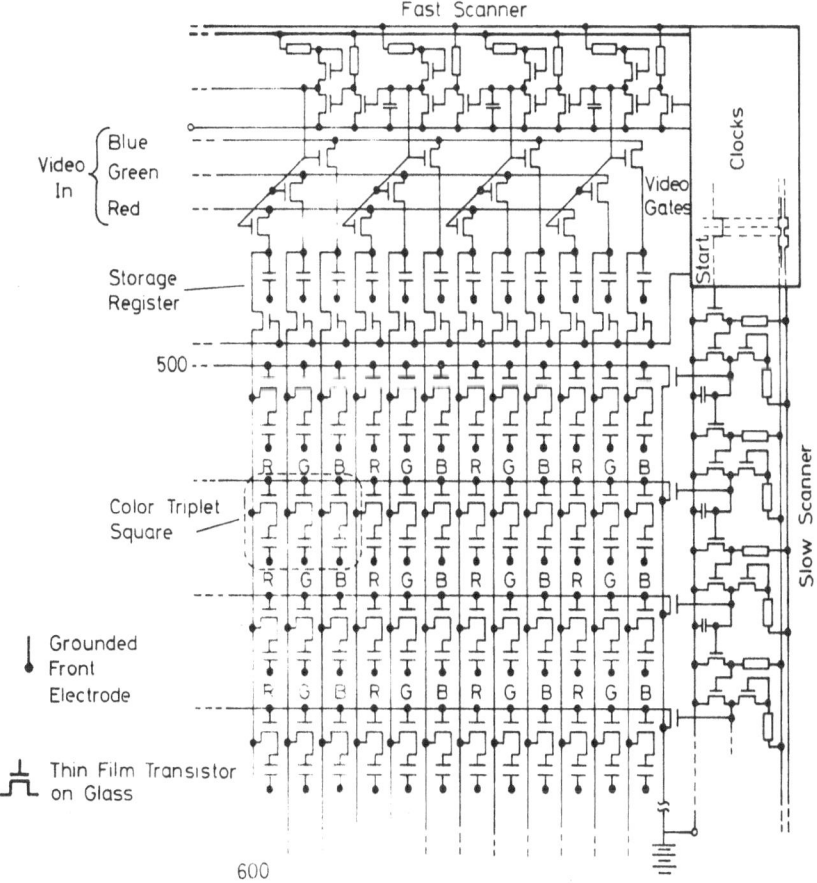

Fig. 1. Twisted nematic liquid crystal layer spread over TFT matrix, between parallel linear polarizers, rear-illumination by white light through registering mosaic color filter. Note that TFT's are used only as on-off switches.

Simultaneously, a broad sealing rim can be deposited around the cell. A glass with matched expansion coefficient must be used. One can even produce a wedge-shaped thickness variation (Fig. 5), e.g. from 5 $\mu$m to 7 $\mu$m, by slowly closing the shutter during deposition.

A new feature is sealing hermetically with indium instead of epoxy. For this, a narrow In rim of 1 $\mu$m thickness is put on either plate, on top of the transparent conducting reactively-sputtered $In_2O_3/10\%\ SnO_2$ common electrode layer and the Janning layer. The two plates are then hermetically sealed together by heating them above the melting point ($\sim 150^\circ$ C) of the metal rims. The TNLC is then injected at about $80^\circ$ C through hypodermic needles which are embedded in the rim at opposite corners of the panel, and which are then closed by pinching them off. When the TNLC cools to room temperature it develops a slight vacuum in the cell, but the multiple spacers prevent collapse of the large panel (Fig. 5).

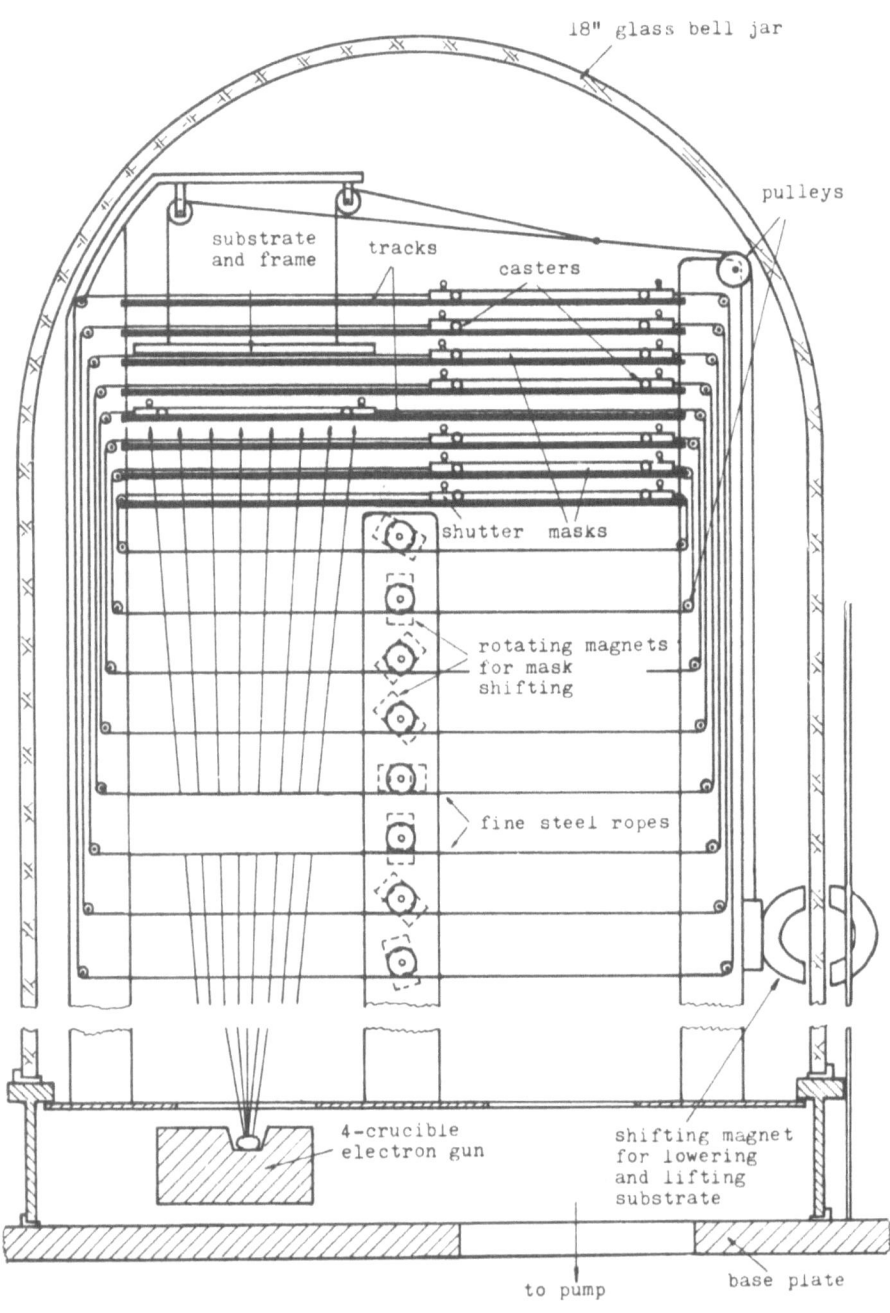

Fig. 2. Vacuum deposition of TFT matrix with mask frames.

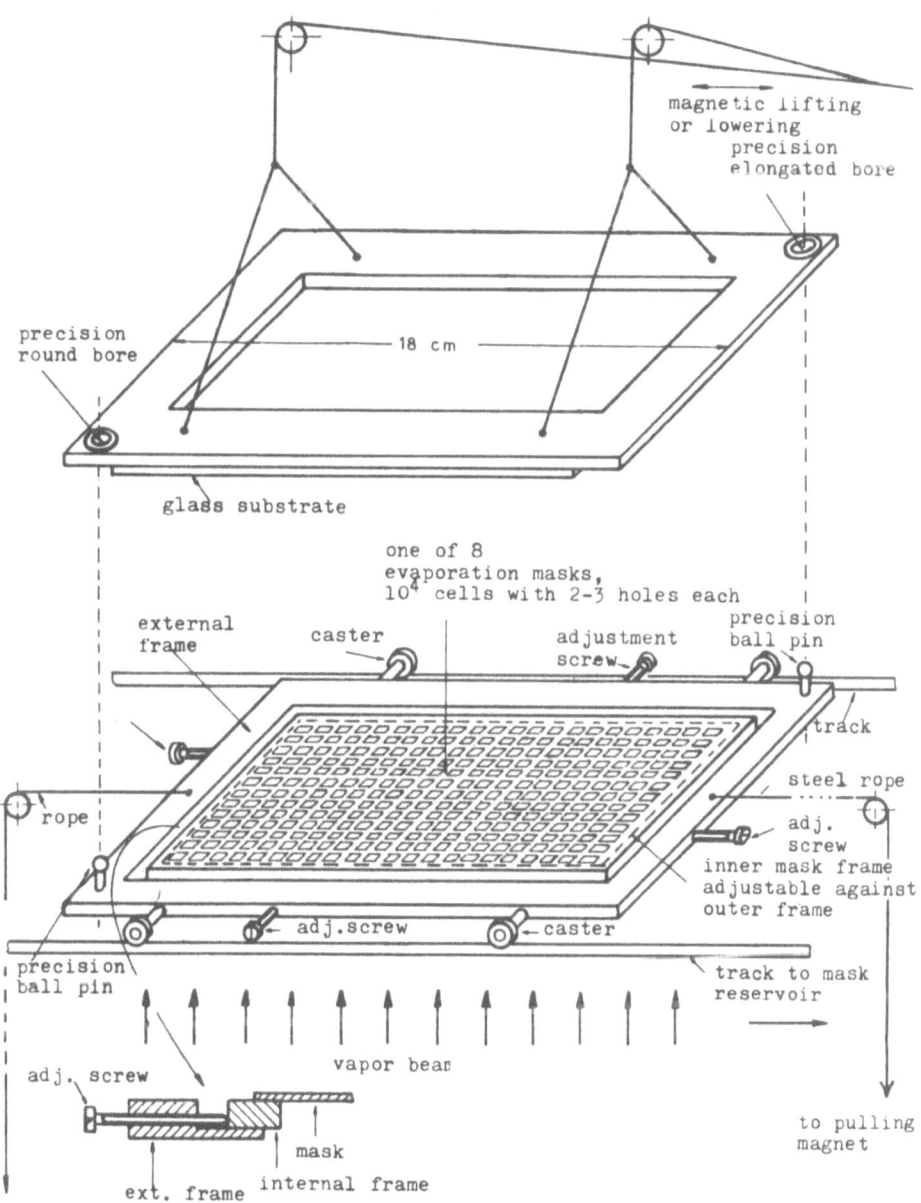

Fig. 3. Principle of mating masks to substrate.

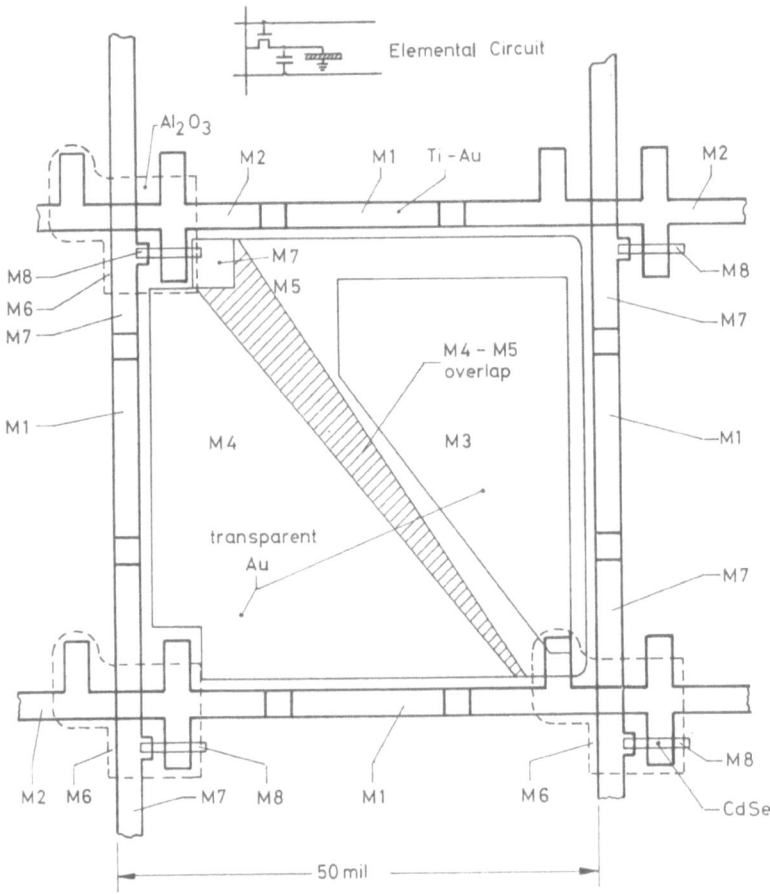

Fig. 4. Mask layout for TFT-addressed TNLC panel with double-gated TFT's.

Rear illumination of the panel with white light (unless sunlight is used) can be accomplished in very compact form by using a white-emitting ac EL powder layer. This EL panel is fabricated by embedding three monograin layers of a specially prepared blue-green-emitting ZnS,Se:Cu,Br powder of long operating life into a thermoplastic high-dielectric resin film of cyano-ethyl starch plasticized with cyanoethyl sucrose which contains dissolved in it fluorescent dyes, such as 'Day-Glo' pigments. They are excited by the EL emission to emit white light.[5] This panel is driven by 100 V, 3kHz sine waves supplied by a simple, battery-operated pulser. Unfortunately, how-ever, the power conversion efficiency of this light source is only about 0.2%.

White light can be generated 100 times more efficiently in fluorescent lamps. Especially suitable are the new Philips 'DeLuxe' lamps which emit

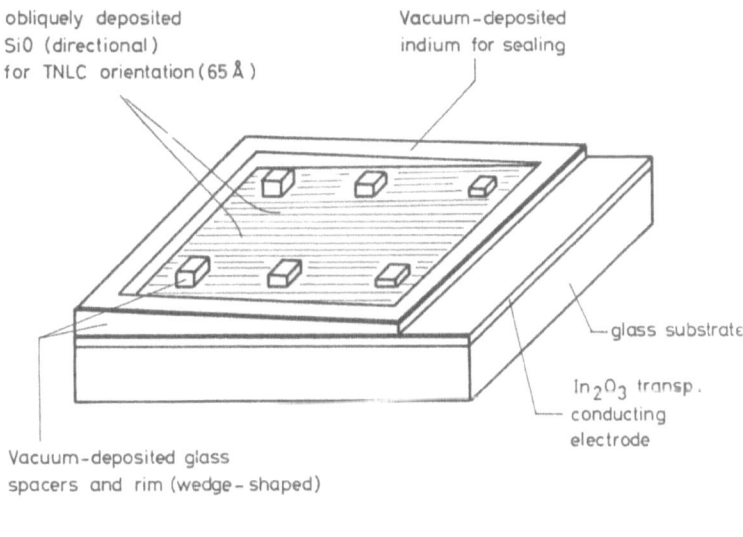

obliquely deposited
SiO (directional)
for TNLC orientation (65 Å)

Vacuum-deposited
indium for sealing

glass substrate

$In_2O_3$ transp.
conducting
electrode

Vacuum-deposited glass
spacers and rim (wedge-shaped)

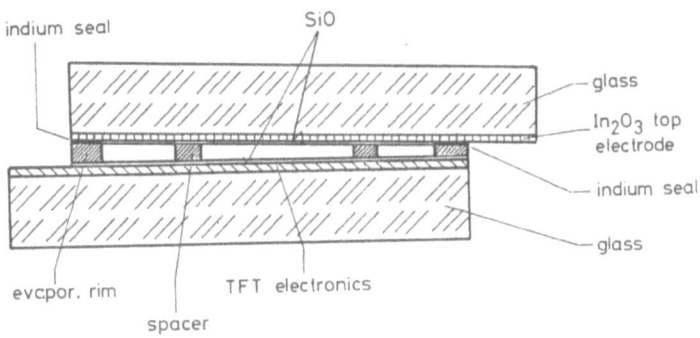

indium seal

SiO

glass

$In_2O_3$ top
electrode

indium seal

glass

evapor. rim

TFT electronics

spacer

Fig. 5. Schematic of TNLC cell construction.

bands centered at the three primary colors[6] at 450, 540 and 610 nm. As suggested before,[1] a plexiglass plate with a sandblasted pattern in registration with the TFT matrix, which is edge-illuminated by a miniature fluorescent lamp, is presently the best solution. The superimposed spatial filter that admits only near-parallel light into the panel (e.g. the 3 M Louvre Filter, or a lenticular plastic foil) must have a pattern in exact registration with the TFT matrix, to avoid the formation of Moire patterns.

Compared to the massive R&D effort which has gone into plasma display panels, the sporadic research effort invested so far in TFT-addressed TNLC panels is minimal. Yet this latter approach promises a lower-cost panel with lower power consumption. Small TNLC displays which last longer than 10 000 hours are on the market already, and the TFT, with new materials and with electron beam evaporation, has now outgrown its infant deficiencies and is comparable in its aging characteristics to MOS transistors,[7] yet much more adaptable to flat display panel needs. The actual operation of an experimental panel (with reduced resolution) with off-the-air signals has already been accomplished.[8] We have here proposed detailed fabrication techniques to facilitate further advances toward flat-panel color TV.

## ACKNOWLEDGEMENTS

This project is supported, in part, by Deutsche Forschungsgemeinschaft. The enthusiastic help of my coworkers K. Koger, J. Knüfer, D. Herbst and E. Adler is greatly appreciated.

## REFERENCES

1    A. G. Fischer, T. P. Brody and W. S. Escott, IEEE Conference on Display Devices, New York, N.Y., October 1972, Conf. Record 72 CH0707-0-ED, 64; A. G. Fischer US Pat. 3 840 695

2    A. G. Fischer, 3rd Internat. Conference on Thin Films, Budapest, August 1975, paper 12-03

3    T. P. Brody, J. A. Asars and G. D. Dixon, IEEE Trans. Electron Devices ED 20 (1973) 995

4    C. J. Gerritsma, S.I.D. Int. Symp. Digest 5 (1974) 164; E. P. Raynes, Electron. Lett. 10 (1974) 141

5    W. Lehmann, US Pat. 2 924 732 (1960)

6    J. M. P. J. Verstegen, D. Radielovic and L. E. Vrenken, J. Electrochem. Soc. 121 (1974) 1627

7    E. Schlam, J. Velasquez and I. Reingold, Proc. 24th IEEE/ECC Conference, Washington, D.C. (1974) 43

8    L. T. Lipton, M. A. Meyer and D. O. Masetti, S.I.D. Int. Symp. Digest 6 (1975) 78

# CONCLUDING REMARKS

C. HILSUM

Royal Radar Establishment, Malvern, Worcestershire, England

A volume as specialized as this is prone to stress the details and ignore the generality, or, in military terms, to concentrate on tactics to the exclusion of strategy. But the strategy in this field is all-important, and we must therefore stand back one or two steps and survey the accomplishments. The term survey is used deliberately, because it is better to think of this contribution in that way, rather than as a summary.

Ten years ago there was no concerted international program on flat-panel displays. The incentive then was no less than it is today, but in some curious way the collective instinct of the device physicists was not aroused by the challenge. These proceedings prove that a most remarkable change has occurred. Far from being hindered by a lack of ideas, we are embarrassed by the plenitude. Even excluding devices which emit light, we are still left with nine physical effects which are serious candidates for discussion. This is both a weakness and a symptom. The many alternative lines of attack dissipate our effort and the available funds. They also reveal that no one solution is so promising that it withers competition. There still remain serious weaknesses in all the methods proposed.

We can justify the statement that the device physicist had an incentive ten years ago to make flat-panel displays. We could not equally substantiate the premise that there was then a need for such devices. We would even have difficulty in proving this today, for we must recognize that alternative solutions have been found for most of the existing markets, and there are few users pressing us to solve their problems. Our drive comes from within us, generated by our own conviction that a topic with such intrinsic fascination must eventually prove economically rewarding. Enthusiasm is laudable, but the new applications which we are sure will develop cannot emerge until we have made more progress.

The problem of justifying research on a novel device is not uncommon in solid-state circles, where we are quite accustomed to convincing accountants that the new markets will definitely appear when engineers can put their hands on this new component. But the display field does have a unique hurdle to overcome — the accountant can see it, and if he is not seduced by its appearance, he will resist all our well-founded arguments. Even in the early stages of development a display should look good.

Here, of course, non-emissive displays face a considerable handicap, because the user generally prefers an emissive display. (Someone might say that people will get used to non-emissive displays, because they are accustomed to reading books. That is true, but remember that once television appeared people stopped reading.) In this context we might wonder why more attention is not paid to the viewer, to the ability of the observer to distinguish accurately the information presented to him, a feature largely linked with what we might call the charm or the eye appeal of the display. This is something which is easy to judge but difficult to measure. We work hard at maximizing the physical effects we want to exploit, and we design the electronics with care, but we have little documented evidence on research to improve the appearance.

We assume for liquid crystals that ac addressing is acceptable, but the circuit engineer would prefer dc techniques. There is no fundamental reason against dc addressing for some systems, and this might make our systems more palatable. In some experiments the problems of dc drive are associated with the electrodes, rather than with the bulk liquid, in that the electrodes are attacked. Should we not try to understand better why we use tin oxide or some variant of this? There are, in fact, many uncertainties about the conductive mechanism in such layers, and studies could lead to new systems that are more suitable to applications.

Another hurdle we must overcome is the reluctance of users to employ liquids, particularly the rather exotic media that we wish to exploit. Most applications engineers think that the only good organic liquids are those you can drink and the only good inorganic ones, those you use for dilution. We cannot dismiss this as unreasonable, because there are few electronic systems based on the liquid state. Semiconductor device engineers can exploit a reputation for reliability — we have to overcome an impression of incipient failure which is only partly merited. Scientists in our field make comments which expose us to criticism; for example, the conclusions of a talk at a recent displays conference read — "There are no theoretical reasons to prevent the development of a good long-life (non-emissive) display, provided all the materials problems which determine operating life can be solved." Moreover, new non-emissive effects have been publicized as though they were complete solutions to the display problem, minimizing the technological difficulties, and devices have been sold despite strong suspicions that they were unreliable.

There are two remedies — to do more research on failure mechanisms in liquid cells, and this entails not just counting elapsed time, but discovering what is causing the degradation; alternatively we can do more work on solids. There is some work on solids proceeding, but does this really explore the effects which are likely to prove most useful? We tend to study lattice perturbations or ionic conduction, the first being a small effect pushed to its limit, and the second an automatic candidate for unreliability studies on electrodes. We should certainly not abandon the search for suitable physical effects, particularly those with time constants faster than one millisecond.

It is in fact very surprising that liquids feature so largely in our field when they play such a limited role in every other form of electrical or electronic device. Individually we must judge whether this is an inevitable corollary to the physical nature of the liquid state, or simply a transitory historical situation. As an erstwhile solid-state physicist I find it both stimulating and annoying to be involved with the vagaries of liquids, and I may be the victim of my past in feeling sure that eventually solids will come into their own again. I might hazard a guess at the contents of a future symposium, held in perhaps five years time. Liquid crystals will still be with us, because of their versatility, but it would be surprising if the other liquid technologies survive on a large scale. We will see new effects in solids, and much more attention to the problems inherent in large area panels.

All modern solid circuits are based on a philosophy of reducing the number of interconnections. Surely the future display panel will not be surrounded by a mass of wires. We must at some stage evolve a real solution to the transducer-decoder interface. This topic is closely linked with that of the last two papers, which always elicits a strange response at display conferences. You may not believe in thin-film transistors, or in glassy semiconductors, but if you have reservations you should be trying to generate alternative solutions. The complex display needs intrinsic circuitry just as much as a suitable visual effect. We are in a field that promises big rewards to those who can find a way of making a good, large flat-panel display. It is the distant horizons we should be scanning, rather than the green fields of the foreground with their limited crop of wristwatches and calculators.

# AUTHOR INDEX

Page numbers in bold-faced type indicate full-length papers and short communications; all other page numbers refer to contributions to discussions.

# SUBJECT INDEX